Valve Radio and Audio Repair Handbook

To my dear wife Jo

Valve Radio and Audio Repair Handbook

Second Edition

Chas E. Miller

Newnes

OXFORD AMSTERDAM BOSTON LONDON NEW YORK PARIS
SAN DIEGO SAN FRANCISCO SINGAPORE SYDNEY TOKYO

Newnes
An imprint of Elsevier Science
Linacre House, Jordan Hill, Oxford OX2 8DP
200 Wheeler Road, Burlington, MA 01803

First published as *Practrical Handbook of Valve Radio Repair* 1982
Second edition 2000
Transferred to digital printing 2003

British Library Cataloguing in Publication Data
A catalogue record for this book is available from the British Library

Library of Congress Cataloguing in Publication Data
A catalogue record for this book is available from the Library of Congress

ISBN 0 7506 3995 4

For information on all Newnes publications
visit our website at www.newnespress.com

Contents

Part 3

Preface

This book has been written for all vintage radio enthusiasts, whether those who have just become interested in the subject and have no previous technical knowledge, or those with a good general grounding in this respect but who wish to learn more about the practical aspects. Part 1 is devoted to the essential basic theory and Part 2 to practical repair work, whilst Part 3 consists of a wide-ranging and valuable collection of valve and other data relating to vintage receivers. Readers who already know the basics may, if they wish, turn straight to Part 2, but even for them a little revision may be useful and interesting.

Essential facts and figures

When I was learning about valve radio in the days when it was contemporary and not 'vintage', I got hold of as many books as I could on the subject, of which there were many in those days, some good, some indifferent and some plain bad. Re-reading them now I can see that some were produced cheaply and quickly to exploit what was then a growing interest, some were written by highly technical people with poor communicative skills and just a few were crafted together by practical radio engineers who knew how to 'get things across'. Obviously, to be able to become reasonably proficient in any subject one needs first to know its basic principles, but far too many books seemed to throw readers in at the deep end and thus 'blind them with science' before they had learned even to paddle, as it were. For myself, a poor mathematician, nothing could have been more daunting than to be faced with pages full of formulae, to understand which the reader would have had to be just as knowledgeable as the author, if not more so. The problem as I saw it was that these writers were envisaging their readers as ending up with degrees in radio engineering, not as simply trying to learn sufficient to enable them to get started in the subject. For this reason I have here restricted the amount of technical matter to the bare minimum necessary for the reader to be able to grasp the whys and wherefores of the practical work. For those who wish to go more deeply into theory, a suggested bibliography is printed at the end of the book.

Vintage radio, vintage terms

To understand and appreciate vintage radio properly one has to project oneself back into the past. For this reason, throughout this book I have retained the use of technical terms that were current in the vintage years of radio, e.g. condenser and not capacitor, cycles per second (c/s) instead of hertz. There is a sound reason for this, not just nostalgia. Anyone engaged on repairing vintage radio sets is going to have to refer to contemporary data published many years ago when the old terms were current, and having mentally to translate them into modern equivalents would only add to the difficulty of understanding circuits and servicing instructions. Likewise, all dimensions, etc., quoted appear in Imperial/American measurements just as they were originally. For instance, resistive mains leads were always specified as having a resistance of either $60\,\Omega$ per foot or $100\,\Omega$ per foot; to attempt to metricate something like this would not only be superfluous, it would be perverse.

The spares position

Some years ago, in a magazine article, I adapted a famous phrase thus: 'the valves are going out all over Europe, and we shall not see them lit again in our time'. It was a decidedly rueful attempt at humour, and no one is happier than I that the prediction has been proven false, for there has been a great resurgence of interest in old valve radios. No doubt some of this stems from the well-known British love of restoring all sorts of old equipment, but where the valve radio scores heavily is that however old it may be it need not be just a museum piece to be brought out on high days and holidays. Once restored it can be placed in everyday service and it will perform reliably and economically.

Every now and again one of those confident know-alls who crops up in all walks of life will say to me, 'of course, you can't get the valves nowadays', and I must confess that I take great pleasure in deflating them by informing them that not only are there millions of valves held in stock by dealers, it is actually easier to obtain some rare types than it was just after the Second World War, when they were being kept 'under the counter' by radio dealers! Thanks to a growing awareness of vintage radio the contents of old shops and of private collectors are frequently offered at auctions, especially those organised by *The Radio-*

phile. As well as valves, components and complete receivers may be obtained from this source, or from specialist dealers, whose expertise and integrity are well proven. The names of some of them will be found in the advertisement supplement at the end of this book.

Thanks

The emphasis of the book is on the practical side of repair work, extended to include mechanical as well as electrical aspects of servicing. Wherever possible original diagrams have been used to illustrate the various features, common or rare, mentioned in the text. As before, there is a large reference section, covering a wide range of information culled from many contemporary sources and I wish to express my gratitude to those persons and organisations who kindly lent material, and to the copyright holders for permission to reproduce.

Finally, I would also like to thank my wife for her unfailing assistance and tolerance, without which this book (or *The Radiophile*) would not have been possible, and Geoff Dixon-Nuttall for his expert proof-reading.

C.E.M.

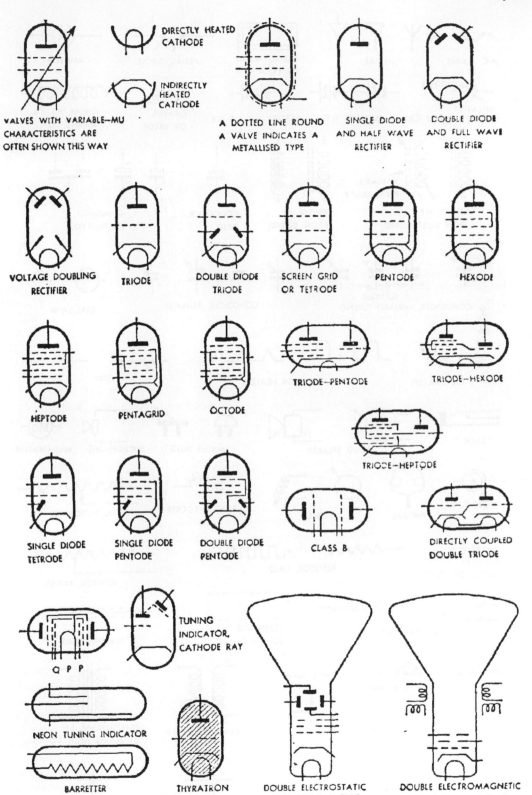

VALVES WITH VARIABLE—MU CHARACTERISTICS ARE OFTEN SHOWN THIS WAY

DIRECTLY HEATED CATHODE

INDIRECTLY HEATED CATHODE

A DOTTED LINE ROUND A VALVE INDICATES A METALLISED TYPE

SINGLE DIODE AND HALF WAVE RECTIFIER

DOUBLE DIODE AND FULL WAVE RECTIFIER

VOLTAGE DOUBLING RECTIFIER

TRIODE

DOUBLE DIODE TRIODE

SCREEN GRID OR TETRODE

PENTODE

HEXODE

HEPTODE

PENTAGRID

OCTODE

TRIODE—PENTODE

TRIODE—HEXODE

TRIODE—HEPTODE

SINGLE DIODE TETRODE

SINGLE DIODE PENTODE

DOUBLE DIODE PENTODE

CLASS B

DIRECTLY COUPLED DOUBLE TRIODE

Q P P

TUNING INDICATOR, CATHODE RAY

NEON TUNING INDICATOR

BARRETTER

THYRATRON

DOUBLE ELECTROSTATIC CATHODE RAY TUBE

DOUBLE ELECTROMAGNETIC CATHODE-RAY TUBE

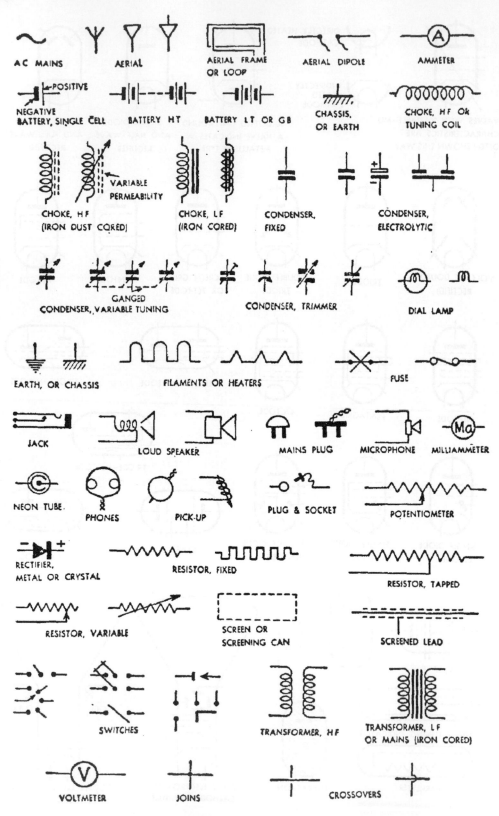

Part 1

Chapter 1

Basic facts you need to know about electricity and magnetism

Electricity from batteries

There are two kinds of electricity, alternating current (AC), the kind that comes from the electricity mains, and direct current (DC), which is the kind that comes from batteries. Every battery has two poles or connections, one positive and one negative. Current drawn from a battery flows continuously in one direction only, hence the old alternative name of 'continuous current'. Batteries may be subdivided into two classes known as primary and secondary types.

Primary batteries produce electricity from chemical action within 'cells' which continues until the chemicals become exhausted and the battery 'runs down'. There are various combinations of chemicals capable of producing electricity at various voltages per cell, but for our purpose we need only consider the 'Leclanché' type of cell used for torch and transistor radio batteries which produces 1.5 V. Two or more cells may be joined in a chain, positive to negative, to obtain any desired voltage for various jobs. Vintage radio sets used batteries of up to 165 V to provide power for the valves. Although they are not 'dry' (except when totally run down!) primary cells and batteries are commonly known as dry types to distinguish them from secondary batteries which employ acid or alkaline solutions.

Secondary batteries also employ chemical action within cells to produce electricity but in this case the chemicals have to be activated ('charged') first by connecting the cell to an external source of DC voltage for a certain time. When charging is complete the cell will deliver voltage until the chemical action ceases, when it may again be recharged. This process may be repeated many times before side effects within the cell cause it to

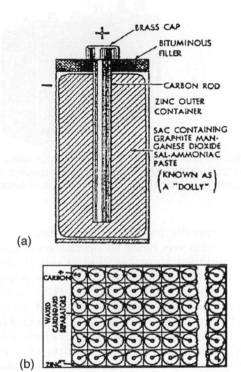

(a)

(b)

Figure 1.1 (a) The construction of a dry cell as used in batteries for vintage radio sets. Modern cells follow much the same pattern. (b) How numbers of dry cells may be connected in series to build up any desired voltage

fail completely. The lead-acid type of cell used in vintage radio sets produces 2 V, and once again two or more may be series connected to provide higher voltages, although most of the sets you are likely to encounter use only a single cell.

Alternative names for a secondary battery are storage battery or accumulator, the latter being the

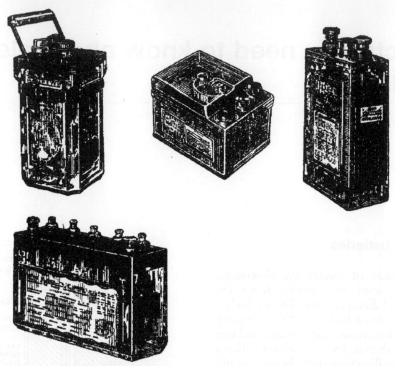

Figure 1.2 Various types of 2 V accumulator used in vintage battery operated receivers. Far left, a typical glass bodied accumulator with metal collar to provide a carrying handle. Centre, a celluloid-case type filled with jellied acid for portable sets. Right, another celluloid-case type but filled with ordinary acid for semi-portable sets

one most generally used in vintage radio.

Vintage radio sets made to operate on batteries usually employed a combination of one or more primary batteries and one accumulator to provide high tension (HT) and low tension (LT) supplies for the valves. Earlier sets also required another source of voltage for the valves to provide grid bias (abbreviated to GB, and of which more later) of up to −16 V. From 1939 the introduction of a new range of valves known as 'all-dry' types made it possible to have receivers working from just two dry batteries without the need for accumulators. However, by far the greatest number of vintage receivers you are likely to encounter were made to operate from mains electricity and to understand how they worked it is first necessary to know the basic facts about how electricity is produced and its connection with magnetism.

Electricity and magnetism

Magnetism was discovered some thousands of years ago when someone tripped over a strange piece of rock, picked it up and found that for some unknown reason it attracted to itself anything made of iron. What was more, if this rock was rubbed along a strip of iron it too started to attract other pieces of iron. Although this was interesting, no practical use for magnetism was found for some time – until around the year 1200 AD, in fact, when someone else discovered that if a magnetised needle was hung on a piece of twine it would spin around and settle into a position where one end pointed north and the other south. Now, this really was useful, because it enabled sailors to travel the world with a sporting chance of reaching their intended destinations and eventually returning home. It also brought about a rather contradictory piece of terminology. Not unnaturally the end, or pole, of the needle that pointed north was called the north pole and the one that pointed south the south pole. However, if two magnets are held close to each other the north pole of one is attracted to the south pole of the other and vice versa, hence the rule that unlike poles attract, like poles repel. Logically, this ought to mean that the pole of a magnet that points north isn't really its north pole but its south, but as things then start to become rather confusing that part is conveniently overlooked.

As for the rock which started the whole thing off, the old name given to it was lodestone or loadstone; lode/load in this instance coming from a old word meaning a way, hence a rock that shows the way. Nowadays it is more prosaically referred to as a form of iron oxide called magnetite.

For the next seven hundred years or so not much else happened with magnets until electricity began to be studied in the early nineteenth century. It didn't take long for scientists to find links between electricity and magnetism and thus unwittingly start the journey that would lead to electric light and power, radio communication, party political broadcasts on television and other advantages which we enjoy today. In fact, it's a fascinating story that is worth investigating in specialist reference books but here we shall have to content ourselves with enough of the gist of it to give the basic principles on which vintage radio sets depend.

What changes an ordinary piece of iron into a magnet is that when stroked with magnetite groups of its molecules called domains, normally pointing in random directions, are lined up to point all the same way. This effect can also be obtained by a different means, as we shall see in a moment.

Magnets produce what is called a magnetic field. An experiment to demonstrate this, once familiar to school boys (perhaps it is still), was to place a magnet under a sheet of thin card, onto which iron filings were scattered, whereupon the latter would obligingly form themselves into exact patterns tracing out what are called the lines of force from the magnet, as in Figure 1.3(c).

The sort of magnets we have been talking about are known as permanent magnets because once they have been magnetised by some means they remain so more or less indefinitely. Now for another type of magnet. If a length of wire is wound around a cardboard tube to make a coil, and its two ends connected to a battery, a magnetic field is set up in

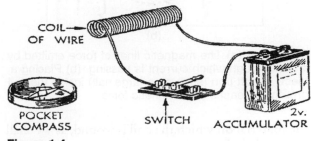

Figure 1.4

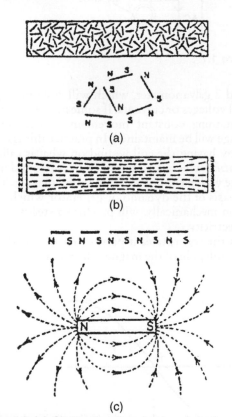

Figure 1.3 (a) Shows the domains in a bar of unmagnetised ferrous metal pointing in random directions. After magnetisation (b) the domains are lined up to point the same way, (c) shows the magnetic lines of force emanating from a bar magnet

and around it. This can be demonstrated by placing a compass alongside the coil, when its needle will be deflected when the battery is connected and will remain so until it is disconnected. If a six-inch nail is placed inside the coil, when the battery is connected it will become magnetised because the lines of force from the coil will have lined up its molecules. The nail will remain magnetised for as long as the current flows through the coil but will revert to just an ordinary piece of iron when the battery is disconnected. This type of magnet is known as an electromagnet and it too produces lines of force that may be plotted with the aid of iron filings.

Because a coil will induce magnetism in this way it is said to possess inductance, the unit of which is the henry, taken from the name of a scientist as are so many electrical and radio terms. Coils of any desired inductance may be produced by varying the number of turns and the gauge of wire used – and

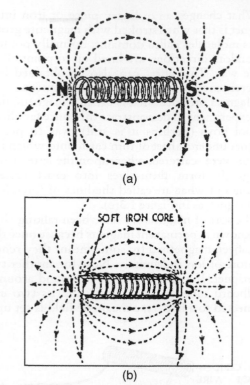

(a)

(b)

Figure 1.5 (a) The magnetic lines of force emitted by a coil through which current is passing. (b) Placing a piece of soft iron (such as a large nail) inside the coil greatly increases the magnetic force

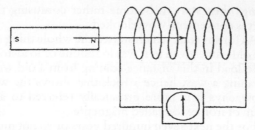

Figure 1.6 Passing a magnet into a coil of wire induces a small voltage measurable on the galvanometer (bottom)

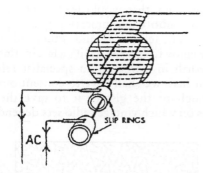

Figure 1.7

the material on which the coil is wound, as we shall see a little later.

That is one aspect of the relationship between electricity and magnetism. Here is another. If that same coil of wire is passed at right angles through the magnetic field of a permanent magnet a voltage of brief duration will be induced into it, sufficient to be indicated on a sensitive electrical instrument called a galvanometer, which will respond to very small voltages or currents. If some means is found of maintaining constant movement of the coil the voltage will be maintained. In practice this is carried out by having the coil wound on what is called an *armature* which is mounted on bearings and free to rotate within the poles of a powerful magnet. This is the basis of the dynamo or generator which, when driven mechanically, will produce a steady current of electricity.

As the coil rotates it passes first the north, then the south pole of the magnet; the voltage induced in

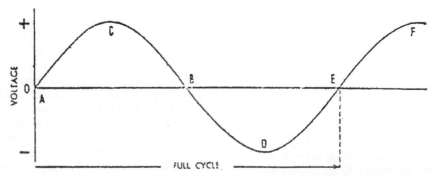

Figure 1.8 Showing how the voltage produced by an alternator swings first positive and then negative to form complete cycles.

it will travel first in one direction, then in the other, thus producing what is called alternating current (AC), the voltage being collected from the coil by means of *slip rings* onto which are pressed carbon *brushes*. The number of complete reversals made by the voltage, or cycles, is called its frequency. In the UK the main electricity supply has a frequency of 50 cycles per second. At this point it should be mentioned that for some years now the name 'hertz' (Hz), derived from the German scientist Heinrich Hertz, has been used as the unit of frequency, and the mains frequency is now called 50 Hz. However, anyone engaged in vintage radio work will be using contemporary reference material such as makers' service manuals published years before Hz came into use, so to avoid confusion in this book we shall stick to cycles per second (c/s) when referring to frequencies. This principle, by the way, will apply also to other vintage terms such as condenser (now capacitor), coil (inductor) and various others as they occur in the text.

Returning to generators, they may also be designed to produce direct current (DC) which flows constantly in one direction only instead of reversing all the time as does AC. If the slip rings are replaced by a *commutator* each voltage pulse is drawn from the coil in the same direction, and if a number of coils are wound onto the armature the pulses from each coil will interleave to produce a steady DC voltage plus a slight 'ripple'. By the 1880s the development of DC generators had reached the stage where it was practicable to use them for electric street lighting and even to drive tram cars by means of electric motors. For technical reasons it was then better to use DC for these purposes, so nearly all the original electricity generating companies supplied it. Another advan-

tage was that it could be used to charge large banks of storage batteries (giant versions of those used in motor cars) to provide back-up power if the main plant had to be closed down for maintenance, etc.

Generally speaking the first places to have electricity, almost invariably DC, for lighting and power were those that had either prosperous communities or those with a good deal of industry. In the latter case the main users of electricity were factories and local electric tramway systems, with few if any domestic consumers. However, as large power stations had to be built to power the trams canny operators began to think of selling any surplus current as a by-product to shop premises and houses along the tram routes where supply cable already had been installed. This is more than just a snippet of history; the growth of DC mains in the 1890s was destined to affect the design of valve radio equipment for something like sixty years. Even today it has vital implications for anyone working on vintage radio sets.

Another aspect of electricity supplies in the late nineteenth century which was to influence radio design was the many different voltages used. In purely residential areas it might have been in the 100 V to 130 V range, which was ample for domestic lighting and small heating apparatus and which could easily be sustained with back-up storage batteries. Industrial users required higher voltages of 400 V to 500 V, and as this was far too much for domestic use the supply companies devised a system whereby the voltage was supplied in two 'halves', that is, a 400 V supply was made up of two 200 V supplies in series so that just the latter voltage could be delivered to houses and shops. Sometimes more than one type of supply system was installed in one town – some had three or more – and the end result was that a domestic consumer might find himself with a supply voltage of anything between 100 V and 250 V. We shall return to this subject later.

Very few country districts had electricity in those days because it was prohibitively expensive to run the long and heavy cables that would have been necessary to carry DC mains from the urban power stations to outlying areas. In a few villages enterprising local businessmen installed generators and supplied electricity to the immediate area, whilst the owners of large country houses often had their own private supply derived from a steam engine driven generator tended by one of the usual employees, such as the head gardener. The plant was situated in a shed far enough away from the house to avoid its noise disturbing the gentry,

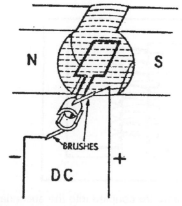

Figure 1.9

whilst back-up batteries took over at night when the man in charge went home. These installations were reliable and long lived (hopefully the employees were as well) and the present writer has personal knowledge of one that was still delivering its 110 V DC in the late 1950s. By and large, though, most country folk had to wait for electricity until the National Grid was developed, which brings us neatly back to the AC generator or *alternator*.

Induction and transformers

Reverting to the subject of the inductance of a coil of wire, let's now see to what practical uses it may be put. One of the oldest, dating from the days of Michael Faraday, is the transformer. We spoke earlier of the way in which a compass needle is affected by the magnetic field set up by passing current through a coil. If we replace the compass by another coil the magnetism set up in the first will induce voltage into this new coil, but only for an instant as the battery is connected and for another as the connection is broken: no voltage is induced whilst steady current is passing through the coil. The reason for this is not difficult to see when considered alongside the generator principle, in which voltage is produced by constant movement of the coil through a magnetic field. In fact, the reverse could also be employed, with the magnet being moved and the coil held steady, the end result being that the magnetic field is constantly moving with respect to the coil. Once that relative movement ceases so does the production of current. When the battery is first connected to the coil and the magnetic field is suddenly created it is the equivalent of movement between magnet

and coil. The same is true when the circuit is broken and the magnetic field collapses, but all the time the battery remains connected the steady magnetic field has no effect.

Now let us see what happens if the coil is connected not to a battery but to a source of AC voltage. The magnetic field will build up in one direction on the first positive half-cycle then die away and build up again in the opposite direction on the negative half-cycle, this process being repeated all the time the current is connected and giving the effect of a constantly moving magnet. The first or primary coil therefore will induce a steady AC voltage into its neighbour, or secondary, coil of a value depending entirely on the number of turns wound on the second coil. If it has a larger number of turns than the first the voltage will be higher, if of fewer turns the voltage will be lower. Since it is impossible to get something for nothing, the actual amount of power in watts being expended in the first coil cannot be exceeded in second coil, so if it is wound to increase the voltage the available current is automatically reduced to give the same wattage (in practice losses would prevent its being 100% of the original but the principle holds good). It is this ability to transform AC up and down to obtain any desired voltage that makes it so much more useful in many ways than DC. It certainly made it possible for mains electricity to spread into almost anywhere in the UK.

The old DC power stations could not produce current at a higher voltage than the maximum supplied to consumers, which as we have seen was about 500 V. We have also mentioned that heavy cables were required to carry current to consumers in order to avoid voltage drop along the way, and how this precluded very long runs. If the total load on the cables was 500 kW at 500 V they had to be

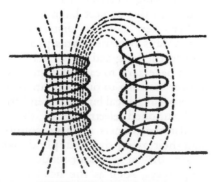 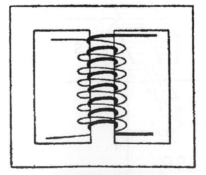

Figure 1.10 Left: how the lines of force produced in one coil (the primary) are coupled into the secondary coil. Right: the basis of practical transformer design

capable of handling 1000 A and were thus enormously thick and expensive. For this reason power stations were built as close as possible to the areas they were to supply, often in the heart of towns, with all the consequent noise and air pollution.

With AC, however, the power stations can produce very high voltages which may be sent out at low current to provide high wattage ratings. For instance, to handle 500 kW at 10 000 V the cables need to carry only 50 A, an enormous saving. The high voltage is transformed down at suitable points along the cable run to provide the usual 230 V for domestic consumers or 440 V for light industrial work. Since voltages of up to 266 000 V are used to carry electricity across the countryside, huge wattages may be handled at comparatively small currents, thus making it economical to supply remote communities. It also makes it possible to locate power stations well away from towns where their impact on the environment is not so great.

The very first power station to supply AC was built in London in the late nineteenth century. The obvious advantages of AC caused it soon to be adopted in many other areas and by the 1920s the numbers of DC and AC power stations were about equal. Once again there was no standardisation of voltage and consumers could expect to receive anything between about 100 V and 250 V, which also caused problems for the designers of vintage radio sets. At the end of the 1920s the government of the day proposed that no more DC stations should be built and that those existing gradually should be converted to AC working. A national electricity grid was to be established whereby current could be passed around from area to area as required, individual undertakings being able to sell surplus current if there was over-production or to buy it in at times of shortage.

The eventual aim of bringing AC mains to every community in the country was worthy but difficult to realise since DC users would have to be compensated if they were compelled to buy new equipment because of the changeover. In any case the Second World War caused the plans to be shelved and it was not until the late 1950s that the change was completed. As we have mentioned before, this dual-current, multi-voltage set-up had influenced vintage radio design in ways which will be discussed in a later chapter.

Practical transformers

In practice, the coils of transformers used for mains frequency are what are called iron cored. In its simplest form each coil is wound on a bobbin through the centres of which a bundle of soft iron wires is passed. These become highly magnetised as the current is passed through the primary and induce a powerful magnetic field into the secondary, raising the overall efficiency of the transformer. Soft iron is used because its ability to retain magnetism is very low, as required for the quick changes from magnetised to unmagnetised state and vice versa. One solid lump of iron would not have this property to anywhere near the same extent, whilst steel would be completely useless. Iron wires were extensively used in the early days of electricity and to improve their efficiency they often were long enough to be bent back over the coils to form almost a loop. This type of construction was known as a 'hedgehog' transformer.

Further improvements in efficiency were brought about by the use of laminations (i.e. thin sheets of soft iron) built up into stacks to form the transformer coil. The classic shapes used for the laminations were and still are 'Es' and 'Is', which is self-explanatory. The coil bobbins are slipped over the centre leg of the Es and the Is are then laid across them to form very nearly a closed magnetic circuit. It is important that this should not, in fact, be quite complete so a thin non-magnetic spacer is placed between the two lots of laminations. Over the years various special alloys have been developed for laminations which have yet further improved transformer efficiency.

Different shapes of cores are used for transformers used in modern electronic equipment, but as far as vintage radio is concerned Es and Is reign supreme.

Auto-transformers

It is possible to use only one winding for a transformer by inserting a third connection or *tap* somewhere along its length which may be in the middle or towards one end according to the

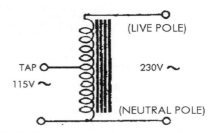

Figure 1.11 The auto transformer may be used to reduce or increase voltage as desired

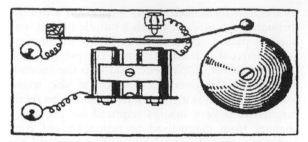

Figure 1.12 Functional diagram of an electric bell

Transformers and DC

Although we have already learned that DC as it stands cannot be used with transformers, methods do exist of achieving this by high speed on/off switching between DC source and primary. To see how this is done we need first to look at the humble electric bell.

The electric bell consists of a small iron-cored electromagnet which, when energised by a battery, pulls towards it an armature consisting of a strip of soft iron mounted on a flat spring so that it will return to its original position when the current is connected. Attached to the armature is a small spherical hammer which moves with it and this a steel gong when the electromagnet operates. Also attached to the armature is an electrical contact in the shape of a small disc about a sixteenth of an inch in diameter and maybe half as thick.

Opposite this is another, pointed contact which is mounted on a small metal post so that when the armature is at rest the two are touching and able to pass electricity. When the battery is connected to the bell the current passes through the contacts to the coil and thence back to the battery, causing the electromagnet to attract the armature and to sound the gong. This breaks the circuit and the armature returns to rest, whereupon the circuit is restored

desired voltage ratings. The entire winding or one section of it may be considered as entire winding or one section of it may be considered as either a step-down or step-up action. One very popular use for auto-transformers is to enable radio and other equipment rated for 110/130 V American mains to be used on British 230 V supplies. In this case the coil is centre tapped with the 230 V being applied across the entire winding and 115 V drawn from one half section. The drawback of the auto-transformer is that there is no electrical insulation between primary and secondary as in a normal transformer, but set against this is higher efficiency with respect to core size, so an auto-transformer may be half the size of an ordinary one for a given wattage rating.

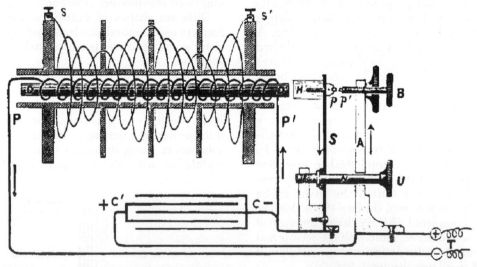

Figure 1.13 Diagram of a Victorian shocking coil P and P′ on the coil are the ends of the primary winding. S and S′ those of the secondary the latter would be connected to a pair of brass handles. The other P and P′ are the points which make and break the connection from battery to primary winding. S is the brass spring armature to which is fastened the soft iron hammer H. B and U are adjusting screws by which the initial setting to the points is achieved to start the device working. It had scientific uses as well as being a source of fun at country house parties

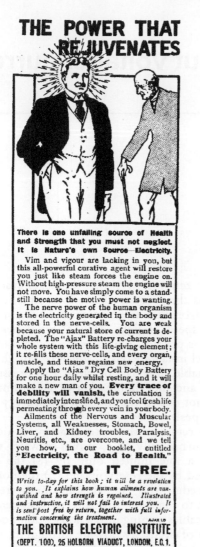

Figure 1.14

In these, very large secondaries were employed to produce high voltages, taken out to a pair of brass tubular handgrips. It was considered frightfully funny to induce some simple soul into holding these one in each hand and then to turn on the current, whereupon the said simple soul received a powerful electric shock. This caused immense hilarity amongst the spectators but not, we suspect, to the hapless victim. Another Victorian manifestation was the 'Electro-medical Apparatus' the purchasers of which were beguiled into believing that electrical discharges were of immense value to health and strength. In these devices the high voltage was applied to various odd-shaped glass electrodes whence it was discharged in a sort of blue haze known technically as a *corona*. The selected electrode was then rubbed upon the part of the body that the owner considered to be in need of rejuvenation. Since one of the claimed benefits was the restoration of virility, heaven only knows what some men suffered in the cause.

A more down-to-earth use of the shocking coil was to provide ignition for the engines of Ford Model 'T' motor cars. As far as vintage radio is concerned adapted versions were used to provide high tension voltage for car radio receivers from the 6 V or 12 V vehicle storage battery, and for enabling domestic receivers designed for use on AC mains to be run on DC supplies. We shall examine these devices in due course.

To sum up

In this chapter we have seen the difference between alternating current (AC) and direct current (DC) and how each may be produced commercially. We have also looked at the various voltages supplied by batteries and electric power stations and the way in which AC may be increased or decreased in voltage as required by the use of transformers. Finally, we have seen how DC too may have its voltage altered by devices based on the action of the common electric bell.

In the next chapter we shall examine the relationship between voltage, current and power, plus a new factor, resistance.

and the process is repeated, for as long as the battery remains connected.

Now, it does not take much imagination to realise that the rapidly rising and falling magnetic field produced by the electromagnet winding could be employed to induce voltage into a secondary winding. In Victorian days this facility was expressed in what were called 'shocking coils'.

Chapter 2

What you need to know about voltage, current, resistance and Ohm's law

The time-honoured method of explaining voltage, current and resistance is to compare electricity with water in a hosepipe. The voltage corresponds with the pressure of water (in fact old text books speak of electricity being delivered at a pressure of so many volts) whilst the current corresponds to the rate at which the water travels through the pipe. The unit of pressure is the *volt* (abbreviated to V) and that of current is the *ampere* (abbreviated to *amp* or A). However, in electrical ratings and formulae current is referred to by the letter I, so you will find, for instance, $I = 10 A$, etc. At one time the letter E was used for voltage and although now rare, you may well come across it in servicing data for vintage radio sets, finding ratings such as $E = 10 V$. There is another term to be remembered: electricity consists of large numbers of infinitessimally small particles called electrons and when an electrical current flows it is regarded as a flow of electrons which is given the name *electromotive force*, or EMF; you can think of it as a quantity of electricity trying to get somewhere to do something.

Let's think a bit more about this hosepipe business. Water comes into the house from a reservoir through just one main pipe, and to start it flowing through the hose all one has to do is to turn on a tap – there does not have to be a direct return path to the reservoir. Electricity, however, always is supplied by the equivalent of two 'pipes', the + and – terminals of a battery or the two main cables from the electrical power station. In either case they may be thought of as flow and return paths, and for electricity to travel there must be a complete circuit of some kind between the two. There is in fact some kind of return path for water

to the reservoir via the drainage system, but whereas the process takes a long time, electricity travels from flow to return via the necessary circuit at the speed of light, 186 000 miles per second. Thus for practical purposes electricity can be said to travel instantaneously, which is why there is no appreciable time lag between the pressing of a switch and the illumination of an ordinary light bulb (not one of the modern low energy fluorescent types!).

Let's return to the water analogy. If a domestic tap is turned fully on a current of water will flow out at a rate limited only by (a) the pressure delivered by the reservoir and (b) the bore of the pipe feeding the tap. If someone were foolish enough to connect a wire directly between the two mains cables coming into a house (don't do it!) the current of electricity flowing through the said wire would be limited only by the pressure, i.e. the voltage delivered by the power station, the thickness of the supply cables and the thickness of the wire joining them together. Two things to be learned from this are that (1) the thicker the wire, the easier it is for electricity to pass through it and (2) the higher the pressure or voltage applied to a given path the greater the current that will flow through it – in other words the less its resistance to electricity. There is always a direct relationship between voltage, current and resistance which we shall look at very soon. Of course, just as we do not want to have a water tap turned full on all the time, neither do we want to have the maximum amount of electricity flowing. With water it is possible with a tap to restrict the flow to a trickle if need be, but with an ordinary electric switch it's a matter of all or nothing. The switch will either stop the current flow altogether or allow it to pass unimpeded. Thus the only restriction in the amount of current that will flow is that presented by the circuit being fed by the switch. We have just mentioned that the thicker the wire, the less its resistance to electricity. By deliberately restricting the thickness of a wire we can increase its resistance to the point where any desired amount of current – and no more – will flow through it.

The unit of electrical resistance is the ohm. usually abbreviated to the upper case Greek letter omega (Ω), and its relationship to voltage and current is indicated by the fact that one ohm connected across a one volt supply will pass a current of one ampere. If the resistance is increased to $10\,\Omega$ the current falls to one-tenth of an ampere or 0.1 A. Increase the resistance to $100\,\Omega$ and the current falls to one-hundredth of an ampere (0.01 A, although by this time the smaller unit of the *milliamp* (one-thousandth of an ampere) would be used and the current stated as 10 mA). The higher the resistance the smaller the current flow at a given voltage.

Suppose the voltage should be increased to 100 V. The $100\,\Omega$ resistance will now pass 1 A, exactly the same as the $1\,\Omega$ resistor at 1 V. Increase the resistance to $1000\,\Omega$ and the current will fall back to 0.1 A; increase it further to $10\,000\,\Omega$ and the current will fall to 10 mA. This relationship between voltage, current and resistance is set down in Ohm's law, which it is the one mathematical formula that will be of constant use to the repairer of vintage radio equipment. Fortunately you don't even have to learn it by heart; all you need to do is to get a piece of paper or card and draw a horizontal line on it. Above the line draw the letter V (for voltage) and below it draw the letters I (for current in amps) and R (for resistance in ohms). You now have a powerful tool with which, as long

as you know any two of the items, you can obtain the third by simply placing a finger or thumb over the letter representing the unknown one and looking at the relative positions of the other two. In repair work one of the commonest needs is to find the value of resistance required to drop a certain voltage at a certain current. For instance, you might need to know what resistance to use to drop 25 V at 20 mA. Knowing these two, cover up the letter R which represents what you need to know, leaving V above the line and I below it. This means that you need to divide the voltage by the current, but hold on a moment and remember that the latter has to be quoted in amps, so alter 20 mA to its equivalent of 0.02 A. 25 divided by 0.02 equals 1250, which is the resistance in ohms required.

As another example you might well need to know how much current would pass through a $5000\,\Omega$ resistance connected across a 200 V supply. Cover up 1 for current and you have V over R, so divide 200 by 5000 to obtain 0.04 or 40 mA.

Again: how much voltage will be dropped by a $4000\,\Omega$ resistance through which 20 mA is passing? Cover up the unknown V and you have I and R side by side, which means that they must be multiplied together to give 0.02 times 4000, equals 80 V.

Unlikely as it may seem to the beginner, you will find yourself using this one handy little tool over and over again when you start to do repair work.

There is an interesting side effect to passing electricity through a resistance which is sometimes useful, but sometimes a wasteful nuisance: the passage of the current causes the resistance to heat up. Generally speaking, the thinner the wire and the more the current passing through it, the greater the heating effect. There is, of course, a limit to the amount of current that may be passed through a given thickness or gauge of wire beyond which it would over-heat and burn out. This effect is exploited in *fuses*, which protect circuits by burning out if the safe current is exceeded.

Let's pause here to consider the materials used for making resistances, or resistors as they are usually referred to in radio servicing data. For all normal electrical wiring purposes the wire used needs to be of what is called low specific resistance, or in other words, will present the least difficult path for the electricity to travel. Silver is the best from this point of view, but since it is too expensive to use the next best, copper, is employed. For reference purposes copper is given a specific resistance figure of 1, to which all other materials

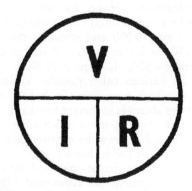

Figure 2.1

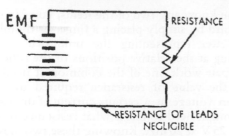

Figure 2.2 Leads from a source of EMF to any electrical circuit need to be thick enought to present negligible resistance to the current flow. These are shown as straight lines. Resistors being designed to impede the current flow are shown as a series of zig-zags

are compared as multiples. Because of its low resistivity copper is not practical for resistors since enormous lengths would have to be used to make up quite small ohmic values. On the other hand, one of the most popular alloys used for winding resistances, a mixture of nickel and chrome called 'nichrome' has, according to the proportions of each metal used, a specific resistance of up to 60 times that of copper. Its melting point is up to 1400°C as against just over 1000°C for copper, so it will stand up much better to heat. One of the most familiar types of resistor made with nichrome, and one which illustrates perfectly the useful aspect of the heating effect of current flowing through a resistor, is the ordinary electric fire element. This is in essence a wire-wound resistor of about 50 ohms which, when connected across a 230 V DC source will pass about 4.6 A. In doing so it will heat up rapidly to the familiar bright red glow and will throw off heat into the room. This brings us to the question of how much power is being used in this or any electrical circuit, which is expressed in *watts* (W) and which may be calculated by an extension of Ohm's law.

The basic calculation for wattage is to multiply the voltage applied by the current passed, so in the case of the fire element just mentioned it would be 230 V × 4.5 A = 1035 W (in practice the element would be rated at 1000 W or 1 *kilowatt* (kW) because it is a tidier figure!). Another familiar application of useful heat produced by a resistor in household use is the ordinary electric lamp (again not one of the low energy types) in which the filament runs at white heat and produces bright light. Take a lamp rated at, say, 60 W at 230 V; we can discover the resistance of its filament by using Ohm's law. Divide 60 by 230 to obtain the current

passed through the lamp, which is 0.26 A. Then divide 230 by 0.26 to obtain the resistance, 846.6 Ω (which in practice would be rounded off to 845 Ω). Although you won't often be calculating details of electric fires and lamps you will find yourself using the same principles over and over again in valve radio repair work.

Now let's consider when the heating effect can be an embarrassment rather than a bonus. It often happens in a vintage radio set that we need to reduce a voltage by means of a resistor, for which we'll take as a practical example a real-life situation that occurred in the Second World War. British radio manufacturers had turned over to producing military equipment for the duration, leading to a chronic shortage of domestic receivers that was met by importing into this country about 100 000 sets from the USA. Having been made for the US mains voltage of between 110 V and 130 V, they had to be adapted for UK mains of between 200 V and 250 V. Typically, this meant dropping 100 V at a current of 0.2 A. Using Ohm's law, the required resistance was calculated by dividing 100 (the voltage to be dropped) by 0.2, the current, giving 500 Ω. The power dissipated in this resistor was 100 V × 0.2 A, equals 20 W. Because this amount of wattage produces a considerable amount of heat resistors used for this kind of work were wound on a heat-proof ceramic former and mounted in a position where there was good ventilation. Although this happened in the mid-1940s, anyone engaged in repairing vintage radio receivers is likely to encounter just the same sort of problem today, and not only in imported sets. We shall be coming back to this in more detail later.

We now come to a couple of handy short cuts. Wattage may also be calculated either by squaring the voltage and dividing the result by the resistance, or squaring the current and multiplying the result by the resistance, both easy tasks with a pocket calculator. Hence, for the resistor just considered we could obtain the wattage by squaring 100 to get 10 000 and dividing this by 500, equals 20 W, or by squaring 0.2 to get 0.04 and multiplying by 500, also equals 20 W.

There are various other possible variations on Ohm's law which may come in handy now and again, but the above is sufficient for the time being. The chief thing to remember is that Ohm's law is a tool which should always be employed in the way best suited for the job in hand.

As far as most calculations used in repairing value radio equipment are concerned Ohm's law may be treated as applying equally to AC or DC voltages. This is not always so but we need not bother ourselves with pure AC calculations at present. What we do need to consider, however, is that AC voltage is always quoted in what is called a *root mean square* (RMS) value. We saw in the last chapter how the voltage derived from an alternator swung up and down in cycles. If the peak voltage reached on each excursion is 300 V, the average or RMS voltage is approximately 70% of this or 210 V, which would be quoted as the output of the machine. Reversing this, to obtain the peak voltage reached by any RMS value the latter is multiplied by 1.414; for example, the peak value of the UK mains standard of 230 V is just over 325 V. As far as wattage calculations are concerned there is an RMS value for current as well, so looking again at the case of a 50 W electric fire element, although the peak current would appear to be 325 V divided by 50 W, equals 6.5 A, the RMS value is 70% of this at 4.55 A, close to the figure arrived at for DC voltage. In fact, almost the only area in which peak as against RMS voltage plays any part in vintage radio work is where receiver power supplies are derived from the mains, and we shall look at this when the time comes.

In the next chapter we shall examine the various practical types of resistor to be found in vintage radio receivers.

Chapter 3

What you need to know about real life resistors

In the last chapter we quoted the case of a voltage dropping resistor which dissipated 20 W. Although resistors of this and even higher wattages do appear in many vintage radio sets, most that are employed in them need to be rated at no more than 1 W, and often as little as one-eighth of a watt. Resistance values, however, span a hung range from as little as 10 ohms up to as high as 10 million ohms (10 *megohms* or MΩ). It is impracticable to make wire-wound resistors of much over 50 000 Ω since the length of wire required would be enormous, the cost prohibitive and the size huge. Fortunately another resistive material is available with which physically small resistors may be mass produced at very low cost. This is carbon, which has a specific resistivity of more than 2000 times that of copper. It is usually sprayed on to small formers made of ceramic or other insulating materials, in thicknesses and lengths according to the value of resistance required.

Typically, resistors used in the vintage years were from about one-eighth to one-quarter of an inch thick and between about half an inch and an inch and a half long, according to wattage. The two main types were those with the carbon sprayed on the outside of a quarter-inch thick former and covered with paint, and those with the carbon on a thin rod protected with a layer of hard material. Connections to the first type were by copper wires wound around each end of the former and soldered to it, whilst in the second the wires entered the protective case axially at each end. As for the paint covering on the first type, it was not there just to make the resistor look pretty but had a definite purpose. Whilst in the early days resistors had their values printed on them in ordinary lettering, mass production demanded

something easier to apply and so colour coding was adopted. With this three different colours could indicate any desired value, as shown in the diagram.

A total of ten colours is used, the values for each being as follows: Black = 0, Brown = 1, Red = 2, Orange = 3, Yellow = 4, Green = 5, Blue = 6, Violet = 7, Grey = 8 and White = 9. The body colour gives the first figure of the resistance value, the end the second figure and the spot the number of zeros following (no apparent spot means no zeros). In Figure 3.1, the body colour red gives the first

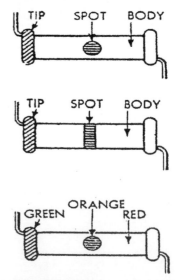

Figure 3.1 Top and centre: two ways of applying the colour coding to resistors of the externally sprayed, twisted wire end type. Bottom: an actual example of the coding for a 25 000 resistor (for explanation, see text)

figure 2, the tip colour green gives the second figure 5 and the orange spot gives three zeros or 000, thus the value is 25 000Ω.

Later a fourth colour was added to indicate another aspect of resistor design, tolerance. Tolerance refers to the margin of error either permissible in circuits using resistance values or to be found in actual resistors and is expressed as a percentage figure. For instance, a 10 kΩ resistor with a tolerance of 10% may in practice have a value of anything between 9 kΩ and 11 kΩ. Some design features in a radio receiver may call for a

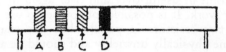

Figure 3.2 An alternative method of colour coding resistors of the enclosed type. A corresponds with the body colour. B with the tip and C with the spot. D is the fourth colour applied to indicate the tolerance

closer tolerance of perhaps as little as 2%, whilst others may require no more that 20%. Wire-wound resistors are usually of fairly close tolerance whilst carbon types may be made, by different techniques, from the 20% just mentioned down to 1%. For most vintage radio purposes 20%, 10% and 5% types sufficed. To indicate the last a gold spot or band was applied to the resistor, whilst on the 10% types it was silver. Twenty per cent types had no fourth colour at all.

Standard values

In theory virtually any value of resistance may be made and for many years radio receiver designers could expect to find 'off the shelf' virtually the exact values called for by their calculations. This called for the stocking of thousand of different values and was plainly uneconomic, so towards the end of the 1930s designers started to use 'rounded-off' values with a tolerance range that included the required value. For instance, had calculations indicated the need for a resistance of 41,300 Ω, unless this was critical it would have been rounded-off to 40,000; even with a 5% tolerance the latter would have had a range of between 38,000 Ω and 42,000 Ω.

Even after the adoption of 'rounded-off' values, until the Second World War over 800 were in general production. The need to streamline production in wartime made this unsupportable and the

close official control on industry enabled drastic pruning to be effected. In 1943 the Inter-Service Component Manufacturers' Council, which coordinated component production throughout the radio industry on behalf of the Government, decreed that henceforth a simplified system of 'standard values' was to be adopted. With this only 255 values were needed to cover the range 10 Ω to 10 MΩ, but there were further restrictions within this band. It was made up of types with 20%, 10% and 5% tolerances with a recommendation that the first should be used wherever possible and the second only when essential. As for the 5% types, a designer would have to obtain special dispensation from the council if he wished to use them, which presumably meant filling in endless numbers of those forms dear to the official mind. As the 20% types numbered only 37 in total and the 10% types 73, for practical purposes the total of values in quantity production had fallen to 110.

Even after wartime controls were swept away, in general the standard values have remained unchanged. This means that no difficulty will be experienced in finding replacements for receivers manufactured after 1943 but for pre-war sets a certain amount of substitution may have to be done. This normally presents no great problem and the subject will be covered later in the book.

Resistors in series and parallel

A series arrangement is when two or more resistors are connected end to end in a row, with the overall resistance measured between the two outer ends. The overall value of resistors in series is R1 + R2 and so on, which is easy enough to grasp. If, for instance, one needed a 20 kΩ resistor but none was to hand, it could be made up by placing two 10 kΩ types in series.

So far, so good. A parallel arrangement is when two or more resistors are connected directly across each other. Most reference books state boldly that the overall value is found by the formula 1 over R1 + 1 over R2 and so on. This is easy enough to work out with a pocket calculator having a reciprocal key. All you have to do is to enter the value of the first resistor (say 330 Ω) and press the 1/x key to obtain the rounded-off answer of 0.003. Enter this in the memory, then clear the screen. Enter the value of the second resistor (say 470 Ω) and press the 1/x key to obtain the rounded-off answer of 0.002. Press the memory + key. Clear the screen and press the memory recall key to obtain the total

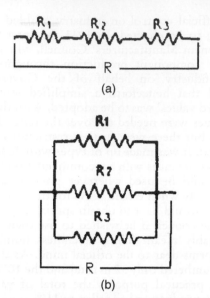

Figure 3.3 (a) Resistors in series. (b) Resistors in parallel

of the two previous figures, rounded to 0.005. Press the 1/x key again and the answer 194.9 appears as the combined value of the two resistors.

(Calculators without a reciprocal key may be used by, for each resistor, dividing 1 by its value and adding the results in the memory, then dividing 1 again by the recalled memory. This takes less time to do than it does to describe.)

It is worth bearing in mind that when two resistors of exactly the same value are wired in parallel the overall value will be exactly half of either. When two or more resistors of dissimilar values are wired in series the overall value will always be 'lower than the lowest'.

In practice, however, the repairer of radio sets is seldom going to be presented with a couple of resistors in parallel and asked to work out their value. What is much more likely is that he will need a certain value which is not to hand and want to know what two resistors might be placed in parallel to obtain it, which is something reference books *never* mention, suggesting that writers of reference books never soil their hands with actual radio repairs. Suppose a certain job called for a 320 Ω. You don't have this but you do have a range of standard value resistors between 390 Ω and 1000 Ω. We have already said that the total value of two or more resistors in parallel will always be lower than the lowest so don't start with the 390 Ω

but with the 1000 Ω. What must be placed in parallel with this to obtain 320 Ω? To find out, first enter 320 in the calculator, press the 1/x key and place the answer in the memory. Clear the screen, enter 1000, press 1/x and use the M-key to subtract the answer from the figure already in the memory. Clear the screen again, recall the memory and press the 1/x key to obtain the answer 470. Thus, wiring the standard values of 470 Ω and 1000 Ω in parallel will give the required 320 Ω. As before, writing this down takes longer than actually doing the calculation.

You will find that this simple use of arithmetic will come in useful over and over again in actual repair work. It is possible, of course, to use more than two resistors in parallel but this starts to become physically unwieldy and should be avoided if at all possible.

Series/parallel resistors and wattage

Another useful aspect of placing resistors in series or parallel is that two low wattage types may be used in place of a single high wattage component. For instance, if a 500 Ω, 20 W resistor should be needed, it could be replaced by two 250 Ω, 5 W types in series or two 1000 Ω, 5 W types in parallel. In this sort of application it is best to use two resistors of the same value since the wattage is then distributed evenly between them. If dissimilar values have to be used care must be exercised to ensure than one or more of them is not over-run. The wattage dissipated in each may be found by the use of Ohm's law as detailed earlier. This is another subject we shall examine again when discussing practical repair work.

What do resistors do in radio sets?

We have already talked about how resistors are used to drop voltages, and although it is not the only job they do it is a very important one. In any radio receiver the valves have to be supplied with different voltages which in a mains set may vary between about 250 V and 10 V, to cater for which the designer starts with an HT line of 250 V and drops it down as required with resistors calculated according to Ohm's law. These resistors do another job as well. To prevent instability it is necessary to keep the various stages of a receiver electrically separated or 'decoupled' from each other, and the resistors help to do this. Decoupling is aided by the use of condensers, of which more shortly.

In some cases resistors may be employed solely for decoupling purposes without any requirement of voltage dropping. They may also be used when it is necessary to provide a return path from certain parts of a valve to the negative side of the supply voltage. In these two instances the values of the resistors are likely to be high, typically from about 250 000 Ω upwards.

Variable resistors

Almost everyone at some time must have used the volume control of a radio set. When you turn the knob you operate the most common type of variable resistor in which the resistive material is in the form of a circular *track*, upon which a small movable contact of metal or carbon (the *wiper*) is held by light spring pressure. This is attached to a metal or hard plastic spindle onto which the volume control knob is screwed. When turned, the knob simply sets the wiper to any point between the two ends of the track, which may be made of resistance wire or of carbon deposited on a thin insulating material. Just as with fixed resistors, wire is used when the wattage is expected to be high, carbon when it will be low. Wire-wound variable resistors are seldom made above about 50 kΩ, whilst carbon types may go up as high as 2 MΩ. This type of variable resistor is known technically as a *potentiometer*, which comes from the days when they were laboratory instruments used in connection with measuring voltages. They have other uses in vintage radio sets other than controlling the volume, one of the most common being for controlling the tone produced by the loudspeaker.

A slightly different type of variable resistor is the *rheostat*, which has only two connections, the wiper and one end of the track. These are used when a resistance value needs to be set fairly accurately to suit a particular job but which then is seldom altered. The tracks of most wire-wound variable resistors have what are called linear characteristics, that is from one end to the other the resistance rises steadily, ohm by ohm. Carbon tracks can be made on the same principle but for certain jobs other types are made, one being the logarithmic or 'log.law' version. In this the resistance rises at first slowly and then by increasingly larger amounts so that the greater amount of the resistance is found within the last few degrees of the wiper's travel. The opposite of this effect is found in the 'anti-log' type. When you are changing a control in a vintage radio set you don't particularly need to know why the designer specified one or the other type but you do need to use the same as a replacement or otherwise the action of the volume or tone control may be unacceptably 'jerky'.

When volume and tone controls can be operated by the owner of a set via control knobs they are often referred to as 'continuously variable'. There is another type of control known as a 'pre-set' which may be operated either by a very small knob or by a screwdriver slot, and which is intended to be adjusted at the factory or by an engineer to provide some kind of fine control over certain adjustments. We shall meet some of these in due course.

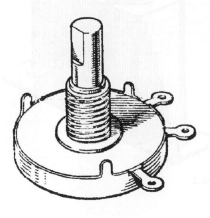

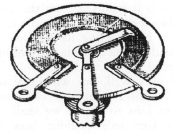

Figure 3.4

Chapter 4

What you need to know about condensers

Condensers of various types are the other main components to be found in vintage radio sets. In essence a condenser consists of two metal plates in close proximity but separated by an insulator (usually called the *dielectric*) which might be air, paper, mica or some other material. If a battery is connected across the plates of a condenser the EMF of the former tries its best to move electrons around but is frustrated by the insulation between the plates. After a small initial movement lasting only a fraction of a second the electrons build up on the plates; at one time this was thought to emulate the action of water vapour condensing on a cool surface, hence the name *condenser*. According to the capacity of the condenser, after a certain time it will accept no more electrons and is said to be fully charged. At this stage what is called a *back EMF* has been built up in the condenser which is equal and opposite to the voltage of the battery. If the latter is now disconnected the electrons remain *in situ* in the condenser, and if its internal insulation is perfect this will continue indefinitely. This may be demonstrated by a practical experiment in which a fair-sized condenser is charged up

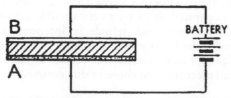

Figure 4.1 Charging a condenser: The battery EME causes electrons to leave plate B and collect on plate A, building up a charge until the latter has a back EME equal to the voltage of the battery

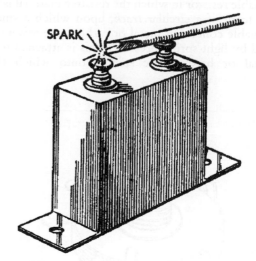

Figure 4.2

from a battery or other DC source giving, say, 100 V. When this is removed from the condenser the EMF within it may be discharged with quite a startling effect by shorting its terminal with a screwdriver blade, when a visible and audible spark will be produced.

This is all very well, but what real use is a condenser in a radio set? In fact, condensers have many different jobs to do, as we shall see in a moment, but first let's examine how real-life condensers are made and how their capacity is rated.

To do this we have to look briefly at another electrical unit called the *coulomb*. This is the amount of electricity that flows in a circuit when a current of 1 A passes for one second. The unit of capacity is defined as the size required to hold one coulomb when charged from a 1 V source. This unit was named the *farad* in honour of the great British scientist Michael Faraday, but whether anyone has ever made a condenser of one farad capacity is open to doubt, since it would have to be of enormous physical size. In practice, therefore, the unit is taken as the microfarad or one-millionth of a farad, for which the Greek alphabet once again provides a handy abbreviation, μ in this application meaning one millionth (we shall meet μ again in various other roles and it is important to be able to identify in which sense it is being used). Farad is usually reduced to fd, so the common form of microfarad is μfd. Even this is too large for many jobs in radio and another unit, a million-millionth of a farad, is used. In the early days of radio this was often written as mmfd, but around 1940 a new term started to come into use, the picofarad or pfd. This is synonymous with the mmfd. Some ten years later another new unit appeared called the nanofarad (nf) which is one ten-thousandth of a microfarad.

In a domestic radio receiver you may expect to find condensers from about 5 pfd up to as much as 100 μfd. In most cases you will find that the capacity or value of a condenser in such sets will either be printed directly onto it or will be found on a label glued around it. Remember that the annotation will differ according to the age of the set. For instance, a condenser made around 1930 may bear the value 0.0002 mfd. In the later 1930s it might be labelled 200 mmfd and in the 1940s 200 pfd. Again, a condenser of 0.001 mfd in 1930 might be 1000 mmfd in the later 1930s, 1000 pfd in the 1940s and 1 nfd in the 1950s. The Philips company was particularly addicted to using multiples of picofarads, even for large value condensers such as 0.1 mfd. The typical Philips condenser of the later 1930s, 1940s and early 1950s was a pitch covered tube with the value expressed in thousands of picofarads, but omitting any reference to the latter, so 0.1 μfd would have been expressed simply as 100 k.

How are condensers made? For many years the most common materials used were silver foil for the plates and paper impregnated with paraffin wax for the dielectric, the latter being sandwiched between two layers of the former. To save space the sandwich would be rolled into a tight cylindrical or curved-end rectangular shape and the whole then sealed in a moulded plastic or metal case. The quality of the paper used had a direct bearing on the longevity of such condensers; whilst cheap types might have

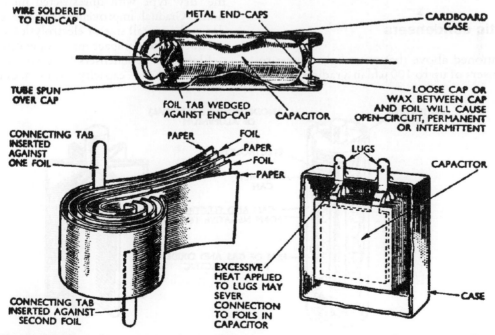

Figure 4.3 The construction of two common forms of condenser used in vintage radio receivers: Top, a cardboard-cased tubular condenser. Bottom left, how layers of foil and paper are interleaved to form the plates of a condenser. Bottom right, a 'Mansbridge' type vertically mounted metal can condenser

lasted only a few years before starting to leak, that is having their insulation break down, it is quite common to find high quality paper condensers still working well after 70 years' use. The range of values covered by paper condensers was from about 0.001 µfd up to about 16 µfd, the larger values usually being of the 'Mansbridge' type depicted at the bottom right of the illustration above.

A better material for insulation is mica, which has a dielectric strength, or resistance to breakdown due to high voltage, of up to eight times that of waxed paper. It is also up to three times as efficient as a dielectric which means that a condenser using it needs less plate area and can thus be made much smaller physically. Mica condensers were made from very low values of perhaps 5 pfd up to around 0.01 µfd. It was chosen for values up to about 1000 pfd mainly because of the size advantage and the fact that the accuracy of the values (the tolerance) was good. Higher values of mica condensers were used more for their excellent insulative properties.

A development of the mica condenser was the replacement of silver foil plates by the spraying of silver onto the dielectric. Silver-mica condensers could be made to fine limits of capacity and tolerance and were much used in the tuning stages of receivers where accuracy is important.

Electrolytic condensers

It was mentioned above that you might expect to find condensers of up to 100 µfd in a radio receiver, but the upper limit for paper condensers was given as about 16 µfd. A different way of making high value condensers was introduced in the late 1920s and soon became popular. The electrolytic condenser uses, in place of a conventional dielectric, a film of oxide produced by passing current through a chemical solution. In essence one plate is the aluminium container in which the electrolyte is held, and the other is a rod of the same metal suspended within it. The case forms a cathode and the rod an anode, and current is able to pass from one to the other when a voltage is applied. The cathode must be connected to negative and the anode to positive.

When the voltage is first applied a considerable current will flow from the cathode through the electrolyte to the anode, but in a short space of time an oxide film is built up on the surface of the anode which acts as an insulator or dielectric. Since this is merely one or two atoms thick, the capacity between the electrolyte itself and the anode is high. By increasing the effective area of the anode, by etching or roughening its surface, or by making it in the form of a spiral, very high capacities may be achieved in condensers of small overall dimensions.

Early electrolytic condensers used a liquid for the electrolyte and thus had to be mounted upright on the chassis, but these were soon superseded by the 'dry' type with unrestricted mounting positions. Gradual improvements in design brought down the overall size of electrolytics, assisting the production of compact mains operated receivers.

Electrolytic condensers for domestic radio purposes range in capacity from about 1 µfd to

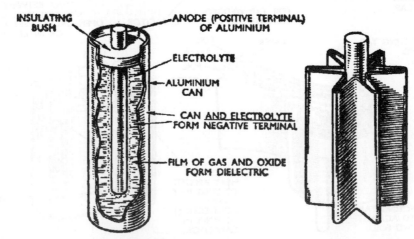

Figure 4.4 Left: the construction of an electrolytic condenser in stylised form GHT: one kind of shape used for the anode to increases its effective area and thus the capacity of the condenser. Many other shapes are used. Early electrolytics used a free chemical solution for the electrolyte but all modern types use a paste

150 µfd, with a wide variety of operating voltages from about 10 V to 500 V. Those used in conjunction with the high voltage supplies used in valve receivers are often marked with two voltages of which one is termed 'working' and the other either 'surge' or 'peak'. This takes account of the fact that when a mains powered receiver is first switched on and until all the valves have 'warmed up' the high voltage may be considerably above its normal value. A typical rating might be '250 V working, 350 V surge'. This is another point to which we shall return in due course.

Leakage current and re-forming

When any electrolytic condenser is in use a small current continues to pass through it to maintain the oxide film. Whilst this is of no account in smoothing and decoupling applications, it precludes the use of electrolytics for inter-valve coupling, when the steady current would cause harm.

If an electrolytic condenser lies unused for a fairly long period the oxide film gradually disappears, and before the condenser can be used again it must be 're-formed', i.e. have the film restored. This is achieved by passing a voltage through it that is very considerably less than the normal working voltage, in conjunction with some kind of current-limiting device and a meter to indicate the amount of current flowing. As the film builds up the current flow falls; the voltage is increased until the current falls again, until the point is reached where just a very small polarising current passes at the normal working voltage.

Time spent on re-forming electrolytics is well spent since it may save the trouble and expense of finding new replacements. Further information on electrolytic condensers will be given later in the chapters on receiver servicing.

Condensers and AC voltages

When AC is applied to a condenser the EMF travels first in one direction then the other, having the effect of repeatedly charging and discharging the plates. This means that current flows in and out of the condenser, although it does not actually pass through the dielectric. If this sounds a little obscure, what it really amounts to is that whilst a condenser in good condition presents a complete block to DC it will allow a limited amount of AC current to pass through it, just how much depending on its capacity and the frequency of the applied AC. The amount of opposition presented by a condenser to AC is stated in ohms, but because only voltage flows and not current no power is developed as in resistors working with DC, so the term resistance is replaced by *reactance*. A condenser cannot be labelled as having any particular reactance because, as stated, this depends on the frequency. Only occasionally will anyone engaged in repairing vintage radio sets need to know reactance figures, because these lie in the set designer's province. There is a formula by which they may be calculated but it will be much easier to look up the sort of tables in reference books that show them related to capacity and frequency. All you need to remember for now is that reactance decreases as capacity or frequency is increased.

Variable condensers

For certain purposes in radio receivers, particularly in connection with tuning (see the next chapter), it is necessary to have *variable condensers*, the capacity of which may be altered at will between certain limits. The most popular type has remained essentially the same right through the valve radio era up to the present day. It consists of two sets of plates, usually of aluminium, occasionally of brass, one of which is held firmly in position on a framework with the other mounted on a movable spindle fitted on the same framework. Individual plates in each set are spaced a short distance apart, making it possible

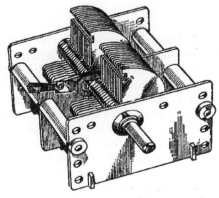

Figure 4.5 An entirely typical variable condenser as found in vintage radio receivers

for the set mounted on the spindle to slide within the fixed set. When the two sets are fully enmeshed the capacity of the condenser is at its highest, and when the two sets are fully apart it is at its lowest. Generally speaking variable condensers used for tuning in stations have a maximum capacity of 500 pfd or 0.0005 μfd (although some may be rather less) and a minimum capacity of a few pfd, for it can never fall quite to zero even with the condenser wide open. In order to achieve accurate settings, in practice the variable tuning condensers in receivers are usually provided with some sort of reduction gearing giving slow-motion drive. Only in the cheapest or most compact sets will you find the tuning knob attached directly to the spindle of the condenser.

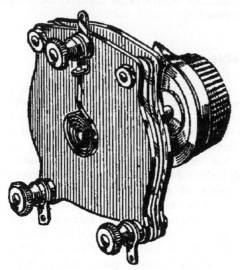

Figure 4.6 A typical solid-dielectric or 'reaction' condenser

The type of variable condenser just described is referred to as air-spaced or air-dielectric. Another type, called solid dielectric, used thin wafers of insulating material between the plates, permitting them to have much closer spacing and also a greater capacity for a given number of plates and overall size. This type was occasionally used for tuning in compact receivers but was more usually employed to control *reaction*, an effect we shall meet a little later. For this reason all solid dielectric variable types tend to be called 'reaction condensers'. They usually have a maximum capacity of about 300 pfd (0.0003 μfd).

Pre-set condensers

For certain fine-adjustment purposes physically small variable condensers of limited range called trimmers are employed. The usual method of construction is to have two sets of plates made of some springy metal, interleaved with wafers of mica, fitted onto a small ceramic base. Typically this would be about the size of a normal postage stamp, from which the once-common name postage-stamp trimmer was derived. The plates and the mica wafers all have a centre hole through which a small bolt may pass. This may screw into a threaded brass bush set into the ceramic base or it may be fixed into the latter, face outwards and have

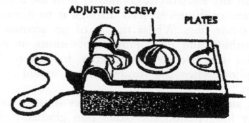

Figure 4.7 'Postage-stamp' trimmer

a nut fitted on the outer end. In either case adjustment consists of tightening the screw or nut to press the plates together to increase the capacity, or loosening them off to reduce it. It was common practice for the adjusting screw or nut to be sealed in position with quick-setting paint once the correct capacity had been obtained. This not only guarded against alteration in capacity due to vibration, etc., it also warned service engineers that someone had been twiddling the trimmer, perhaps without the necessary knowledge!

Yes, but what do condensers *do*?

A good question, and the answer to which is, lots of things. First of all, let's look again at what we have just discussed, the ability of a condenser to block DC but to pass AC. In any and every receiver the need arises for just this property, as in coupling and decoupling. The most common form of coupling is when the output of one valve in a receiver has to be passed on to another. For reasons which will be explained in another chapter this consists of a mixture of AC which needs to travel as unimpeded as possible and DC which must be blocked to prevent its damaging the next valve. Clearly, a condenser is ideal for this work. We have already mentioned decoupling in Chapter 3 in connection with the use of resistors to drop voltages fed to valves and the need to separate the sections of a receiver to prevent instability. The latter is caused if AC voltages in one part of the set enter another, so the places where they are likely to occur, such as the lower ends of voltage dropping resistors, are connected via condensers to the metal chassis on which the set is built. These effectively short unwanted AC voltages out whilst not affecting the DC voltages. There are other uses of decoupling which will be discussed when we start to look a receiver design in general.

The ability of a condenser to hold a charge is exploited in the power supply sections of mains operated receivers in which it is necessary to change the incoming AC from the supply line to DC suitable for the valves by means of a rectifier (of which again more later). Large electrolytic condensers are used to 'smooth' the DC output from the rectifier to make it as nearly as possible the same as current obtained from batteries.

A condenser may be called upon to do two jobs at once. If we look again at Figure 1.13 a rudimentary symbol for a condenser, marked C and C', will be seen at the bottom, connected across the points P and P'. Without this condenser,

each time the points open the self-inductance of the primary coil would cause it to build up an EMF great enough to cause heavy sparking at the points. This EMF would also be built up when the points close again, slowing down the build-up of current through the primary. Both these effects would reduce the amount of EMF induced into the secondary and its output would be lowered. With the condenser across the points the EMF that would otherwise have caused the sparking is absorbed as a charge, which when complete is discharged back into the primary in the opposite direction. This in turn assists the induction EMF from primary to secondary and results not only in suppressing sparking at the points but also increases the voltage output from the secondary. Note that because the condenser can charge only when the points open, the high voltage at the secondary is obtained when they 'break' and not when they 'make'. Motorists with long experience of the traditional type of coil ignition will be aware, perhaps only too well, that if the condenser fitted across the points in a distributor fails the engine just won't run!

Having touched on the subject of condensers being used in conjunction with the coils we come to the effects which are at the heart of radio transmission and reception, which are so important that they deserve a chapter to themselves.

Chapter 5

What you need to know about tuning

Coils have something in common with condensers in that they too present opposition to the flow of AC, again called reactance, although, unlike condensers, they present little or no resistance to DC. The reactance of a coil also differs from that of a condenser in that it *increases* with frequency. This leads to the interesting and useful fact that if a coil and a condenser are connected in parallel the reactance of each will be equal at only one particular frequency. At that point maximum voltage is developed across the two and the *resonant frequency* of the circuit is attained.

We now have to consider the question of phase. It is a massive simplification of a complex subject to say that it approximates to harmony and that things that are in phase assist each other whilst those out of phase oppose each other, but this is all that need concern anyone wishing merely to learn how to repair vintage radio sets. Once again there are numerous reference books available for anyone who wishes to pursue the subject further.

When an AC voltage *f* is applied to a coil L and condenser C in a parallel circuit the voltages developed across each are exactly in phase and equal, but the currents are out of phase and cancel each other out at the resonant frequency. *At this frequency and at no other, maximum voltage is developed yet no current flows through the circuit.*

The parallel circuit just described is called a *rejector circuit* because it rejects all frequencies save those chosen by the operator. If, however, the coil and condenser are wired in series a different effect is obtained. This time the voltages occurring at resonance will be of opposite phase and thus oppose each other. The result of this is that there is

zero voltage drop across the circuit which then behaves as a virtual short circuit limited only by the resistance of the wire making up the coil. This means that *signals at the resonant frequency will pass straight through the circuit but all others will be blocked*. This type of circuit is called an acceptor because it accepts just the selected frequency and no other.

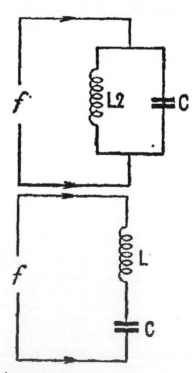

Figure 5.1

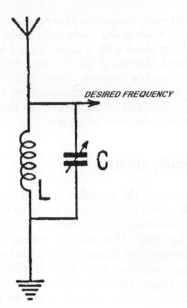

Figure 5.2

If we replace the fixed condenser in the circuit by a variable type the resonant frequency may be varied over a range determined by the maximum and minimum capacity of the condenser plus the inductance value of the coil. For instance, a variable condenser of 500 pfd maximum in parallel with a coil of 200 microhenries (200 µH) will cover the medium wave broadcast band of 200 m to 550 m (1500 kc/s to 545 kc/s), and enable the circuit to resonate at any desired frequency within it.

At this point we need to look at one of those anomalies that can so easily confuse anyone learning about vintage radio. When you tune a radio set it is natural to assume that you are selecting a particular station and rejecting others, so you might well imagine that the tuning knob operates the variable condenser of an acceptor circuit. Not so; tuning is almost in every case carried out with a rejector circuit, so what you are doing is accepting all the stations you don't want and rejecting the one you do! This needs to be explained.

Imagine the rejector circuit being connected to an aerial and earth system. At any time of the day hundreds of different frequencies put out by hundreds of broadcasting stations will arrive simultaneously. To all those differing from the resonant frequency of the circuit the latter will act as a short circuit and they will be returned to earth, leaving maximum voltage at the resonant frequency available for use at the point marked on the diagram.

A use for the acceptor circuit

Sometimes it is handy to be able to select just one particular frequency and to suppress it rather than make use of it. This happens when a radio set is sited near a very powerful transmitter which threatens to 'swamp' it, that is, to break through onto other stations. A case in point occurred when the BBC opened a high power transmitting station at Droitwich in the mid-1930s; many sets of the day located fairly close to the site were unable completely to tune it out and it appeared as a constant background to other stations. To overcome this new receivers were equipped with a 'Droitwich rejector' and you will probably come across sets with this feature. The 'rejector' in fact usually consisted of an acceptor circuit that could be tuned accurately to the Droitwich frequency and short circuit it out. You will find as we go along that apparently contradictions in terms are not uncommon in vintage radio.

Frame aerials

It is possible to wind the tuning coil of a receiver on a large former which itself can pick up signals from radio stations without the need for an outside aerial or earth. They were, of course, widely used in portable and transportable receivers, very often being wound on a wooden frame just big enough to fit snugly inside the cabinet. Frame aerials have directional properties, maximum signal strength being received when the side of the frame is pointing towards the station. Apart from portability, they are also useful for separating two competing stations on nearby wavelengths.

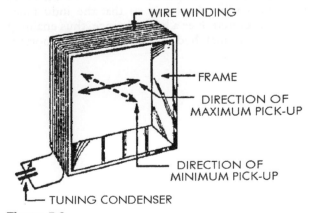

Figure 5.3

Air cores and iron dust cores

So far all the tuning coils we have considered have been 'air-cored', which does not necessarily mean that they are wound on thin air! Some short wave coils are wound with thick wire which needs no former, but these are called 'self-supporting' coils. It is coils wound on some kind of non-metallic former, such as a hollow paxolin tube or a wooden dowel, that are known as air-cored. It is not possible to use a soft-iron core to increase the

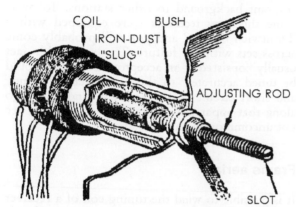

Figure 5.4 A typical arrangement for a coil fitted with an iron-dust core having a threaded brass rod for adjustment purposes

inductance because it would introduce too many losses, but early in the 1930s it was discovered that powdered iron bound with some glue-like medium can be moulded into highly effective cores for coils. Iron-dust cores. as they are called, increase inductance values and enable coils to be smaller in physical size yet to be more efficient than air-cored types. A later development was to make the cores movable within the cores so that the inductance could be altered over a limited range, thus enabling coils to be matched closely, a very important

property in radio receivers. There are two popular methods of making the cores adjustable, the first being to form them with a screw thread which fits into internal threading in the coil former; the second is to have smooth cores into which are moulded thin brass rods, threaded to fit into a fixed nut at one end of the former, so that turning the rod moves the core to or fro.

Permeability tuning

An extension of the movable core principle is to reverse the usual tuned circuit by having a coil of variable inductance in association with a condenser of fixed capacity. This sort of tuning was very popular in receivers featuring push button tuning, employing a selection of coils each with a limited frequency range, e.g. 200–300 m, 300–400 m and so on. It was also occasionally used in the main tuning in compact sets, particularly automobile receivers where a conventional tuning condenser would have required too much space.

Ferrite rods

Appearing towards the end of the valve era was a remarkable non-metallic magnetic material called ferrite which, like iron dust, can be moulded into any convenient shape but which has far greater powers of enhancing inductance. Miniscule coils employing ferrite cores can be made, but the best-known use of the material is the ferrite rod aerial, which took the place of the conventional frame aerial in portable and transportable receivers.

We have now seen how it is possible to tune in desired radio stations. We now need to know the principles on which they work, which in turn will enable us to understand how radio receivers work. First, though, we need to look at the device that made broadcasting possible, the valve.

Chapter 6

What you need to know about valves (1)

The simplest kind of valve, and the one that inspired the name, is the two-electrode type or diode (from the Greek dis, twice + electrode). In essence this consists of a filament, not unlike that of an electric lamp, around which is mounted a metal plate in the form of a cylinder or flat box, the whole being sealed into a glass bulb and evacuated. When the filament is lighted by means of a battery or a source of low voltage AC it emits negatively charged particles of electricity called electrons. If another battery of, say, about 30 V is connected with its negative terminal to one side of the filament and its positive to the metal plate a current will pass between them; however, if the battery

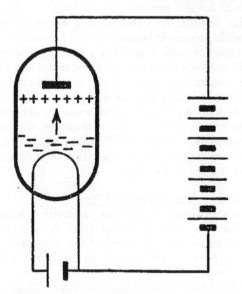

Figure 6.1

connections are reversed no current will flow. This one-way action mimics that of a mechanical one-way valve, hence the name given to the diode, which at one time was the only type available.

The correct name for the two parts of the diode are cathode (from the Greek, a way up) for the filament and anode (also from the Greek, a way down) for the plate. In practice, though, the term cathode is not used for valves with simple filaments, otherwise known as directly heated valves, which require a pure DC to power them. In practice this usually meant accumulators, so early filaments tended to be made suitable for 2 V, 4 V or 6 V for operation either from one, two or three cells. By the 1930s the standard voltage was 2 V and nearly all the battery radio sets likely to be encountered were for use with a single accumulator. This is known as the low tension (LT) battery.

AC is not suitable for filaments because they would dim 100 times per second at the zero point of each cycle. Although this is imperceptible to the eye, the effect would be to impose a 50 c/s hum on the output from the valve. Nor is the type of current that used to be obtained from DC mains altogether suitable since, as we saw earlier, it contains a certain amount of AC ripple which again would impose hum on the output. After a number of false starts the problem finally was solved in the late 1920s when valves specifically designed for operating on low voltage AC transformed down from the house mains were developed. In these valves the filament or heater is contained in a metal tube, from which it is electrically insulated, made of a material which emits electrons when heated and is unaffected by the 50 c/s variation in voltage. The term 'cathode'

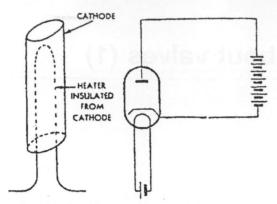

Figure 6.2 Above left: diagrammatic representation of a cathode. Above right: how the cathode takes the place of the filament in a circuit

was brought back into use for this new electrode. The same principle was used later for valves intended for use on DC mains and subsequently for either AC or DC operation. We shall look at mains valves more closely later but for the moment we will confine ourselves to directly heated valves.

The term anode instead of plate also was not generally adopted in Britain for many years, and for a long time valves and valve holders were marked with a letter 'P' to indicate Plate. This term survived right through the valve age in the USA. Therefore, when you read any literature to do with valves you may assume that anode and plate are synonymous.

If an AC voltage is applied to the anode of a diode only the positive half-cycles will be able to pass through it, producing at the filament a positive DC voltage pulsating at a rate equivalent to the incoming frequency. When the latter is high the half-cycles are so closely packed that the effect is almost equivalent to pure DC. Note though that the DC cathode voltage is equivalent to the peak voltage of the AC input and thus is 1.414 times that of its RMS value (or at least in theory; in practice the actual voltage may be somewhat less).

Although the diode is a useful device it lacks the ability to amplify. This problem was solved when the three-electrode valve or triode (from the Greek treis = three) was developed. In this a new electrode consisting of a fine wire mesh called a grid is inserted between the filament and anode. Another battery, this time of between about 50 V and 150 V and known as the high tension (HT) battery, is connected with its negative terminal to one side of the filament and its positive to the anode (AC is unsuitable in both battery and mains triodes). This attracts to the anode electrons

emitted from the filament, which have to pass through the grid to reach it. If a small negative voltage is applied to the latter with respect to the filament the electron flow is impeded. This in turn reduces the anode current flowing through the valve, and increasing the negative voltage on the grid will reduce the current still more until it reaches zero, or what is called the cut-off point. Conversely, making the grid positive with respect to the filament increases the anode current. If an AC voltage is applied to the grid the anode current will swing up and down in sympathy with the positive and negative cycles of the input but – and this is the important part – only a very small change in grid voltage is required to give a relatively large change in anode current. If the voltage to the anode is supplied via a resistor there will be a varying drop across the latter in accord with the variations in anode current, and this voltage change will follow that at the grid but will be a much magnified version; in other words, the triode is able to amplify voltages. It is this property that made possible the introduction of broadcasting and of radio receivers as we know them.

Valve characteristics

Each valve has a set of what are called characteristics which determine how it performs and for what job it is best suited. In the early days, characteristics were likely to be more a matter of luck than judgement and it would have been difficult to find any two valves that were identical in this respect. As valve making became a more scientific process it became possible to evolve a set of standards which enabled characteristics to be predicted with accuracy and to be maintained over long production runs. Note that although the filament or heater voltage and current of a valve, which may be in various combinations, are quoted for reference, they are not included in the list of characteristics. Those which are of importance to anyone engaged in repairing rather than designing radio sets are as follows:

Anode voltage (symbol V_a).
Anode current (I_a) at that particular voltage.
Grid bias voltage ($-V_g$) (this is a standing negative voltage applied to the grid, the purpose of which will be explained a little later).
Mutual conductance (g_m) or 'slope'.
Anode resistance (r_a).

Amplification factor, for which the Greek letter μ was chosen as an abbreviation. It should not be confused with the first two letters of mutual conductance! The other use of μ as a submultiple is regrettable in a way, but the context in which it is used will normally prevent confusion.

Optimum load (R_a – note upper case to distinguish it from anode resistance). This is the value of impedance or resistance to be placed in the anode circuit to obtain the best all-round output.

The mutual conductance of any valve is a direct measure of its sensitivity and is calculated by dividing the change in I_a brought about by a certain change in V_g. It may be displayed as a graph called a characteristic curve in which the change of anode current at a given voltage is plotted with respect to various voltages on the grid. In Figure 6.3 the anode voltage is taken as

6 mA. In practice care is taken to avoid the grid becoming positive since the sustained high anode current that would result would damage the valve.

Look again at the section of the graph which lies between −3 V and 0 V on the grid and between 1 mA and 4 mA I_a, and indicated by a dotted line. This is called the 'straight-line' part of the curve and when a valve is working it gives its best performance when the I_a is restricted to these limits. For this reason a standing grid voltage or bias is applied to maintain the I_a at the mid-point of the curve when the valve is at rest. When we speak of the input voltage to the grid as swinging between negative and positive the true variation is between more negative and less negative.

The rate at which the I_a increases with a reduction of negative voltage on the grid determines the angle of the $I_a V_g$ curve. In a valve of low sensitivity the curve will rise only slowly, whilst in a highly sensitive valve it will climb sharply. For this reason a valve possessing good sensitivity, or high mutual conductance, is sometimes referred to as a 'steep-slope' type.

It is important to note that mutual conductance is in itself *not* a measure of the amplification factor of the valve. This is determined by another set of figures. If the bias voltage on the grid is made less negative by a very small amount the I_a will increase slightly (for this test it is assumed that there is zero resistance between the HT supply and the anode). If the V_a is then reduced gradually a point will be reached where the I_a is the same as before the grid voltage was changed. Dividing the change in V_a

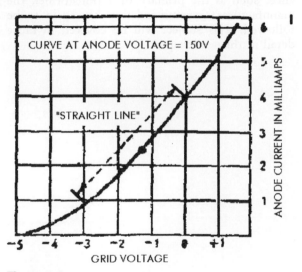

Figure 6.3

150 V and the grid voltage as between −5 V and +2 V. Starting at the bottom left-hand corner with a grid voltage of −5 V, the anode current is zero, or at the cut-off point of the valve and no further increase in negative voltage will make any difference.

However, as the grid voltage is reduced to −4 V a small anode current of about 0.25 mA begins to flow. A further reduction to −3 V increases the anode current to 1 mA. From this point a smooth increase in I_a is recorded until the grid voltage reaches 0 V; as it goes into positive values the I_a rises more quickly and at +2 V reaches more than

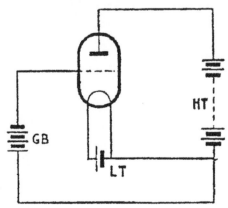

Figure 6.4 Showing how a battery is used to place a standing negative 'bias' on a valve. (GB = grid bias, LT = low tension, HT = high tension, tension itself means the same as voltage)

necessary to achieve this by the change in $-V_g$ gives the amplification factor or μ.

Depending on the type of valve μ may be anything from between about 5 up to 125. Although this gives us an idea of by how much the valve is likely to amplify a signal applied to the grid, it is *not* what will actually be obtained in practice. It was mentioned above that μ is calculated with no resistance between HT+ and anode, but in real life there always is some. When V_g is changed and the I_a alters it will bring about a change in V_a as well, which upsets the calculations. It was thus customary for valve manufacturers to work out a realistic stage gain figure that might be expected under typical operating conditions.

Anode resistance, by which is meant the actual resistance in the valve to passing current, is not easily pinned down. It can be calculated as a DC value of ohms by dividing the V_a by I_a, but this will only hold good for one particular combination of the two. Valves are normally operated with alternating voltages on the grid so a more useful figure is obtained by taking the difference in V_a as against I_a on the straight-line part of the characteristic curve. This figure, written as r_a, is referred to as the 'AC resistance' or impedance of the valve. As a generalisation, high impedance valves take less anode current but are likely to have a high μ, whilst medium and low impedance types take more anode current and probably have a fairly low μ. Note, though, that either may have large or small values of g_m. There is a certain amount of correlation between g_m, μ and r_a, so that if any two are known the third may be calculated. For instance, multiplying g_m by r_a gives μ, whilst dividing μ by r_a gives g_m.

Optimum load. When a valve is operated with an anode load consisting of a resistor the optimum R_a is between two and three times the r_a; there is no easy way to calculate this exactly and it is best to use the value suggested in the valve maker's data book.

If the anode load takes the form of an impedance, such as the primary of a transformer, the manufacturer's recommendation has again to be followed. This subject will be covered in more detail in the next chapter.

Chapter 7

What you need to know about valves (2)

High, medium and low impedance valves each have certain advantages over the others which make them suitable for specific jobs in a radio receiver. Taking them in order, high impedance types are best suited to what are called *resistance-coupled* amplifiers, used to build up weak voltages or signals. The circuit of a typical example, with batteries to supply the filament, the anode voltage and the grid bias, is shown in Figure 7.1. As mentioned in the previous chapter, the optimum value for the load resistor in the anode circuit of the triode is usually calculated to be about 2 or 3 times that of the r_a of the valve and is typically of between about 50 kΩ and 1 MΩ. Also, as mentioned earlier, the voltage from the battery used to supply the anode is dropped across the load resistor according to the current passing through the valve, which again is dependent on the grid voltage. A varying voltage on the grid, such as might be obtained from a microphone, causes the anode current to move up and down, thus altering the amount of voltage drop across the load resistor.

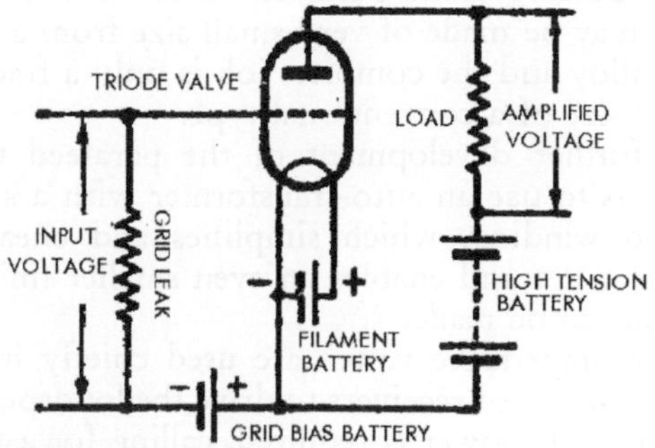

Once again, the voltage swing here is much greater than that at the grid. It is also in the opposite phase, an important point which gives the valve its ability to oscillate, of which more later.

Note that in the diagram the incoming signal to the grid is between the latter and the GB battery, whilst the output is between anode and HT battery. For practical purposes the impedance of new batteries may be taken as very low, so the incoming and outgoing signals are effectively both with respect to filament (as batteries age the impedance rises, so they are usually by-passed by a large value condenser to offset this).

Normally the amplified signals at the anode are coupled by a condenser to the grid of another valve used either to provide further voltage amplification

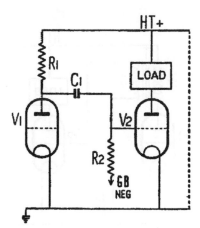

Figure 7.2 How one amplifier valve may be used to drive another. Typical values for the components: R1, 100 K; R2, 1 M; C1, 0.01 F.

Figure 7.1

or to drive a loudspeaker (we shall look at this application shortly) with the value of the condenser chosen to have a low reactance at the lowest frequency expected to be handled by the amplifier.

In all radio receivers there is what is called an 'earth line' to which the negative of the HT and LT batteries and the positive of the GB battery are connected. In the vast majority of sets likely to be encountered by the reader the earth line will be the metal chassis on which the set is built, therefore it is customary to refer to these as inputs and outputs, and voltages to be referred to as 'with respect to chassis'. These terms will be used extensively through this book. In this respect, in the diagram the resistor in the grid labelled 'grid leak' is provided to allow any electrons on the grid (which eventually would build up as a negative voltage) to leak away back to the filament. In each and every valve application there must be a return path between grid and filament or chassis.

Another type of amplifier, and one which was used extensively in the 1920s and early 1930s, employs an inter-valve or audio-frequency (AF) transformer in the anode circuit instead of a resistor. The DC resistance of the transformer primary, through which the anode voltage passes, is relatively low and little voltage drop takes place. However, its AC impedance may be quite high, in the neighbourhood of several thousand ohms and for optimum performance is best 'matched' to a medium impedance valve.

In this type of amplifier the change of voltage at the anode is very small although the I_a fluctuates considerably in accordance with the signal input to

the grid. This effectively results in a steady DC I_a plus an AC component, which induces voltage into the secondary of the transformer. Typically this would give a step-up ratio of between 1:3 and 1:5 (although as much as 1:7 might occasionally be encountered), providing a high value of 'drive' at the grid of the next valve.

Before we move on to low impedance valves it is worth looking at two variations on transformer coupling which were both very popular in valve radio sets. One drawback of the steady DC passing through the primary of the transformer was that it tended to magnetise the laminations and inhibit its performance. To get over this problem the parallel-feed ('parafeed') transformer circuit was developed. In effect it is a combination of resistance and

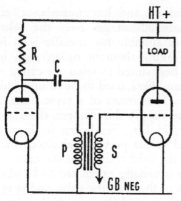

Figure 7.4 The circuit of a parallel-fed or 'parafeed' transformer coupling in which no HT current passes through the primary. Typical values for R: 50 K; for C, 0.25 F

transformer coupling, with the primary of the latter being fed via a condenser from the anode of the valve. Thus only the AC component passes through the primary and no magnetising takes place. Because of this parafeed transformer laminations may be made of very small size from a light steel alloy and the complete job is only a fraction of the size of a conventional type.

A further development of the parafeed technique is to use an auto-transformer with a single tapped winding, which simplifies and cheapens manufacture and enables an even smaller finished product to be made.

Low impedance valves are used chiefly in the output stages of receivers to drive the loudspeaker. For this job power is required, calling for a valve which will pass a fairly heavy anode current, and

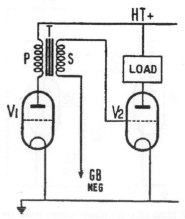

Figure 7.3 The circuit of a transformer-coupled amplifier

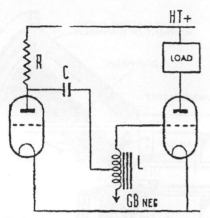

Figure 7.5 Parafeed coupling using an auto-transformer

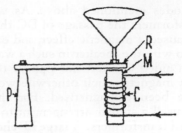

Figure 7.7 Diagrammatic layout of a moving-iron loudspeaker. Reed 'R' to which the cone is attached, is mounted on pivot 'P' a very small distance from magnet 'M' on which is wound coil 'C'

typical battery output triodes of the 1920s had an impedance of between about 2000 Ω and 4000 Ω, and passed a current of between about 10 mA and 20 mA.

Before going into this application further we need to look at the different types of speaker. The earliest were effectively just a large earphone attached to a horn which amplified the relatively weak sounds it produced.

like armature was held close to a small permanent magnet on which was wound a coil corresponding to one of those in an earphone. Attached to the armature was a paper cone flexibly mounted on a metal frame to permit it to move to and fro. Variations in current through the coil caused the armature and thus the cone to vibrate and thus reproduce the speech or music from the set. Moving-iron speakers were much better than horns and despite the upper and lower frequency response still being limited, can still sound surprisingly good if heard in isolation, and not compared with modern types. They lasted well into the 1930s and the reader has a fair chance of coming across one.

The impedance of the coil in a moving-iron speaker was probably in the region of 2000 Ω, suitable to provide a reasonable match if wired directly into the anode circuit of one of the battery

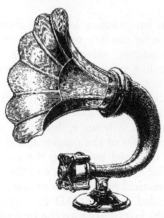

Figure 7.6 A typical horn loudspeaker of the early 1920s (Amplion)

The end result was a surprisingly large amount of volume but with an extremely restricted frequency response, despite the claims in advertisements for 'life-like tone'!

By the late 1920s horn speakers were obsolete and the reader is very unlikely to come across one except as a collector's item. They were superseded by the moving-iron or 'cone' type in which a reed-

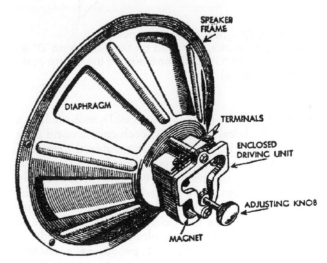

Figure 7.8 An actual moving-iron loudspeaker in chassis form. The adjusting knob alters the spacing between reed and magnet

output triodes mentioned above. As with inter-valve transformers the passage of DC through the winding caused a magnetic effect, and care had to be taken to wire the speaker in such a way that the magnetism aided and did not oppose that of the permanent magnet, which otherwise would gradually have been demagnetised. For this reason some sets used a similar arrangement to that used for parafeed transformers. A large LF choke with a suitable impedance took the place of the loudspeaker winding, with the latter fed via a condenser of about 2 µfd, which permitted the AC component to pass but blocked the DC current. This system was particularly favoured in the early mains powered sets (q.v.) which have much more powerful output valves.

By far the best loudspeaker design to appear in the 1920s was the moving-coil type. Despite being nearly 75 years old, its principle is still being used today in countless millions of radio sets and amplifiers. Figure 7.9 shows how it works. At the

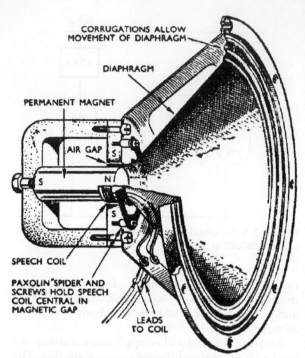

Figure 7.10 A typical moving-coil loudspeaker of the mid-1930s. Note the output transformer mounted on the chassis. A common practice

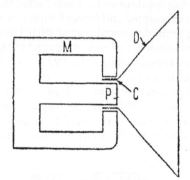

Figure 7.9 Diagram of how a moving-coil loudspeaker works. Mounted on the diaphragm 'D' is the speech coil 'C' positioned around the centre pole 'P' of the magnet 'M'

centre of a light cone, very flexibly mounted, is a small cylindrical coil (the speech coil) wound on a thin former. The speech coil sits within the poles of a circular magnet, and is coupled to the output valve. When the varying voltages produced by speech and music are passed through the coil the electro magnetic effect which is produced either assists or opposes the existing magnetic field and causes the coil to move backwards or forwards, in turn moving the cone and producing a faithful reproduction of the original sound.

The speech coil consists of only a few turns of fairly thick wire and thus has an impedance of only

a few ohms (in practice between about $2\,\Omega$ and $16\,\Omega$). This is far too low to match even the lowest impedance output valve and so a step-down transformer has to be used. This brings us to another valve characteristic which is always quoted with respect to output valves. It is termed the optimum load, abbreviated to R_a with upper-case R to distinguish it from r_a for valve impedance. The required ratio of the output transformer, as it is called, is obtained by taking the square root of the R_a divided by the impedance of the speech coil. To take a practical example, to match a valve with a recommended R_a of $9000\,\Omega$ to a speech coil of $3\,\Omega$, divide the last figure into the first to obtain the answer 3000. The square root of this is just under 55, so the transformer ratio would be 55:1.

Note that all early examples of the moving-coil speaker used an electromagnet, called a *field magnet*, rather than a permanent type as in the 1920s and early 1930s – it was not easily possible to make the latter of sufficient power. Typically the magnet was energised by a 6 V storage battery, which made such speakers expensive to buy and own. A short-lived development was the field coil which could be energised from the mains and thus gave rise to the name 'mains energised loud-speaker'. This term is all too often misapplied to later types with fields powered by the HT supply of mains operated receivers, properly known as 'HT energised speakers'. Always be suspicious of any speaker described as 'mains energised' because there is a 99% chance that it won't be!

The development of new alloys in the later 1930s made possible small and efficient magnets (and in turn smaller speakers for portable sets), but energised types remained popular for certain jobs right into the 1950s and they will often be encountered.

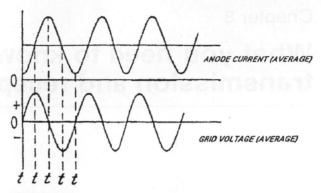

Figure 7.12

The triode as oscillator

In the circuit of Figure 7.11 is shown a simple oscillator circuit working on the feedback principle. L1 and L2 are respectively the primary and secondary of a high frequency transformer: in addition L1 and C1 form a tuned circuit. The action of the circuit is as follows: assuming that the LT battery is already connected, as soon as the HT supply is switched on (t_0) anode current will start to flow in the valve (t_1). As the current passes through L1 the resulting magnetic effect causes a voltage to be built up across L2, which is wound so that the voltage is positive on the grid with respect to the

filament. As we have seen, making the grid of a triode positive increases the anode current and thus increases the positive grid voltage still further. This action continues until the peak current of which the valve is capable is reached (t_2), the time taken to attain this being determined by the resonant frequency of L1 and C1. At this point the anode current begins to fall back (t_3), causing an opposite voltage to be built up across L2 and making the grid negative with respect to filament and thus further reducing the anode current. Eventually cut-off point (t_4) is reached, and with anode current ceasing, the negative voltage on the grid collapses, permitting the valve to start drawing anode current again and repeating the sequence of events over and over again as long as the HT supply is connected. The result is a continuous oscillatory voltage built up across L1 and C1 at their resonant frequency. By choosing suitable values of inductance and capacity for L1 and C1 any desired frequency of oscillation may be obtained.

In the next chapter we shall look at how the two abilities of the triode to amplify and to oscillate are combined to make possible the broadcasting of speech and music as electromagnetic waves. This in turn will enable us to discuss how the latter may be received and converted back into audible sounds.

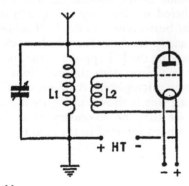

Figure 7.11

Chapter 8

What you need to know about the principles of transmission and reception

In the previous chapter we looked at a basic valve oscillator circuit. It is possible to use this as a simple radio transmitter. If an aerial were to be connected to the top of the coil L1 and an earth to its bottom end, the oscillations produced would be radiated as electromagnetic waves, but not very effectively because the power would be low. However, high frequency amplifying stages can be used to build up the strength of the oscillations, culminating in a power output stage capable of putting out anything from a few watts to hundreds of thousands of watts into the air.

What would be sent out from such a transmitter would be a continuous series of magnetic waves which by itself does not contain any information in the way of speech or music but which is able to carry such sounds if they are imposed upon it. For this reason it is known as a carrier wave. The process of imposing speech and music on it is called modulation and the end result is known as a modulated carrier wave. The modulation system used for broadcasting in the long, medium and short wave bands is called amplitude modulation, discovered as long ago as 1913 and still unchanged in principle. The term means exactly what it says: the amplitude of the continuous waves is modulated – altered in size – by the speech or music. One of the best-known methods of achieving this is called anode modulation, which works on the basis of varying the HT applied to the anode of the oscillator in unison with the variations in sound picked up by microphone. A simple circuit capable of achieving this is shown in Figure 8.2, in which you will see that the secondary of an LF transformer has been inserted in the feed from the HT

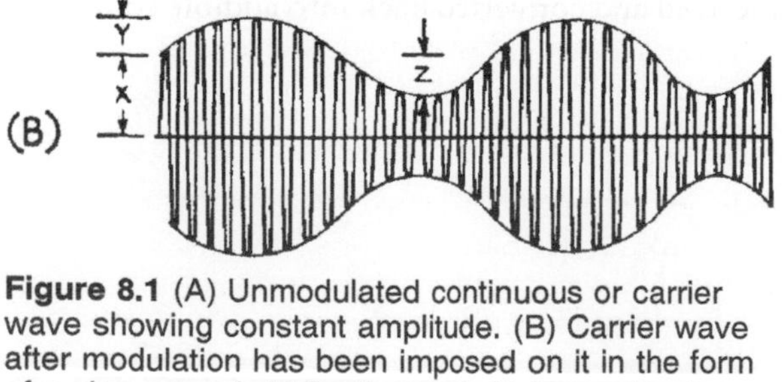

Figure 8.1 (A) Unmodulated continuous or carrier wave showing constant amplitude. (B) Carrier wave after modulation has been imposed on it in the form of a sine wave (middle diagram). Note amplitude varying in concert with the sine wave

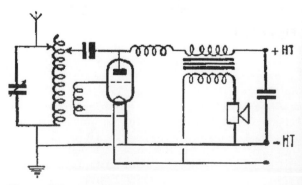

Figure 8.2

battery to the HF choke. The microphone shown in the primary circuit of the transformer is a carbon type, which needs to be supplied with an energising voltage, in this case obtained from the LT supply for the valve heater. Sounds reaching the microphone are converted into fluctuating DC currents which, flowing through the transformer primary, induce voltages in the secondary which either aid or oppose the steady DC voltage from the HT battery, thus causing the anode voltage on the valve to rise and fall in sympathy. This in turn alters the amplitude of the oscillations generated by the triode. The amount by which their amplitude is affected by the modulating device is called depth of modulation. This has to be set carefully, as too little would result in the transmissions having background noise, whilst too much would cause distortion of the sounds. The simple transmitter shown here would certainly work to a certain extent but not very well because the depth of modulation would not be very great. This could be much improved by introducing another triode as an amplifier for the microphone, as shown in Figure 8.3. It will be seen that the anode of the amplifier valve is connected to the HT feed to the oscillator and also to an LF choke inserted in the lead to the HT supply. As we have seen in the previous chapter, the anode current of the amplifier will rise and fall with the variations in sound at the microphone, thus altering the voltage drop across the choke and imposing a good depth of modulation on the oscillator. Just as the strength of the latter can be built up by amplifiers, so can the power of the former, the aim always being to match them correctly. Anode modulation, as it is called, was for a long time the principle method used, but others, such as grid modulation, have

joined it. Whatever the means of modulation the end result is the same and if a 1913 speech transmitter could be found and made to work it could be received on a modern AM receiver. It is this continuity of basic principles that gives vintage radio a unique advantage over other hobbies: no matter how old the set, it can still be made to give out speech and music.

Sidebands

As well as modulating the amplitude of the carrier wave, the sounds picked up by the microphone have another effect, of causing it to 'spread out' on either side of the basic frequency of oscillation. If we accept that the range of human hearing is from about 20 c/s up to 16 kc/s, to reproduce them all properly they would have to be imposed on the carrier wave. Supposing that this were 1000 kc/s, what are called sidebands of 16 kc/s would be set up on either side consisting of 1000 kc/s − 16 kc/s and 1000 kc/s + 16 kc/s, that is from 984 kc/s to 1016 kc/s. Thus the total 'airspace' taken up by the transmitter would 32 kc/s. This would be fine if there were but a few transmitters on the air but in real life there are hundreds on the medium wave band alone. This band, covering as it does from 550 kc/s to 1600 kc/s, has a total usable 'space' of 1005 k/cs, which could accommodate only 31 stations. The situation would be even worse on long waves where the total space is only about 150 kc/s and fewer than five stations could be squeezed in. In practice radio stations are separated by no more than 9 kc/s so the sidebands must be limited to no more than 4.5 kc/s. This places a severe limitation on the range of musical frequencies that may be

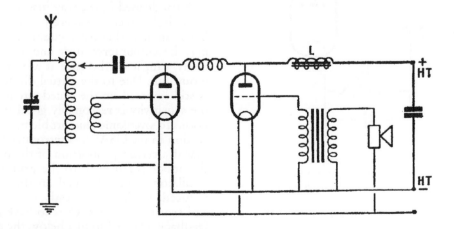

Figure 8.3

broadcast, but the tolerance of the human ear is such that the end result at radio receivers is reasonably acceptable, especially where speech and modern 'non-music' are concerned.

We shall look again at sidebands shortly when discussing the way in which they affect the design of radio receivers.

Receiver principles

We saw in Chapter 5 how the combination of a coil and a variable condenser makes it possible to select any particular frequency from the hundreds arriving at an aerial system. If this chosen frequency is fed to the anode of a diode it will pass only the positive half-cycles, so closely spaced in radio transmission that the result at the cathode will be a steady DC voltage plus varying audio frequencies mimicking exactly those used to modulate the transmitter. In simple receivers the carrier wave has no further use and will be discarded, whilst the AF signals may be fed directly to a pair of earphones or to an amplifier.

The good points of a diode as a detector are that it will accept large inputs without overloading and its output is free from distortion. However, its sensitivity is low and it imposes a loading effect on the preceding tuned circuit which impairs it selectivity.

The amplifier following a diode may be a triode. However, it is possible to make the latter act as both detector and amplifier by taking the incoming signal to its grid via a small condenser. The grid

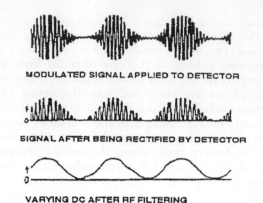

MODULATED SIGNAL APPLIED TO DETECTOR

SIGNAL AFTER BEING RECTIFIED BY DETECTOR

VARYING DC AFTER RF FILTERING

Figure 8.5

then acts as the anode of a virtual diode consisting of itself and the filament (or cathode). In this case the steady DC resulting from the carrier is built up across the condenser, with the filament positive and the grid negative, the AF signal also imposed on the grid. If the negative voltage were allowed to build up indefinitely it would reach a level where it would 'cut off' the anode current, so a resistor is connected between grid and filament or cathode through which the voltage can leak away – hence the term 'grid leak'. Its values may be from about $1\,\text{m}\Omega$ up to $5\,\text{m}\Omega$ and it may be connected either directly between grid and filament or indirectly by shunting it across the grid condenser. The AF signals on the grid modulate the electron stream as usual and result in a varying anode current carrying the information which may be reproduced in earphones or passed on by resistance capacity or transformer coupling to another amplifier and ultimately to an output stage and a loudspeaker.

A triode used in the way just described is called a grid leak detector or merely a grid detector in the USA (an upmarket term 'cumulative grid detector' may be encountered now and again but it never really caught on because it was too much of a mouthful). Due to the virtual diode effect loading is still imposed on the tuned circuit and although the sensitivity is reasonably good the selectivity is poor. To overcome this problem, reaction (regeneration in the USA) was introduced. It was found that if a certain proportion of the output signal at the anode is fed back to the grid, much as in an oscillator circuit, the load on the tuned circuit is reduced, resulting in greatly enhanced selectivity and sensitivity. The trick is to have the amount of feedback adjusted to just below the point at which

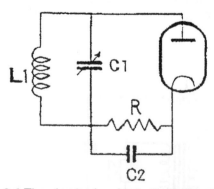

Figure 8.4 The circuit of a simple diode detector L1 and C1 are respectively the tuning coil and the tuning condenser. *R* is the diode load resistor, on which the AF output from the diode is developed. C2 is a by-pass condenser which filters out the remaining RF

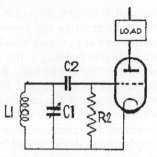 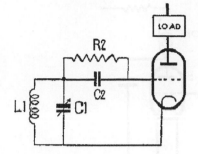

Figure 8.6 Two versions of grid leak detection

oscillation would occur, otherwise the triode would behave as a low power transmitter capable of causing severe interference to neighbouring receivers. In the early days of broadcasting carelessly adjusted reaction and oscillating detectors were a major menace and the old British Broadcasting Company and the radio press carried on a constant publicity campaign to persuade listeners to avoid offending in this way.

going part of an incoming signal will have any effect now since it will make the grid less negative and cause anode current to flow; on the other hand the negative going signal will only increase the negative bias and maintain the cut-off condition. If an RF signal from a tuned circuit is applied to the grid the result will be a varying positive voltage derived from the original AF modulation which will then be reproduced as

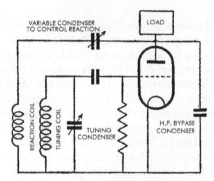

Figure 8.7 Adding reaction to a grid leak detector L2. The 'reaction coil' is wound on the same former as L1 and induces positive feedback into it. C2, the 'reaction condenser' enables the degree of feedback to be adjusted. C3 by-passes any residual RF to the earth line

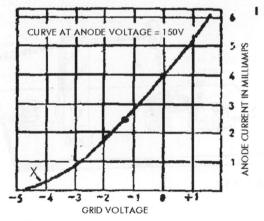

Figure 8.8

An alternative way to use a triode for detection is to make use of its characteristic curve. Cast your mind back to the way in which an amplifier is operated at the centre of the straight part of the curve (Y) in order that both positive and negative portions of the AF signals may be reproduced equally. Now consider what will happen if the grid bias is adjusted to reduce the anode current to near the cut-off point (X). Only the positive

amplified AF signals at the anode. Because of its use of the bend in the characteristic curve this effect is known as anode-bend detection (simply anode detection in the USA).

The anode-bend detector imposes little loss on the tuned circuit and good selectivity is obtainable without the need for reaction. However, the incoming signal does have to be fairly strong to make it work properly and for this reason it was seldom employed without an RF amplifying stage preceding it.

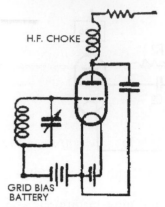

Figure 8.9 Basic anode-bend detector circuit

Before leaving the subject of detectors we should look at the simplest of all, and one that was used in thousands of receivers through the 1920s and even later. This is the so-called crystal detector which was usually composed of a small chunk of lead sulphide, otherwise known as galena. This has the useful property of permitting electricity to travel

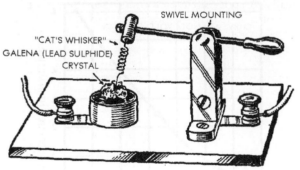

Figure 8.10

through it in one direction only, just like a valve diode. In practice the crystal would be held in a small metal cup, above which was mounted a small movable arm holding a tiny spring-like coil of wire called the 'cat's-whisker'. Randomly disposed on the surface of the crystal are 'sensitive spots'; the trick is to adjust the cat's-whisker to make contact with one of these, at which point diode action is obtained. An aerial and an earth are taken to a simple tuning device which is connected to one side of the detector, and a pair of headphones is connected to the other. When a good signal is

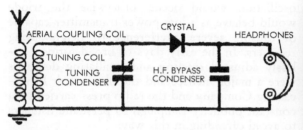

Figure 8.11

received from a transmitter, and the cat's-whisker is adjusted properly, enough rectified signal passes through the headphones to enable them to reproduce the original AF. The only optional refinement is a small condenser shunted across the headphones to remove any residual HF, so the complete receiver consists of no more than four or five components. No batteries are required, hence the attraction of the crystal set to listeners on low incomes.

With the principles of reception now established we can look in the next chapter at some typical early designs for receivers.

Chapter 9

Practical receiver design (1): battery operated TRFs

The basic single valve grid leak detector with reaction receiver described in the previous chapter is capable on its own of giving acceptable results for anyone willing to listen only with headphones. However, since most of us prefer to use loudspeakers, more valves have to be employed to amplify the output of the detector suitably; this subject has already been covered in Chapter 7. In theory it would be possible to add as many stages of amplification as necessary to make even the weakest of stations audible through the loudspeaker, but in practice this would be unsatisfactory. An ordinary grid detector, even with reaction, still has relatively poor selectivity, that is, the ability to pick out a desired station cleanly from all the others arriving at the aerial. The result on weak stations is that they may well be interfered with by more powerful ones on neighbouring wavelengths. To overcome this the tuning has to be 'sharpened' and a relatively simple method of achieving this with a tuned circuit

preceding a detector is to make it the secondary of an HF transformer, with the primary used to couple the aerial inductively instead of directly. An extension of this is to use two sets of HF transformers, one in the aerial circuit coupled to another feeding the detector and with each having its own separate tuning condenser. Properly designed this arrangement would select only a relatively narrow band of frequencies, hence its generic name of 'band-pass' tuning. Better-class receivers continued to use it right through the vintage radio era.

Whilst improving the tuning enables particular stations to be selected more easily it is not a complete answer to receiving those situated at a long distance from a receiver, for any remaining interference will be amplified along with the required AF. A far more effective and satisfactory idea is to amplify the RF signals from the transmitter before detection. This could be achieved by using separate RF transformers to couple one or

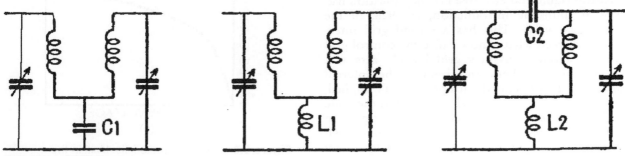

Figure 9.1 Three forms of band-pass tuning using (from left) capacity coupling, inductive coupling and a combination of the two

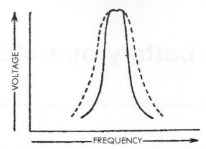

Figure 9.2 Response curve of a band-pass tuner (solid line) as compared with that of a single coil and condenser (dotted line)

more triode amplifiers in sequence and then to the detector; as each transformer would have its own tuning condenser the selectivity would also be greatly improved. Sets which use one or more RF amplifiers before the detector are known generically as tuned radio frequency (TRF) receivers, or sometimes colloquially as 'straight' sets.

Unfortunately triodes will not work satisfactorily as HF amplifiers above about 100 kc/s, after which they become unstable. As when any metal objects are held in fairly close juxtaposition, a certain amount of capacity exists between the various electrodes of a valve. These internal capacities are small enough to be negligible at fairly low frequencies but above the 100 kc/s mark they become more and more significant until they start to introduce instability and even self-oscillation into an amplifying stage. A part-answer to the problem was to fit small variable condensers between the anode and grid of HF amplifiers to oppose and cancel out the internal capacities. This process, called neutralisation, was used with varying degrees of success (often dependent on the skill of the listener) for some years, but even at best a triode can give only a relatively small overall gain at HF.

Really stable and effective HF amplification had to wait until the introduction, in 1926, of the screen-grid valve. This has a second grid interposed between what is now called the control grid and the anode. This new grid literally acts as a screen between them and reduces the internal capacities. It is operated with positive HT applied, the voltage being between about half and the same as that on the anode. Extremely large stage gain figures may be realised with screen-grid RF amplifiers, a single one being sufficient for many receivers. Because it has four electrodes the screen grid is properly called a tetrode, but this term was

not widely used in vintage radio literature; far more popular was the abbreviation s.g. valve.

It is fair to say that the s.g. valve revolutionised TRF receiver design but it is not without its faults. In use the anode itself emits electrons caused by its bombardment from the filament or cathode. This secondary emission, as it is called, tends to collect on the screen grid and brings about a strange non-linearity in the current drawn by the latter and by the anode. As Figure 9.3(a) shows, there is a pronounced kink in both at a certain combination of voltage and current which has the effect of making the s.g. valve unsuitable for other than RF amplifiers. Subsequently yet another grid was added, this time between the screen grid and the anode, which had the effect of suppressing the secondary emission from the latter, thus giving it its name of suppressor grid. In operation the suppressor is connected to the filament or cathode, to which the secondary emission is returned harmlessly. This new five-electrode valve or pentode is extremely suitable for both RF and AF amplification and as a power output valve. It offers good sensitivity and linearity and rapidly became almost supreme in these fields. Its only serious

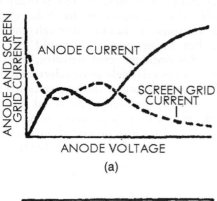

(a)

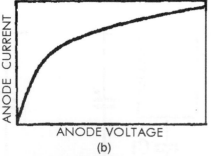

(b)

Figure 9.3 (a) The characteristic 'kink' in the screen-grid valve's current vs. voltage curve (b). However, the pentode has a nice linear curve

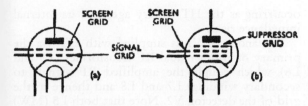

Figure 9.4 Theoretical symbols for (a) screen-grid and (b) pentode valves

challenger was a development of the screen-grid valve in which the spacing between the electrodes is critically adjusted to cause the electron stream from the filament or cathode to travel in sharply defined beams rather than in random formation. Known as the beam tetrode or kinkless tetrode, this type of valve has characteristics similar to those of the pentode and in many cases they are interchangeable in receivers.

Sharp cut-off versus variable mu

A major problem with the s.g. valve, the pentode and the beam tetrode was that in RF amplifiers their high gain could, in certain circumstances, be a drawback rather than an advantage. It was extremely useful for receiving weak stations but when the set was then tuned to powerful locals severe over-loading took place. The gain could not be reduced by putting extra negative bias on the control grid because all these types of valve have what is called a short grid base or sharp cut-off whereby the anode current drops to zero at very little more than the normal value of operating grid bias. It was found that by winding the control grid in a certain way the grid base could be lengthened very considerably, with the cut-off bias being

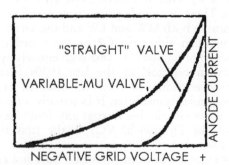

Figure 9.5 Comparing the slopes of typical 'straight' and variable mu valves

increased to as much as −40 V below the operating voltage. By varying the grid bias between these points the gain of the valve could be controlled very accurately, giving rise to the term variable-mu valve (remote cut-off in the USA). To distinguish them, non-variable-mu types are called straight valves.

Extending the tuning range

Early receivers, of which there are still thousands in circulation amongst radio collectors, employed separate coils with different numbers of turns to cover different wavebands. Anyone engaged in repairing radio sets is almost certain to encounter one of these at some time or other. Coil changing did not appeal to non-technical listeners and a

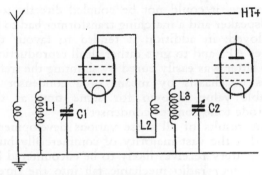

Figure 9.6 Basic two-stage RF amplifier using pentodes, for one band only

much more convenient development was to wind several different coils on one former, each for a certain band, and to select them as required by means of a switch to obtain coverage of a desired band. This feature was universal in commercially built domestic receivers by 1930. Initially only medium and long waves were covered: note that some sets in this class had the waveband switch marked 'short' and 'long', which could be confusing to the beginner. From about 1935 growing interest in true short waves led to many firms including an SW band in even cheap receivers, the usual coverage being from about 16.5 m to 52 m.

Also by this time the pentode was beginning generally to take over from the triode in output stages, where it offered good sensitivity and output power with reduced HT current consumption. Its only drawbacks were that, having a high internal

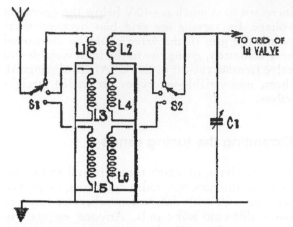

Figure 9.7 The circuit of a triple wound RF transformer for short (L1/L2), medium (L3/L4) and long (L5/L6) waves

resistance, it could not be coupled directly to a loudspeaker and a matching transformer had to be employed: in addition it tended to favour the higher AFs and to give rather shrill reproduction. This latter was easily cured by shunting the treble notes to chassis by means of a condenser of suitable value, known for some time as the 'pentode tone control condenser'.

The results of all these various developments was that the vast majority of commercially built TRF battery receives likely to be encountered by the amateur radio mechanic fall into the three-valve class with an s.g. or pentode as RF amplifier, a triode (sometimes a straight pentode) as detector and a pentode as the output stage. Two representative circuits are shown in Figures 9.8 and 9.9; they are entirely typical and their design points should be studied.

The first employs a straight s.g. valve as RF amplifier, its control grid fed from the secondary of a dual RF transformer covering medium and long waves (L1/L2 and L3/L4). Note that L3 and L4 are shorted out by S1 and S2 when the set is switched to medium waves. Three different aerial input sockets are provided, A1 going to the top of L1, A2 going to a tapping on it and A3 also going to the tapping but via a small condenser (typically 100 pfd). These are to suit aerials of different length and efficiency. VC1 is the first tuning condenser.

Note that the screen grid of V1 is supplied with voltage from a separate tapping on the HT battery. Condenser C5, typically 0.1 μfd, decouples the screen grid (i.e. provides a low impedance path) to chassis to obviate the possibility of instability

occurring as the HT battery ages and its internal impedance rises.

The anode of V1 is supplied with HT via the primary of another dual RF transformer (L5 and L6) which couple the amplified RF signals to secondary windings L7 and L8 and thence to the grid of the detector, V2. Note that both L5 (MW) and L6 (LW) are in circuit on both bands whilst S3 shorts out L8 on MW. VC2 is the second tuning condenser.

Typical values for C2, the grid condenser, and R1, the grid leak, would be 100 pfd and 1 mΩ respectively. Any residual RF remaining after detection is by-passed from V2 anode to chassis by C3, typically again 100 pfd, and prevented from reaching the primary of the inter-valve transformer, L11, by the HF choke L10. Also connected to the anode of V2 is the reaction winding L9, which is common to both MW and LW. VC3 is the reaction condenser and placing it at the bottom end of L9 enables its moving plates and control spindle to be connected to chassis, enhancing stability. The top end of the the inter-valve transformer secondary, L12, is connected to the control grid of V3 whilst the lower end goes to a suitable tapping on the grid bias battery. L13/L14 form the output transformer coupling the signals from V3 anode to the speech coil of the moving-coil loudspeaker. C4 is the pentode tone control condenser, a typical value being 0.01 μfd. HT for both anode and screen grid of V3 and for the anodes of V1 and V2 is obtained from another tapping on the HT battery. C5, typically 1 μfd, decouples the latter to chassis.

Our next circuit (Figure 9.9) may appear superficially very similar to the first but closer inspection will reveal many differences.

The RF amplifier is now a variable-mu pentode, with input to its control grid from the dual secondary of an RF transformer (L3 for MW and L4 for LW) with S1 shorting out L4 on medium waves. There is a single primary or aerial winding common to both MW and LW and the aerial input is brought in either via C2 or the tuned circuit comprised of L1 and C1. This is an entirely typical 'Droitwich rejector' of the late 1930s fitted to reduce overloading in receivers installed near to that particular transmitter. It is actually, of course, an acceptor circuit able to pass any frequency but that of the Droitwich. VC1 is the first tuning condenser.

Note that the bottom end of L4 does not return to chassis but goes to R1, typically 500 kΩ. This in turn is connected to the slider of potentiometer

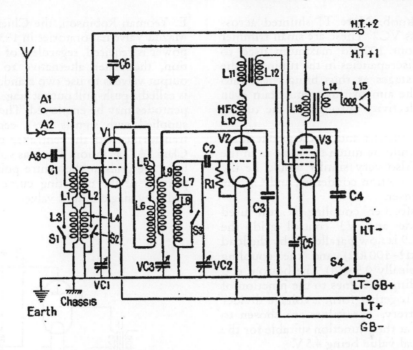

Figure 9.8

VR1 which is connected between GB− and chassis. In this case the GB battery would be a 9 V type, sufficient fully to control an average battery RF pentode. C3 is a by-pass condenser, typically 0.1 μfd, used to give L4 a low impedance RF path to chassis.

Choke-capacity coupling transfers the amplified RF signals at V1 anode to the control grid of V2, with L6 and L7 as grid tuning coils. S2 shorts out L7 on MW. VC2 is the second tuning condenser and the dotted line joining it to VC1 indicates that the two are ganged or operated simultaneously by

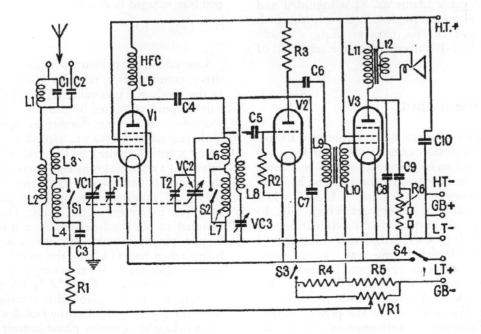

Figure 9.9

a single control knob. Note T1 shunted across VC1 and T2 across VC2. These are small trimmer condensers of about 100 pfd maximum used to cancel out small discrepancies in the tuning of the RF and detector stages as thus bring them into close alignment, the aim of this to maintain even sensitivity and selectivity throughout the tuning ranges.

C6 is the grid condenser and R2 the grid leak for V2; their values would be much the same as in the previous circuit. Also very similar would be the value of VC3, the reaction condenser, and C7, the RF by-pass condenser.

Although transformer coupling is still used between V2 anode and V3 control grid, the primary winding L9 is now parallel fed by the load resistor R3 (50 kΩ–100 kΩ) and the coupling condenser C6 (typically 0.5 μfd). The bottom end of secondary winding L10 goes to the junction of R4 and R6 which together form a voltage divider across the GB battery. The values are chosen to give a bias voltage at their junction suitable for the grid of V3, a typical value being 4.5 V.

C8 (around 0.01 μfd) is the fixed pentode tone control condenser while C9 (around 0.02 μfd) and R6 (around 22 kΩ) from an extra semi-variable tone control, a plug and socket arrangement, enabling R6 to be shorted out at will to cut even more the high frequency AFs. C10 is an HT decoupler of about 1 μfd capacity.

S4 is the main on/off switch, breaking the LT+ supply to the valve filaments. S1 is included and ganged with S4 to break simultaneously the path between R4 and chassis, thus avoiding a steady discharge of the GB battery when the set is out of use.

Improving the output

As regards power output, battery receivers always were subject to design constraints whereby good performance had to be balanced against economy, for HT batteries never were cheap. "The heart of many a conscientious set designer is broken when he tries to develop a battery set which will do justice to his ideals of quality of reproduction, for whilst engineering knowledge tells him that a power output of about one watt is necessary, common sense dictates that no HT battery is obtainable at a reasonable cost will supply this power. Even with a high-efficiency pentode the drain on the HT battery for the power output valve would be about 20 milliamperes", wrote Mr.

E. Yeoman Robinson, the Chief Engineer of the Mazda Valve Laboratories in 1932. If more output power is required, regardless of battery consumption, the better alternative to using one large output valve is to use two standard valves in what is called a push-pull output stage. Either triodes or pentodes may be employed. The control grids are supplied with signals by a centre-tapped input transformer with the centre tap taken to GB–. For Class "A" operation. The bias voltage is selected to hold the grids at the centre point of the straight portion of the operating curve, just as with an ordinary single output valve.

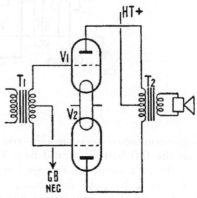

Figure 9.10 Push-pull output stage using triodes for either class 'A' or 'B' operation-only the value of the grid bias voltages is altered

One end of the primary winding of the input or driver transformer is taken to HT+ and the other to the anode of a preceding AF amplifier or driver valve. Signals amplified by this valve appear on the secondary of the transformer in opposite phase with respect to the centre tap, that is, a positive going signal is applied to the grid of one output valve and a negative going signal to the other. Thus the anode current of one valve rises as the other fails and vice-versa as opposite phase inputs continue to arrive at the grids. The output is coupled to the loudspeaker by a special centre-tapped output transformer, the centre tap itself being taken to HT+ and the outer ends to the anodes. The total effective output power developed by two valves in Class "A" push-pull is likely to rather more than double that of a single valve of the same type. In addition the fact that the anodes are working in opposite phase cancels out certain forms of distortion so the quality of the sound is usually better than with a single valve. Yet another advantage is that the anode currents flow in opposition to one another and reduce the magnetising effect on the soft iron core of the transformer. This in turn means that the transformer may be physically smaller that otherwise would be the case.

There is, of course, a price to be paid for higher output in the amount of current drawn from the HT battery. One attempt to ameliorate this was the Class 'B' push-pull stage, in which the grid bias for the output valves is arranged to hold the anode current near to cut-off point in the absence of incoming signals. These are again supplied by a centre-tapped driver transformer but each valve can respond only to the positive half-cycles of the input, when the grids are driven positive and the anode current increases. On the negative half-cycles the grids are driven even more negative to way past the cut-off point. In practice the power output of a Class 'B' stage may be quite impressive - typically 1.25 W – but the saving in HT current is not as great as might be thought. Although the no-signal current may be low it may soar to as much as 45 mA with some valves on peak positive inputs. The average would probably be around the 15 mA mark, but added to this a high level of drive power is required for the grids and the preceding stage may have to employ a power triode taking around 3 mA if the full output is to be realised.

True economy combined with high output power in battery sets had to wait until the development of Quiescent Push-Pull (QPP) in the Mazda Valve Co.'s laboratories during 1932. For QPP operation two output pentodes are arranged in a similar way to that used for Class A push-pull, but with the grid bias selected to take each valve to a very low point on its characteristic curve at which the anode current is extremely low; in the case of the entirely typical Mazda Pen220A it is about 1.8 mA. For a pair of valves the total drain, including the screen grids is around 4 mA, giving a very significant saving in HT current under no signal conditions. When signals arrive the two valves again operate alternately (they respond only to the positive swings) and the anode current is exactly proportional to the amplitude of the signal. Operated under these conditions the theoretical efficiency of a pentode is raised in theory by as much as 80% and in practice by about 50%. What this means in performance is well illustrated by data recorded

in the Mazda Laboratories. A pair of Pen 220As were found to draw a maximum of 16 mA with a continuous high amplitude signal corresponding to transmitter modulation of around 90–100%. However, as the average modulation level of a radio stations is (or at least was in 1932) no more than 15–20% the potential saving in HT current is evident. Used as they would be in an ordinary domestic receiver, a pair of Pen220A working with a 120 V HT battery and with a grid bias of 10.6 V delivered a genuine 1.3 W to a loudspeaker for an average anode current of about 6–7 mA, a most remarkable achievement.

Initially two separate pentodes were used for QPP but quite soon the major valve manu-

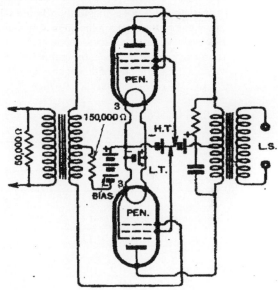

Figure 9.11 The original Mazda circuit for QPP. Note the separate tappings on the HT battery for each screen grid

facturers had developed double-pentodes especially intended for QPP work. Final matching of the two sections had to be carried out in the individual receiver by adjusting the screen grid voltages according to code letters stamped on the valve envelope. The bias, too, had to be set with care. By about 1936 a new generation of QPP valves arrived in which the need for separate screen grid voltages was eliminated and a standard type of 120 V HT battery could be employed.

Automatic Grid Bias

It was also possible to use automatic grid bias which adjusted itself effortlessly as the HT battery ran down. The bias is obtained by inserting a resistor between the negative input from the HT battery and chassis so that a small voltage drop occurs across it as the steady HT current flows. This means that HT– is now slightly negative with respect to chassis, and this voltage can be used to provide grid bias for the valves. By using two or more resistors in series and choosing their values carefully different suitable voltages may be obtained for all the various stages in a receiver. Interestingly, automatic grid bias was invented in the early 1920s but apparently a patent prevented its being used generally for a dozen or more years.

Chapter 10

Mains valves and power supplies

As mentioned in Chapter 6, valves for mains operation have an indirectly heated cathode as the emissive electrode. Apart from this they operate in the same way as battery valves and may be substituted for them in all the applications we have looked at in the previous two chapters. In fact, the two TRF circuits shown and described could very easily be converted to mains operation with the addition of a power supply unit and a few other components. The great difference is that since power from the mains is cheap and virtually unlimited the valves can be made to perform much more effectively; in particular very large power outputs may be obtained.

Cathode bias

Mains valves need negative grid bias just as their battery equivalents, but thanks to their cathodes no battery is necessary. Since the HT current

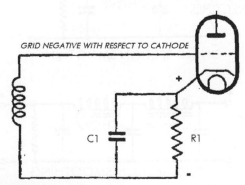

GRID NEGATIVE WITH RESPECT TO CATHODE

C1   R1

Figure 10.1

passing through the valve returns to HT– (normally the chassis) via the cathode, if a resistor is inserted in the circuit a certain amount of voltage drop will occur across it. If the grid is returned to chassis it will effectively be at a negative potential with respect to it, thus providing the necessary bias. The value of the cathode or bias resistor (R1 in Figure 10.1) is found in the normal way with Ohm's law by dividing the required negative voltage by the total anode and screen current drawn by the valve. To provide an AC path for RF or AF currents the bias resistor is decoupled by a suitable value of condenser (C1), typically 0.1 μfd for RF and 25 μfd to 100 μfd for AF.

Types of mains valves

Mains valves as a whole fall into two broad classes, those with relatively low voltage, high current heaters for parallel operation, and those with high voltage, low current heaters to be operated in series. Generically these are known respectively as AC-only and AC/DC valves, terms which need explanation.

AC-only valves are types which are intended to have their heaters powered by a low voltage secondary winding on a mains transformer in the power supply unit. The actual voltage will normally be, in UK receivers, 4 V or 6.3 V; 4 V valves were in common use from about 1928 to 1939 whilst 6.3 V types were introduced around 1937 and by 1940 (the year that domestic radio production virtually ceased) were set to take over almost completely. In the USA 2.5 V heaters were widely used until around 1935, when 6.3 V types were introduced and soon ousted them. The reason for the choice of 6.3 V was quite simple: it permits the valves also to be used on a 6 V storage battery for automobile receivers. The heater current taken by normal domestic type AC-only valves ranges from about 0.2 A up to 3 A, whilst a few intended primarily for automobile use were rated at 0.15 A.

We should also mention that a very few AC-only output triodes and pentodes had directly heated filaments rated at 2 V or 4 V. The triodes in particular were mostly employed in expensive receivers and radio-gramophones of the 1930s and immediate post-war period by A.C. Cossor Ltd, Decca, EMI (HMV and Marconiphone) and the Radio-Gramophone Development Company (RGD). They are unlikely to be found in other makes.

The low voltage or LT winding on the mains transformer is sometimes centre tapped with the tap connected to chassis, either directly or via a small potentiometer called a hum-dinger. This is adjusted to minimise any hum from the loud-speaker of a set due to slight heater-cathode leakage in a valve or valves. This device is normally only found in older receivers and those employing directly heated output valves. Otherwise it is usual simply to connect one side of the LT winding to chassis.

AC-only valves as used in the average domestic radio are intended to work with anode and screen-grid voltages of between about 200 V and 250 V, whilst high power output valves employed in large radio-gramophones and in audio ampli-fiers may take up to 500 V. These voltages nor-mally are obtained from another winding on the mains transformer, but as this is of course AC it has to be converted into DC – rectified – and then smoothed, that is, have any remaining AC ripple removed. Two main types of rectification are employed, half-wave and full-wave. As the terms imply, the first rectifies only one half of each cycle whilst the second rectifies both halves. The rectifiers themselves consist in the vast majority of vintage receivers of our old friend the diode, exploiting its ability to pass voltage in one direction but not in the other. For half-wave circuits only one diode is required, in conjunction with a single HT winding on the transformer. (A) in Figure 10.2 shows how one half of each cycle applied to the anode of the diode results in a DC pulse at the cathode. When a full-wave rectifier is employed with a centre-tapped HT winding and two diodes twice the number of DC pulses

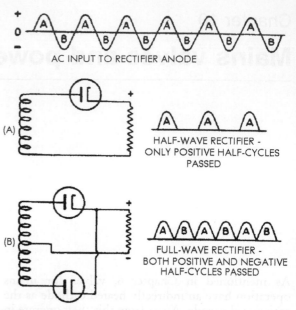

Figure 10.2

appear on their cathodes, which are coupled together. The closer spacing of these pulses (B) make it inherently easier to smooth them effec-tively, so full-wave rectifiers are used in the majority of AC-only sets with only a few, gen-erally cheaper makes using half-wave types.

The standard method of smoothing the HT is carried out by a combination of large value condensers and one or more LF chokes, called in this application *smoothing* chokes. The voltage from the rectifier filament or cathode is first applied to a *reservoir* condenser which evens out to

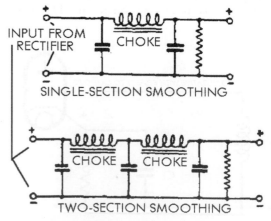

Figure 10.3

a considerable degree the DC pulses. The result is DC which is almost, but not quite, steady since it still contains a small AC ripple. The purpose of the smoothing choke is to filter this out, which it does by presenting a low resistance to DC but a high impedance to AC. On the other side of the choke another *smoothing* condenser finally removes whatever tiny ripple remains. Most ordinary domestic receivers needed only this basic choke plus two condensers arrangement but some large radio-gramophones and amplifiers employed more than two chokes and three condensers.

A very popular arrangement during the 1930s and 1940s was to have the smoothing choke in the form of a field coil in a loudspeaker employing an electromagnet instead of a permanent type. This was doubly useful as the electromagnet generally

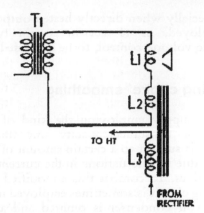

Figure 10.5

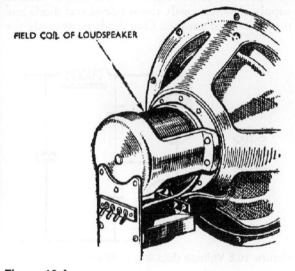

Figure 10.4

was superior to the permanent magnets of the day and its efficiency did not wane with age, whilst it also saved the cost of a choke. Some sets employing double smoothing used the field plus an ordinary smoothing choke.

In Figure 10.5 L3 is the field coil of the loudspeaker and L1 its speech coil, fed via L2 from the secondary of the output transformer T1. There is a danger that any remaining ripple voltage in the field would be picked up by the speech coil and reproduced as a hum; it is the job of L2 to prevent this. Known as the *humbucking coil*, it consists of a few turns of wire wound on the same former as the field coil but in anti-phase to it. It picks up the

same ripple voltages and by feeding them to the voice coil cancels out the others.

The introduction of high capacity; small size electrolytic condensers prompted some firms, notably Philips/Mullard, to adopt resistance smoothing in which the choke or loudspeaker field is replaced by a high wattage resistor. The smoothing properties of the latter are very limited and there is very great dependence on very large values for the condensers.

Whatever smoothing arrangement is employed in a receiver, a certain amount of voltage is bound to be dropped across it. Because of this the HT winding on the mains transformer is designed to supply a higher voltage to the rectifier than will be employed as HT in the receiver, etc. Typically, a set with an HT line rated at 250 V will employ a 350–0–350 V HT winding, allowing for 100 V or so to be dropped along the way.

Negative smoothing

The vast majority of sets deploy the smoothing choke or speaker field in the positive HT line with the centre tap of the transformer and the negative terminals of the smoothing condensers taken to chassis, which acts as the HT– line. This is called positive smoothing, but a few sets placed the smoothing components in the negative line. This feature is mainly to be found in sets from the middle 1930s, especially those made by EMI. The negative terminal of the reservoir condenser is isolated from chassis and is negative with respect to the latter by the amount of voltage dropped by the choke or field. This negative voltage may be used to bias the valves as an alternative to cathode bias, especially when directly heated output types are employed, or to act in certain types of automatic volume control, to be discussed later.

'Swinging choke' smoothing

Any HT supply employing the kind of choke-capacity or resistance-capacity smoothing just described is subject to a certain amount of voltage variation due to fluctuations in the current drawn by the valves. To combat this a modified type of smoothing circuit is sometimes employed in which the reservoir condenser is omitted and a special type of smoothing choke is fitted, the impedance of which is directly affected by the load current. The absence of the reservoir reduces the initial no-load output at the rectifier by perhaps as much as 100 V, but the subsequent variation due to current drain is greatly reduced. As far as the writer knows this device was used in only one UK domestic receiver in the late 1940s, but even then it was speedily replaced by conventional smoothing. It is as well to be aware of the swinging choke, since it may well be encountered in high power amplifiers.

Directly or indirectly heated?

Although Figure 10.2 shows two separate diodes in the full-wave circuit, in practice they are almost always combined in one envelope. From the point of view of their output there is no need for them to be indirectly heated but this may be an advantage in actual operation. Directly heated rectifiers start to work almost immediately a set is switched on, some time before the indirectly heated valves begin to draw HT current. Note that the theoretical DC output voltage is equal to the peak value of the AC input, i.e. 1.41 times the RMS value. Although this is not usually realised fully in practice the DC voltage may well be significantly higher than the AC input, particularly during the initial warm-up period of the other valves, and may cause harm. For this reason indirectly heated rectifiers were developed having similar warm-up times to the others in the set.

Normally both types of rectifier have their filaments or heaters supplied by another LT winding on the mains transformer, well insulated since it too is up to 500 V DC above chassis. The vast majority of indirectly heated rectifiers have the cathode internally connected to one side of the

heater, but a very few were made with separate, highly insulated cathodes enabling them to work from the same LT winding as the other valves. Their use is principally confined to automobile receivers but they did occasionally appear in domestic sets. Their use is to be avoided if possible since a breakdown of insulation between cathode and heater, with one side of the latter earthed, results in a dead short from HT to chassis and the probable burning-out of the mains transformer. One particular well-known radio manufacturer discovered this to its cost in the early 1950s.

The voltage doubler

A third type of rectifier circuit, found almost exclusively in very early AC-only sets, is the voltage doubler. A single HT winding is used to supply simultaneously the anode of one diode and the cathode of another. One diode passes only the

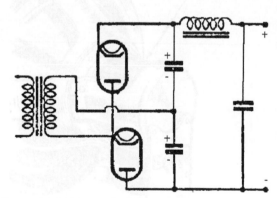

Figure 10.6 Voltage doubling rectifier

positive halves of each cycle, the other diode the negative halves, the resulting voltages being developed across two reservoir condensers wired in series. Each of these receives a charge notionally equivalent to 1.41 times the RMS input to the anode or cathode, and as they are added together to give the HT output to the receiver the combined voltage should be 2.82 times that of the AC input. In practice this is seldom obtained except at light loads, but there is a useful step-up.

Voltage doublers are associated mainly with the alternative to the diode, the 'metal' rectifier which takes advantage of the fact that certain combinations of metals, such as copper and copper oxide, will pass voltage in one direction only. The principle was exploited mainly by the Westing-house Brake & Saxby Signal Co. Ltd, which was the major manufacturer of metal rectifiers during the 1920, 1930s, 1940s and 1950s. It is said that voltage-doubling circuits were favoured with metal rectifiers since each of the two used had only to work at half the total HT voltage, thus reducing stresses on them.

The conventional metal rectifier employed small copper washers oxidised on one surface sandwiched between lead washers of similar size and much larger thin metal plates used only to dissipate the heat set up in the rectifying process. The number of washers and cooling plates employed depended on the working voltage rating; rectifiers for use in radio sets tended to be large and rather cumbersome. More compact types with smaller, thicker discs and plates appeared in the late 1940s, whilst in the 1950s a radical new version was developed. This was the so-called *contact-cooled* rectifier in which the discs were contained in a thin aluminium box, the latter being bolted down firmly to the chassis of the receiver through which the heat was dissipated.

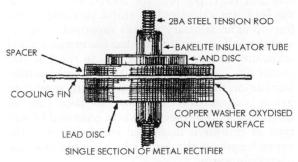

Figure 10.7

Contact-cooled rectifiers were responsible for a brief revival of a fourth type of rectifying circuit called the bridge. This uses four diodes to give the same kind of output as a full-wave rectifier but with only a single HT winding. As will be seen from Figure 10.8 the diodes are arranged in two pairs, each of which is wired with the anode of one connected to the cathode of the other. The four are then arranged so that the remaining anodes are joined together and the remaining cathodes are treated similarly. The AC input is applied to the two junctions of anodes and cathodes so that during each cycle one of each pair of diodes must be conducting in one direction or the other. The positive side of the output is taken from the combined cathodes and the negative from the

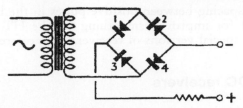

Figure 10.8 Bridge circuit employing metal rectifiers, represented by symbols 1, 2, 3 and 4. The flat bar is the cathode and the arrowhead the anode

combined anodes. The bridge circuit had long been employed in battery chargers, using conventional disc type rectifiers, and continues to be popular in this field. Its reign in radio receivers was limited to a few makes for only two or three years, and it is most likely to be found in imported continental sets.

Isolated chassis

The mains transformer in a true AC-only set isolates the 'works' from the mains itself and consequently the chassis may be earthed for safety as with other metal electrical appliances.

Different mains supplies

In Chapter 1 it was shown how many early power stations supplied their customers with DC. In fact in 1926, when the National Grid was authorised by Parliament, there were 223 different DC-only power stations as against 156 supplying AC only. A further 103 stations supplied both DC and AC. The range of voltages was astonishing: 24 different DC voltages and no fewer than 35 for AC of between about 100 V and 250 V. Most AC supplies were at 50 c/s but 26 stations supplied others ranging from 25 c/s to 100 c/s.

It will be seen from these figures that AC-only receivers made for general sale throughout the country had to be readily adaptable to suit individual mains supplies. The voltage problem was catered for by having the primary windings of mains transformers tapped for different inputs, sometimes every 10 V between 200 V and 250 V. Some early sets also had tappings for 100/130 V but usually listeners on these low voltage supplies had to order special versions of a receiver. Most transformers accepted between 50 c/s and 100 c/s quite happily but it was often necessary to have special versions for 25 c/s. In any case, the greater time spacing between positive pulses in the latter called for improved smoothing for the HT, and again special versions of receivers were offered.

AC/DC receivers

By 1930 there was near parity between the number of AC and DC power stations and by the end of the decade the ratio was about 5:1 in favour of AC. However, it has to be remembered that it had mostly been the more affluent communities which were first supplied with electricity, and that the vast majority of these were DC. They remained as a very significant potential source of sales income for radio firms. Initially DC-only receivers were made but were swiftly outmoded by sets capable of working on either AC or DC without adjustment, save perhaps that of a voltage tapping. The valves used in AC/DC or Universal receivers have

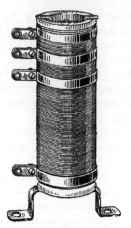

Figure 10.10 Typical mains dropper

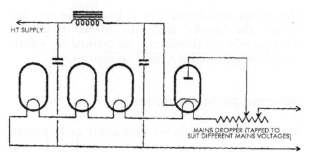

Figure 10.9 Heater chain in AC/DC receiver showing also half-wave rectifier and HT smoothing

their heaters wired in series – a *chain* – and although each may have a different voltage rating they all must pass the same current. There was no general consensus amongst valve makers on the latter, which may range from 0.1 A to 0.3 A. The heaters are supplied from the mains via a voltage dropping resistor usually known simply as a *dropper*. The value of the latter depends both on the total voltage of the valve heaters in the chain, their current and the mains voltage.

Droppers produce heat which has to be dissipated into the air space within a receiver cabinet. Whilst this is of no great significance in the average table set it can be highly inconvenient if not downright dangerous in a small cabinet. This problem was solved by the use of a special kind of mains lead (*line cord* in the USA) incorporating a resistive element. The latter consisted of thin

resistance wire wound on an asbestos core and surrounded by a woven asbestos sheath. When it plus one, two, or occasionally more conductors were enclosed in cotton braiding it was no thicker than a conventional mains lead. The resistance value was generally 60 Ω per foot for 0.3 A rated lead and 100 Ω per foot for 0.15 A. The length of the lead was calculated to suit a particular valve line-up at an arbitrary voltage supposed to represent the half-way mark between the minimum and maximum mains voltages in use. In the UK this meant 225 V and in the USA 117.5 V, and a certain latitude on the part of the valves with respect to heater voltage was supposed to take care of discrepancies. The inevitable heat generated by any type of resistor is dissipated harmlessly over the entire length of the lead, which should run just pleasantly warm.

It has been the writer's ill-fortune to have read in recent years a tremendous amount of nonsense written by ill-informed persons regarding the alleged dangers of resistive mains leads, from its heating effect of the fact that it contains asbestos. Yes, a resistive lead *will* run too hot if too much current flows through it, but this is not the fault of the lead *per se*. It can usually be traced to a bad repair or to an owner shortening the lead for the sake of tidiness. Again, it is conceivable that if one were regularly to chew the resistive element over a period of many years the asbestos might cause harm, but few rational people are likely to indulge in this habit. The truth is that a resistive mains lead, properly installed, poses no more danger than any other type, although naturally, as with any other cable, it needs to be examined for worn insulation and exposed conductors.

Barretters

The barretter is a special kind of electric lamp having an iron filament in a hydrogen-filled bulb. It has the curious but useful characteristic of passing the same current at a wide range of applied voltages, and if placed in series with an AC/DC valve chain it automatically adjusts for different mains voltages. In view of this convenience it is rather surprising that the popularity of the barretter was limited, probably because it cost more than a conventional dropper and also required a special holder.

Rectifiers in AC/DC receivers

Only a half-wave rectifier can be used in AC/DC sets, wired as shown in Figure 10.9. On AC supplies it delivers pulsating DC to the reservoir condenser to be smoothed as in an AC-only set, whilst on DC it acts simply as a low value resistance. Good heater to cathode insulation is, of course, of prime importance (the cathodes in some early AC/DC rectifiers were so thick as to take as much as three minutes to warm up from cold).

Since the HT voltage in AC/DC sets is likely to be somewhat less than that of the mains input, in general UK valves were designed to work with anode and screen voltages of around 200 V maximum. In the USA, where mains voltages were in the 110/120 V range, valves were designed to work efficiently on under 100 V HT, but some could also accept up to 200 V when used in UK receivers. Standard choke-capacity smoothing may be found in better quality AC/DC sets, especially those designed for 200/250 V mains where some voltage drop may be tolerated. Energised loudspeakers too occasionally were employed. Cheap 200/250 V sets and most 110/120 V midgets had the HT for the output valve anode taken directly from the rectifier cathode, with a resistor of a few thousand ohms feeding the rest of the set, in conjunction with large value reservoir and smoothing condensers. The loudspeakers in such sets seldom had a wide enough frequency response to reproduce 50 c/s hum!

An alleged improvement on the above arrangement came into vogue in the 1950s. It featured a special output transformer with an overwind on the primary through which the HT current from the rectifier cathode passed, supposedly to smooth it. It is not unusual for the overwind to fail, and if an ordinary transformer is substituted it is almost impossible to detect any difference, so perhaps this represents a defeat of hope by experience.

Apart from the differences in the power supply section the circuitry of AC-only and AC/DC receivers does not differ basically, although there are, of course, wide variations between makes and models.

'Live' chassis

Most UK manufacturers continued to use the chassis in AC/DC sets for the negative HT line, which meant that it had to be connected to one side of the mains input. In UK and in most other countries, one side of the mains – the neutral – is permanently connected to earth, with the other side – the live – at 230 V above it. This means that should the live side of the mains be taken to the receiver chassis it will be at 230 V above earth and therefore dangerous to touch. Radio manufacturers were well aware of this and with a few dishonourable exceptions took elaborate precautions to prevent owners from coming into contact with any metalwork whilst operating their sets. These included close fitting cabinet backs, covers over chassis fixing bolts and wax fillings over the grub screw used to fasten control knobs onto spindles; these must always be replaced carefully by anyone who has cause to dismantle an AC/DC set. In addition care should be taken to wire the mains lead so that the neutral is taken to chassis.

The better class of American midget sets had the HT– line isolated from chassis, although this was still employed as a return for RF currents. The result was that the chassis could not become live whichever way round the mains was connected.

The writer has also seen a great deal of rubbish expounded by know-all alarmists about the alleged inherent dangers of AC/DC sets. The fact, borne out by statistics, is that AC/DC receivers are no more dangerous than any other electrical appliance and are very probably a good deal safer than some when installed and handled properly.

Part AC-only sets

This oxymoron applies to certain sets in which the valve heaters are supplied by a transformer, making them suitable for AC mains only, but with HT drawn directly from the mains as in an AC/DC set. As a general rule this arrangement was confined to small cheap sets and to firms intent on cutting costs to the bone, possibly as a prelude to going out of business. The same remarks regarding live chassis apply.

Appendix

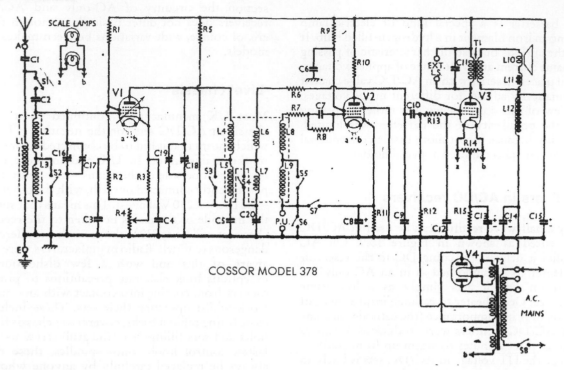

COSSOR MODEL 378

Shown above is the circuit of an actual 'straight three' AC-only receiver, the Cossor model 378 which appeared in 1936. It has some interesting features which illustrate many of the aspects of design mentioned in the text.

The aerial input is either to the primary of the RF transformer formed by L1, L2 and L3, or when S1 is closed to the top of the secondary, which is the MW tuning coil. The switch would be labelled 'local' and 'distant' for different reception areas. S2 shorts out the LW coil on MW. The dotted line enclosing the coils indicates that they are mounted in a metal screening can. C16 is the first tuning condenser and C17 its trimmer. V1 is the variable-mu pentode RF amplifier. Its screen grid is supplied by the voltage divider comprised of R1 and R2, and is decoupled by C3. R2 returns to chassis via R4, a variable resistor in the cathode circuit of V1 which acts as a gain control by varying its grid bias. C4 is the cathode decoupling resistor.

Tuned-anode coupling is used to pass the amplified RF signals on from the anode of V1. This is relatively rare since it involves the fixed plates of the tuning condenser C19 and of the trimmer C18 to be at HT with respect to their moving plates. R5 and C5 decouple the HT feed to V1 anode.

L4/L5, the anode coupling coils, form the primary of an RF transformer with L8/L9 as the secondary. Switches S3 and S5 short out the appropriate LW windings on MW. These coils plus the reaction windings L6/L7 also are contained in a screening can. R6 and R7 are 'stopper' resistors incorporated to improve the stability of the circuit. C7 is the grid condenser and R8 the grid leak for the detector (V2), a straight RF pentode. Note that the bottom of L9 does not return directly to chassis but via S6, which on radio is closed, together with S7. The latter shorts out the cathode bias resistor R11 and by-pass condenser C8 for V2, bias not being required for a grid leak detector. It is needed when the valve is operating as an ordinary amplifier, however, so for gramophone record reproduction (with a pick-up plugged into the sockets marked P.U.) S6 and S7 are opened, bringing R11 and C8 into play and returning the grid of V2 to chassis via the pick-up, which would have been a magnetic type with a DC resistance of about 2000 Ω.

R9 is the screen grid feed resistor for V2, decoupled by C6. R10 is the anode load and C9 the RF by-pass condenser. C10 is the AF coupling condenser, R12 the grid return for V3 and R13 another 'stopper'. C12 is second RF by-pass condenser, an unusual addition. Note that V3 is a directly heated pentode so a humdinger (R14) is wired across the filament. The centre tap of this is returned to chassis via R15, causing the filament to act as a cathode and to provide negative bias for the grid. C13 is the by-pass condenser for R15.

The output transformer in the anode circuit of V3 has a fixed tone control condenser wired across it. The sockets marked EXT LS are for an extension loudspeaker, not a good idea since any wiring to the latter will be at around 250 V DC above earth.

L10 is the speech coil of the moving coil loudspeaker. The way in which L10 and L11 are shown wound on a common iron core indicates that the former is a humbucking coil and the latter a field winding. It does duty as a smoothing choke for the HT supplied by the full-wave rectifier V4, in conjunction with reservoir condenser C14 and smoothing condenser C15.

Note that the wiring to the heaters of V1, V2 and V3 is not shown but indicated by the letters a and b adjacent to them and to the LT winding on the mains transformer. The letters also show that the two scale lamps are fed from the same source.

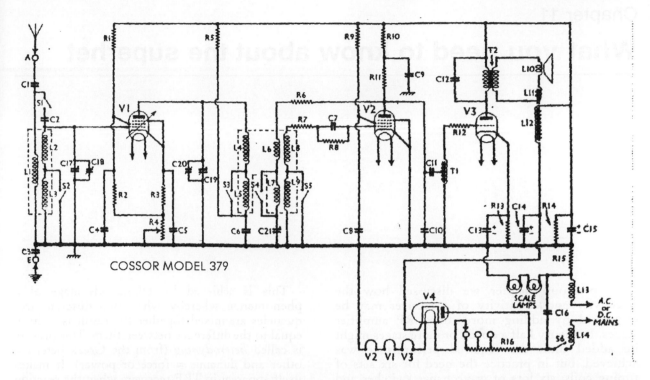

COSSOR MODEL 379

The circuit of the Cossor Model 379 shown above provides an excellent example of how an AC-only set may be adapted for AC/DC working with a minimum of circuit changes. In fact, other than the valves having 13 V 0.2 A heaters there is no difference in the design as far as V2 grid and screen grid apart from the omission of the pick-up sockets. An addition in the anode circuit is a decoupling resistor R10 and condenser C9, the purpose of which is to remove any very slight AC ripple on the HT line remaining after the smoothing process. C10 is the RF by-pass condenser and C11 the coupling condenser feeding the auto-transformer T1 which boosts the AF signals travelling to the grid of V3 via the 'stopper' R12, due to the low HT.

Note that V3 is again an unusual type in being an indirectly heated power triode. R13 is the cathode bias resistor and C13 its by-pass condenser. The circuitry in the anode of V3 is almost the same as in Model 378 except that the extension loudspeaker sockets are omitted.

V4 is the indirectly heated half-wave rectifier with the HT taken from its cathode and the anode supplied from the live side of the mains. S6 is the on/off switch and noise suppressor choke L14. C16 is fitted to suppress 'modulation hum', an obtrusive noise which appears to be tuned in with strong stations and is particular prevalent in AC/DC receivers. L13 is another noise suppressor choke; it was common practice to have them wound on a common former 'back to back' so that electrical noise impulses in the one cancelled out those in the other, and vice versa.

Note that the heaters of V1, V2, V3 and V4 are all wired in series and fed from the mains via the dropper R16, which has three tapping for different voltages. Note also that the heater of V2 is last in the chain before the chassis. It is a convention in all AC/DC sets to place the detector valve in this position in the chain as a precaution against hum caused by heater/cathode leakage.

The scale lamps are wired into the lead from neutral main to chassis, thus carrying both the heater and HT current. They are shunted by R15, which helps to protect the lamps from surge voltages when the set is switched on from cold and the resistance of the valve heaters is lower than normal. It also permits the set to remain working even if one or both of the lamps should go open circuit, which otherwise would break the heater and HT circuits. Positioning the lamps here has other advantages: the lampholders are at low potential with respect to chassis and the passage of both HT and heater current through them assists in keeping them bright when the valves have fully warmed up.

Chapter 11

What you need to know about the superhet

In a previous chapter we discussed how the selectivity and sensitivity of a receiver may be improved by adding more tuned RF amplifier stages. In theory half a dozen or more stages might be added until the required performance was achieved, but in practice the need for six sets of tuning coils, six lots of wavechange switches and six tuning condensers, plus the necessity to keep them in perfect alignment over perhaps three or more wavebands by means of a vast number of trimming condensers, would make the exercise impossibly complicated. The *superhet* receiver makes it possible to have eight or more tuned circuits with no more RF tuning components than in a three-valve TRF. It does this by converting all incoming frequencies to a single one which is passed to pre-tuned amplifier stages that need no adjustment on the part of the listener.

This is achieved by taking advantage of a phenomenon whereby when two different frequencies are mixed together the result is a third, equal to the difference between them. This process is called *heterodyning* (from the Greek hetero = other and dunamis = force or power). It makes itself apparent in TRF receivers when the reaction control is advanced too far and the detector begins to oscillate at close to the frequency of the incoming station. If this is at 1000 kc/s and the oscillation at 999 kc/s the result will be a difference frequency of 1 kc, manifesting itself as a piercing whistle from the loudspeaker.

The full name of the superhet is the *supersonic heterodyne* receiver, which indicates that the heterodyne frequencies employed are above the range of human hearing; in practice they range from about 100 kc/s to 500 kc/s. This is called the *intermediate*

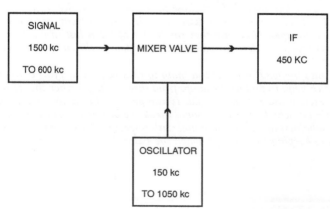

Figure 11.1 Block diagram of the frequency changer stage in a superhet

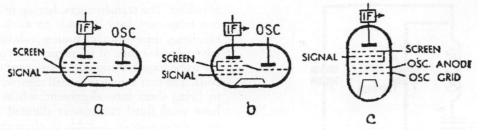

Figure 11.2 Three popular types of frequency changer valve: (a) The triode-pentode; (b) The triode-hexode; (c) The heptode or 'pentagrid'

frequency (IF). To simplify the workings of a superhet let's consider one covering just the MW band from 600 kc/s to 500 kc/s and employing an IF of 450 kc/s. Its aerial input and tuning circuitry feeding the grid of the first valve would be just like that of the RF amplifier in a three-valve TRF, but the valve involved does more than simply amplify. Associated with it is a local oscillator, the job of which is to generate oscillations over a band of frequencies exactly 450 kc/s higher than that of the incoming frequency; for instance, if a station on 1000 kc/s is to be received the local oscillator will produce 1000 kc/s + 450 kc/s = 1450 kc/s. When this is mixed with the incoming frequency the result is a heterodyne at 450 kc/s. The tuning condenser for the local oscillator is ganged with that of the RF tuning so that if the latter is adjusted to tune in 1100 kc/s the local oscillator will produce 1550 kc/s, and so on.

The local oscillator itself may employ a separate valve or the RF valve may be made to self-oscillate.

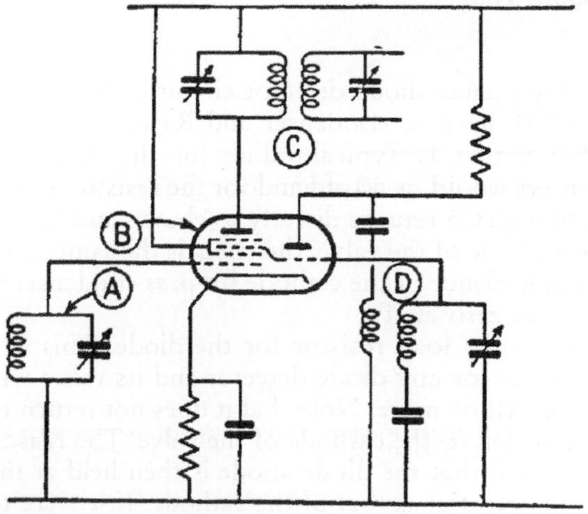

However, by far the most popular arrangement is to use a special *frequency-changer* (f.c.) valve with yet more electrodes than a pentode. One popular valve has no fewer than five grids and is therefore called a heptode or *pentagrid*. Numbering these conventionally from that nearest the cathode, G1 acts as control grid of a triode with G2 as its virtual anode; these are employed as a local oscillator. G3 and G5 are a pair of screen grids connected together and spaced either side of G4, the main control grid.

The oscillations produced by G1 and G2 modulate the electron stream from the cathode on its way to G4 and thus frequencies received from the aerial mix with those of the oscillator to produce at the anode the IF.

Another valve working on this principle is the octode, in which an extra grid (G6) is placed between G5 and anode, where it acts like the suppressor grid in a pentode.

An alternative type of frequency changer is in fact a double valve incorporating a triode section to act as local oscillator. It shares a common cathode with the receiving section, which in early examples was a pentode, where the mixing of oscillations takes place via the common cathode. A later improvement was the triode-hexode, the latter section having four grids. Of these, G2 and G4 are screen grids. According to individual design either G1 or G3 may be the control grid, leaving the other to act as injector grid; connected internally to the grid of the triode, it modulates the electron stream as in a heptode or octode. The triode-heptode has a fifth grid, G5, which acts like the suppressor grid in a pentode.

Keeping the local oscillator in step

The most popular method is to use a standard two-gang variable condenser for tuning both RF and

Figure 11.3 Basic frequency changer stage employing a triode-hexode valve

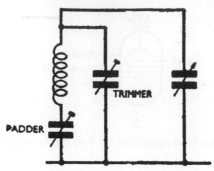

Figure 11.4 Simplified diagram of how trimmer and padder condensers are wired in a local oscillator. The trimmer is always adjusted at the high frequency end of a band and the padder at the low frequency end

local oscillator coils, with series and parallel condenser added to the latter to effect adjustments at the low and high frequency ends of each band. An alternative is the use of a two-gang tuning condenser with one normal set of plates for RF tuning, and for the oscillator a smaller, specially shaped set which automatically keep the two in step. *Tracked* tuning condensers, as they are called, appeared in only a few sets, probably because they were dearer to produce. Whatever type of frequency changer is employed, the end result is always an IF appearing at the anode of the mixer section. This is coupled by special *intermediate-frequency transformers* (IFTs) to the grid of an IF

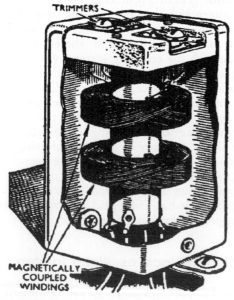

Figure 11.5 Air-cored IF transformer

amplifier. The transformers, having to handle only one frequency, may be made to work with higher efficiency than ordinary tuning coils which have to cover a wide range of frequencies. Two main types are to be found, using either air-cored or iron-dust coils. The first type have small trimmer condensers to bring them into alignment, whilst the second have small fixed condensers shunted across them with the dust core movable to effect the tuning. In either case the response curve may be made to approach the ideal for good station separation consistent with adequate sideband coverage.

Another IFT couples the IF amplifier to either a second amplifier in expensive receivers or, more commonly, to the detector. Although early superhets used a triode in either grid-leak or anode-bend mode, they were very soon supplanted by the diode, with its ability to rectify strong signals without distortion. The diode may be a separate valve or, more likely, it will be incorporated into another, often a triode used as first AF amplifier.

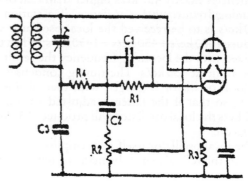

Figure 11.6

A specimen diode detector circuit is shown. C3 is an IF by-pass condenser and R4 acts as an IF filter with C1. Typical values for the two condensers would be 50 pfd and for the resistor 47 kΩ. Although C3 returns directly to chassis and C1 via the cathode of the valve, the effect is the same since the impedance of the cathode by-pass condenser is close to zero at IF.

R1 is the load resistor for the diode. This is a necessity for any diode detector and its value will be 500 kΩ or more. Note that it does not return to chassis but to the cathode of the valve. The reason for this is that the diode anode is then held at the same potential as that of the cathode. If it were to return to chassis it would receive an effective negative bias equivalent to the voltage appearing across R3. This in turn would prevent it from conducting until the voltage of the incoming signal exceeded that of the bias. The effect would be to restrict reception to strong stations only, with all others suppressed.

However, bias *is* necessary for the triode section of the valve, so its grid must return to chassis, which it does via the potentiometer R2, which also acts as volume control by enabling any desired proportion of the total AF voltage across it to be passed to the grid. C2 permits the AF signals from the diode to travel freely to the top of R2 but blocks the DC voltage from the cathode.

The circuitry of a superhet following the detector and first AF amplifier is precisely the same as in a TRF and needs no further discussion.

Automatic volume control

Even a simple superhet with one stage of IF amplification following a mixer can have greatly improved sensitivity over a TRF. This could make it rather uncomfortable for the listener when tuning along a wave band as, if the volume control were turned up to receive weak stations, strong ones would be deafeningly loud. Thus tuning had to take place with another hand ready to operate the volume control, not a particularly convenient or attractive proposition as a sales point. Another problem was that distant, weaker, stations tended to fade, that is their signal strength varied, especially at night on MW. With *automatic volume control* (AVC) the gain of the receiver is controlled by the strength of the signal received and thus automatically adjusts itself to weak or strong stations or to fading. In the 1940s there was a tendency to replace the term AVC by *automatic gain control* (AGC) and both may be found in vintage radio literature; they may be taken as synonymous. Once again the humble diode carries out the work. We have seen previously how the effect of rectifying a signal results in a steady DC voltage plus the AF modulation; in TRF sets only the latter is required and the DC voltage dumped. However, since this voltage is both negative with respect to chassis and is directly proportional to the strength of the incoming signal it can be used as grid bias for the mixer and IF amplifier valves when these are variable-mu types. In fact, this is exactly what was done in many cheap sets, AVC bias simply being tapped off from the signal diode load resistor. Simple AVC, as it is called, works well enough on local stations but it is far from ideal for long distance reception. The reason for this is the fact that any level of signal at the diode will start to generate AVC bias and thus reduce the gain of the set, even though this is undesirable for weak stations. *Delayed AVC* is designed to prevent this happening by preventing AVC bias from being developed until and unless the signal strength of a station is adequate to give full loudspeaker volume. This entails the use of a separate diode to rectify the carrier wave purely for bias purposes; known as the **AVC rectifier or rectifier**, it is again usually incorporated in the envelope of the first AF amplifier, as shown in Figure 11.8.

The anode of the AVC diode is fed with the IF signal via C1, typically 100 pfd. At one time it was common practice to obtain the signal from the detector diode anode, as shown, but research showed that this tended to introduce AF distortion, so the preferred method is to couple the AVC diode to the anode of the IF amplifier. R2 is the usual cathode bias resistor for the triode section, whilst R1 is the load resistor for the AVC diode

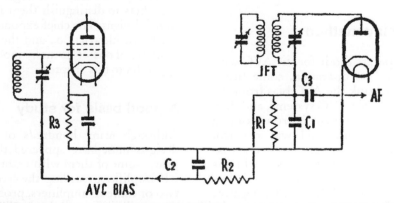

Figure 11.7 Simple AVC derived from the negative DC voltage appearing on the detector diode load resistor

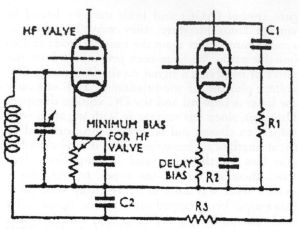

Figure 11.8

anode, typically about 1 mΩ As the resistor returns to chassis, the diode anode is biased negatively by the amount of voltage existing across R2. The set designer knows how much signal voltage is required to obtain full volume at the loudspeaker, so he arranges for the AVC rectifier anode to be biased to the same amount; e.g. if the required signal voltage is 3 V the bias also will be 3 V. Thus the value of R2 is chosen to give that amount of cathode bias. In some cases a high degree of delay is required which cannot be obtained in this way, so it is 'borrowed' from elsewhere in the set, such as a negative smoothing circuit (see previous chapter). We shall look at advanced AVC systems in a later chapter. R3, typically the same value as the load resistor, filters out any remaining IF on the diode anode, with C2, typically 0.1 μfd, as a decoupling condenser. In this circuit only the IF amplifier is shown as receiving AVC bias but in practice it would usually be fed to the frequency changer and RF amplifier, if used.

An alternative delay method

Instead of using a separate diode for the AVC with bias applied to 'delay' its operation, some firms, especially in the USA, took the AVC bias from the detector diode, as in simple AVC systems, and then used a 'clamp' diode to prevent it from coming into effect before a certain level was reached. Basically, the clamp diode is simply shunted across the AVC line and a small amount of positive bias is applied to its anode to make it conduct and thus effectively to short out the AVC until its negative voltage exceeds that of the positive voltage on the

diode anode. When this occurs the diode ceases to conduct and the AVC takes control of the valves. The positive bias for the clamp diode is usually obtained from a voltage divider across the HT line, with the resistor connected to HT+ being of high value, typically between 4.7 MΩ and 22 mΩ. One of the advantages of this system is that the delay bias, no longer depending on the cathode voltage of a double-diode-triode, readily can be set at any voltage the designer wishes.

A further refinement of the method was designed to save the use of an extra diode to provide the clamp action. It was found that the suppressor grid of an RF pentode in the IF amplifier stage could be used as a virtual diode, and one instance of this being exploited appears to have been in the well-known 'Wartime Civilian Receiver' (1944). A year or two later Mullard produced a miniature RF pentode incorporating a single diode (EAF42/UAF42) expressly designed for this kind of work, the diode being used as detector and to provide AVC and the pentode suppressor grid used to provide the delay.

The 'short' superhet

By the mid 1930s advances in valve technology had made possible the production of steep-slope mains output pentodes having Gm figures of up to 10 mA/V. Their high sensitivity enabled them to be driven directly from the detector in a superhet without an intermediary AF amplifier (or, at least, this was the intention, not always realised as we shall see later). A natural extension of this was to incorporate the detector diode, plus an AVC rectifier, into the same envelope, thus producing the double-diode-output pentode (DDP). Superhets using just three valves, frequency changer, IF amplifier and DDP, plus the necessary rectifier, soon began to be called short superhets to distinguish them from more conventional designs. The chief exponent of the genre was the Ultra Electric Co., and the circuit of a typical example of one of its many models is shown in the appendix to this chapter.

A good basis for study

Although many hundreds of different superhet designs must have appeared through the vintage years, some of them with extra features such as an RF amplifier preceding the frequency changer, or two or more IF amplifiers, probably about 85% of them will differ only in detail from the two basic types discussed above. As long as you have a good grasp of the principles involved you should be able to work your way without difficulty through almost any circuit.

Reflex amplifiers

Despite the claimed ability of double-diode-pentodes to work directly from the detector, the lack of an AF amplifier did make itself apparent in some receivers. To make amends for this a device called *reflexing* was adopted. This makes use of the ability of a valve to amplify both HF and AF simultaneously without mutual interference. The usual arrangement is to have the AF signals from the detector fed back to the grid of the IF amplifier via the secondary of the first IFT, along with the AVC bias. Amplified AF signals may be taken either from the anode of the IF valve or from its screen grid acting as a virtual anode. When the latter method is used the value of the decoupling condenser on the screen grid has to be chosen carefully to ensure that it bypasses currents at IF effectively without affecting the AF signals.

Part superhets

This description refers to sets which have a frequency-changer stage as in a normal superhet, but instead of an IF amplifier have a grid-leak detector usually with reaction. They do not have AVC and use the same sort of volume control system as in TRF receivers. It was a cheap way to be able to make a set that could be sold as a superhet but which cost little more than a TRF, with a marginal improvement in performance. A.C. Cossor Ltd was the main exponent of this genre, some acting as a superhet on long, medium and short waves but with others operating as a superhet only on short wave and as a TRF on medium and long. The part superhet appeared first around 1936 and continued to crop up now and again over the next 20 years, virtually to the end of the valve era. It is most likely to be found in cheap 'midget' sets (see T43DA at the end of chapter).

Battery superhets

Equivalent battery valves were produced for all the mains types discussed above (with the exception of the high slope double-diode-pentode) and most receiver manufacturers produced battery powered superhets. They follow exactly the same principles

as their mains equivalents except, of course, that the LT, HT and GB supplies conform to those already discussed for battery TRF sets. There is thus no need to enlarge upon their design in the RF amplifier, frequency changer, IF amplifier and output stages, but we need to look at the detector and AGC stages, where some unusual valves may be encountered.

From about 1935 onwards the vast majority of battery superhets used diodes for detection and AGC, mostly in the same double-diode-triode package as found in mains sets, except, of course, that they lacked a cathode. The circuitry is very similar apart from the methods of biasing the triode grid and the diode anodes. The Ever-Ready 5030 affords a typical example. The grid is returned, via the bottom end of the volume control R22, to a battery tapping 1.5 V negative with regard to chassis. The signal diode is isolated from the volume control and this voltage, by C12 and R10, returns it to LT+. The AGC diode returns to chassis via R14 and R15 and so receives a modest amount of bias by virtue of the filament being slightly positive. This would delay the AGC action to a small extent, but not enough to satisfy some designers. For their benefit some valve manufacturers introduced an almost unique device – a battery powered double diode which was indirectly heated.

Two alternative circuits taking advantage of this valve are those of the Cossor 376B and the Decca PT/ML/B (or PT/B). Cossor used a simple potential divider across the HT supply to provide approximately 6 V on the cathode of their 220DD valve. The signal diode was returned to the latter via, again, the volume control, and the AGC diode to chassis via a 1 MΩ resistor. The delay voltage was thus that of the cathode, permitting better reception of weak stations.

Decca plumped for cathode bias derived from an ordinary grid bias battery, the connections of which are a little confusing at first sight. It is, in effect, split into two sections, the chassis being connected to the −6 V socket and the cathode to the positive. The −9 V tapping goes to a potential divider network consisting of R15, R16 and R17, which provide different levels of bias for the various valves. The diode cathode returns to the junction of R15 and R16, so the actual delay voltage is that found at the junction, some 0.75 V added to the 6 V from the battery, giving rather more delay than in the Cossor receiver. The valve used in the Decca was the Mullard 2D2, equivalent to the 220DD, but with a filament current of only 0.09 A as against 0.2, this being kinder to the LT supply.

Appendix

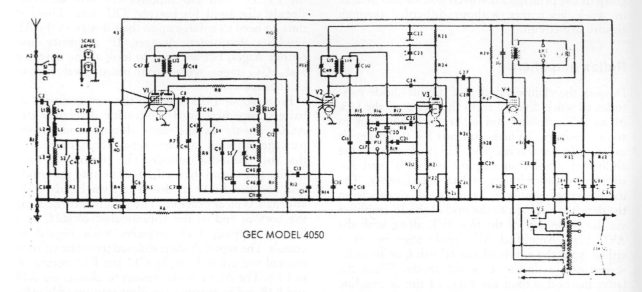

GEC MODEL 4050

The GEC Model BC4050 was released in March 1939. It is an absolutely typical circuit of a four valve plus rectifier AC-only superhet of the latter half of the 1930s, a design which continued right through until the end of the valve era 20 years later and as such it contains many of the features discussed in the text. It covers long, medium and short waves; note that the aerial input and RF tuning coils are little different from those of a TRF, with the exception that the bottom of the long wave coil does not return to chassis but to the AVC line.

The frequency changer is a triode-hexode and the local oscillator has been simplified by the use of a single anode coil for all three bands. Note the padding and trimming condensers associated with the oscillator grid coils.

The IF is 456 kc/s, and the pentode IF amplifier is followed by a double-diode-triode action as detector, AF amplifier and AVC rectifier. Note the very comprehensive IF filtering in the diode load circuit. Further filtering takes place at the anode of the triode, which is resistance-capacity coupled to the grid of the pentode output valve.

There is a fixed tone correcting condenser across the primary of the output transformer and a continuously variable tone control consisting of C32 and R31. The power supply section uses a conventional full-wave rectifier with a mixture of choke and resistance smoothing. Note that one side of the LT winding is connected to chassis.

All the valves have octal bases, which first appeared in the USA in 1935 and swiftly spread across the Atlantic. This type of base nominally has 8 pins for electrode connections (although not all installed for the simpler valves), set around a locating spigot which ensures that it is inserted into its holder correctly. The receiving valves have 6.3 V heaters and that of the rectifier is rated at 5 V.

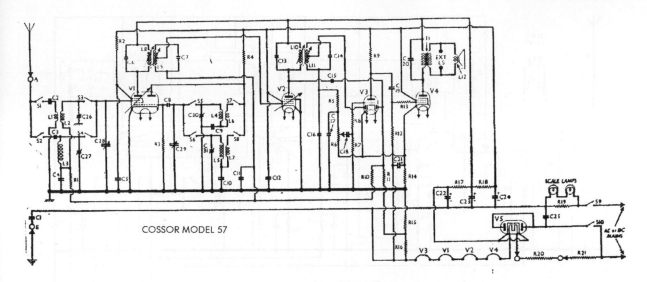

COSSOR MODEL 57

The Cossor Model 57 appeared in 1942, which was rather surprising since not only had domestic radio production virtually ceased by that time, A.C. Cossor Ltd was a major contractor to the Government for radar equipment. The 57 is a four-valve plus rectifier AC/DC having many of the features discussed in the text, and may be considered as very typical of this class of receiver, with the sole exception that as long wave broadcasting by the BBC had ceased on the outbreak of war the set covered only medium and short waves.

Following the growing trend for transportable receivers, the medium wave RF tuning coil was actually wound as a frame aerial, attached to the back cover for the cabinet. A triode-hexode frequency changer is used and the IF is 465 kc/s. The IF amplifier is an RF pentode.

A double-diode-triode is used as detector, AF amplifier and AVC rectifier: note that its cathode is strapped to that of the output pentode. This was a fairly common device to have both valve's HT current combined to produce a larger voltage drop than would have been the case with just one. This voltage is fed back via the detector load resistor to the diode anode, thus preventing it from being negatively biased.

The delay bias for the AVC rectifier is obtained in a very interesting manner. The two resistors R15 and R16 are in series between the HT negative line and the chassis, and carry the entire HT current drawn by the valves. The resulting voltage drop across the two resistors is used to bias the grid of the output pentode, whilst the reduced negative voltage at their junction is fed back to the AVC diode anode and also along the AVC line to the grids of V1 and V2. This means that cathode bias is unnecessary and that their cathodes may be returned directly to chassis, saving the cost of two resistors and decoupling condensers.

Although V5, the rectifier, has two sets of anodes and cathodes these are strapped and the valve operates in the usual half-wave mode associated with AC/DC sets. Note the two-stage resistance smoothing for the HT. R21, the main dropping resistor for the heaters is in this set incorporated in the mains lead. By itself it is suitable for mains inputs of 200/225 V while the addition of R20 permits supplies of 225/250 V to be used.

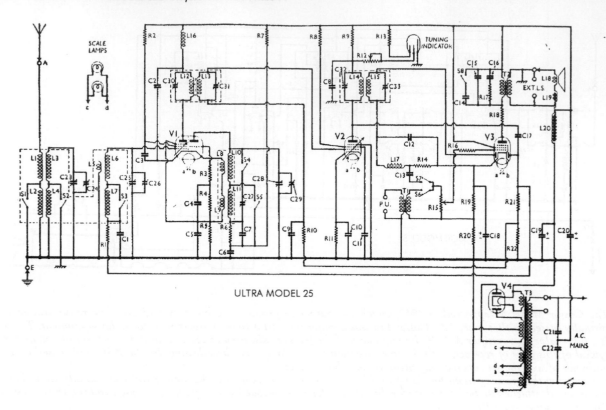

ULTRA MODEL 25

The Ultra model 25 is absolutely typical of the 'short' superhet and of Ultra's products in general, for the firm used this type of circuitry extensively. The sets were not usually very exciting but they were well designed and made.

Note the use of band-pass tuning preceding the triode-pentode frequency changer, V1. The local oscillator is well designed with iron-dust cored coils and fixed padders (C6 and C7). The IF is 456 kc/s and air-cored IFTs are used. The IF amplifier (V2) is an RF pentode; ignoring the usual practice its suppressor grid is returned to chassis and not to cathode. Resistors R9, R12 and R13 in the anode circuit feed the neon lamp tuning indicator.

The detector diode load includes an HF choke to filter out any residual IF, and is returned to the cathode of the pentode section to avoid its anode receiving negative bias. Note that the cathode resistor is made up of two (R1 and R20) in series; this is because the bias voltage required for the pentode grid is lower than that needed for the delayed AVC. Thus the grid is returned via the volume control to the junction of the two resistors whilst the AVC diode load returns to chassis. Ultra was ahead of many other makers in feeding the anode of the AVC diode from that of the IF amplifier, a much better idea than feeding it from the detector diode.

Due to the omission of an AF amplifier before the output stage 'short' superhets were always plagued with low gain when a gramophone pick-up was used. In this set a step-up transformer (T1) is fitted to mitigate this problem.

The load for the AVC diode also consists of two resistors (R21 and R22) in series, with the bias for V1 being taken from the top and that for V2 tapped off at their junction. The effect of this is that V2 receives about three-quarters of the bias voltage whilst V1 has the full amount.

Note the use of two 'top cut' condensers (C15 and C16, the latter in series with R17) across the primary of the output transformer (T2), with a third condenser (C14) able to be brought in by closing S8 if desired by the listener to decrease still further the treble response.

An energised loudspeaker is used, the field of which smooths the HT provided by the indirectly heated full-wave rectifier. An unusual feature was the inclusion of a separate LT winding on the mains transformer for the dial lamps. Condensers C21 and C22 in series across the mains transformer primary, with their junction taken to chassis, supress mains-borne interference and 'modulation hum'.

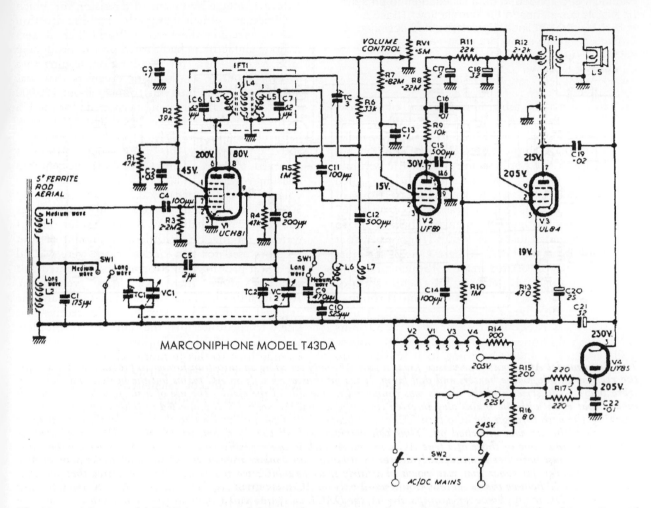

MARCONIPHONE MODEL T43DA

The Marconiphone model T43DA was one of the very last receivers to be made by EMI before it closed its radio and television division. A part-superhet transportable AC/DC set, it was arguably the worst design ever produced by the firm, a sad and uncharacteristic ending to a glorious era. However, it stands as an example of many of the design features mentioned in the text, and of how not to make a radio set.

A ferrite rod aerial is used, carrying the medium and long wave tuning coils which are coupled by C4 to the control grid of the triode-heptode frequency changer: its grid is returned to chassis by R3. The tuning knob is attached directly to the spindle of the variable condenser VC1/VC2 without benefit of slow-motion drive and no dial lamp is fitted.

The triode section of this valve is employed as the local oscillator designed on the cheapest lines possible – just one HF transformer basically for medium waves but with the grid coil shunted by C9 to give LW coverage. V1 has to operate without any bias at all, there being neither cathode resistor nor AVC.

The single IF transformer feeds signals to V2, an RF pentode used as a grid leak detector with reaction, TC3 being a preset control for the latter. The resulting AF at its anode is resistance-capacity coupled to the grid of the output pentode, V3, with RF filtering by C15 and C14. No proper AF volume control is fitted: instead, using technology from a by-gone era, this is effected by varying the HT voltage on, simultaneously, V1 anode, screen grid and oscillator anode, and on the screen grid of V2. It is a minor mystery how the volume control potentiometer, RV1, managed to carry all this current without over-heating.

V4 is the half-wave rectifier with HT smoothed by the over-wind on the output transformer TR1, resistors R12 and R11 and condensers C21, C18 and C17.

The big question about this set is, why would anybody do it on purpose?

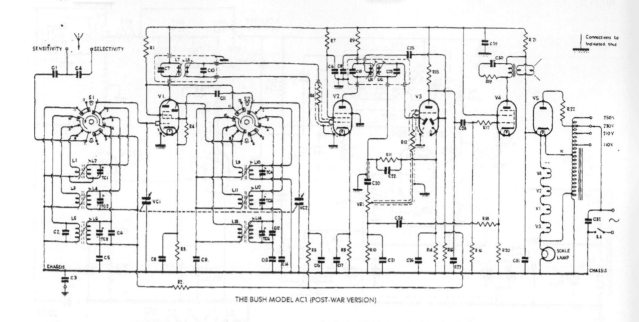

THE BUSH MODEL AC1 (POST-WAR VERSION)

The Bush model AC1 (post-war version) was a part AC-only set using an auto-transformer to provide the voltage for the rectifier anode, valve heaters and dial lamp. It would thus work only on AC mains but the heaters and dial lamp were operated in series and the chassis was connected to one side of the mains. The use of a single-pole on/off switch meant that even with the mains plug inserted correctly, so that the neutral was connected to chassis when the set was working, the path through the auto-transformer and heaters back from the live main to chassis meant the latter became 'live' when the set was switched off. Thus this receiver had all the disadvantages of an AC/DC set without its advantages, and none of those of a true AC-only set. As such it was arguably an exercise in futility because the auto-transformer must have been considerably dearer to produce than a mains dropper. It did run much cooler than an AC/DC set, however, for what that was worth in a fairly spacious table cabinet. Perhaps Bush considered that potential customers would believe that a set that would work only on AC was intrinsically better than an AC/DC type. To cater for people on DC mains, however, a companion set, the DAC1, was fitted with a mains dropper and thus became suitable for either AC or DC. Strange are the workings of salesmen's minds. The two alternative power supply stages are shown side by side (the frequency-changer stage is omitted for clarity). In either case only resistance-capacity HT smoothing was employed.

Bush exploited this bizarre fantasy about 'AC-only' and AC/DC models in a series of other receivers, so presumably it must have been commercially successful.

Another feature of the AC1/DAC1 was something Bush called 'bi-focal tone', a device which was supposed to give the same fidelity of reproduction at high or low settings of the volume control. It was in fact negative feedback from the cathode of the output valve (note the absence of a by-pass condenser) to the bottom of the volume control via R16 and C24. It was an interesting idea but it was soon quietly dropped.

Chapter 12

Some more special features found in superhets

Tuning indicators

The greatly improved selectivity of the superhet caused problems for some listeners who had recently converted to them from antiquated TRFs. In the latter it was usually possible to get reasonably good reception even if the tuning control(s) were not set accurately; indeed, in the absence of proper volume controls it was common practice deliberately to de-tune a set to reduce the output. With a superhet, however, mistuning whether unintentional or deliberate results in very harsh reception due to only part of one sideband being reproduced (sixty-five years of experience have not eradicated completely the problem!).

To assist listeners to tune their sets properly manufacturers began to fit visual tuning indicators. The most popular early type took the form of a small milliameter wired into the HT supply of one of the valves – usually the IF amplifier – controlled by the AVC. At the point of correct tune the AVC bias would be at its peak and thus the anode and screen-grid currents of the valve at minimum. Rather than give the meters just a plain dial, designers often placed a small lamp behind them so that they gave their indication by casting a variable shadow onto a small celluloid screen near to the tuning scale of the receiver. One of the best examples was EMI's 'fluid light tuning'.

Another indicator based on light was employed by Pye in several models. A small LF transformer was placed in the HT supply to the IF amplifier so that its anode current flowed through the primary. There were two secondary windings wired in series with each other and also with a dial lamp supplied from the 4 V AC LT winding on the

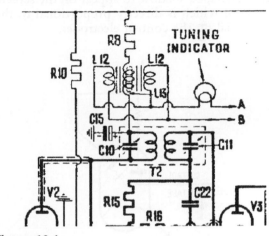

Figure 12.1

mains transformer. The amount of DC current passing through the primary altered its magnetic saturation which in turn varied the impedance of the two secondaries to the 4 V AC and caused the brightness of the lamp to alter in sympathy with the AVC bias applied to the IF valve.

An alternative light-type indicator that appeared in some makes of receiver consisted of a special kind of tubular neon lamp about four inches long and half an inch in diameter. It had three electrodes, two to produce the initial glow and one to control the length of the glow along the lamp. The control electrode was again connected into the anode and screen-grid circuit of the IF amplifier, which in turn was supplied from the main HT line via a dropping resistor. At the point of exact tune the AVC reduced the current passing through this resistor and thus the amount of voltage dropped across it, increasing the voltage on the control electrode and the length of the glow. It was a picturesque and reasonably effective means of obtaining exact tune but unfortunately the neon lamps eventually suffered from blackening of the glass inside the bulb.

The definitive tuning indicator came with the invention of the 'magic eye' around 1936. Really a miniature cathode ray tube, it has the shape of a conventional valve but at the top of the bulb is a small screen about an inch in diameter which fluoresces green when connected to the HT line and bombarded with electrons from the cathode. The latter is mounted to protrude through a circular hole in the middle of the screen but concealed by a metal shield. Also hidden by this shield is a blade-like control electrode which causes a V-shaped shadow to appear on the screen, the angle of which is directly proportional to the HT potential on the control electrode.

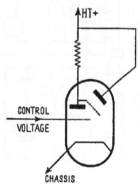

Figure 12.2 Theoretical diagram of a 'magic eye' tuning indicator

Mounted below the screen assembly but sharing its cathode is a small low-mu triode, the anode of which is connected directly to the control electrode and to the HT line via a high value resistor. The grid is connected to the detector diode load resistor so that it receives negative bias as a signal is tuned in. Under no-signal conditions with zero bias the triode takes its maximum anode current, there is considerable voltage drop across its feed resistor and the control electrode receives little HT and the shadow is at its widest. When a station is tuned in the negative bias produced causes the anode current of the triode to drop, increasing the anode and control electrode voltage and the shadow narrows, exactly in proportion to the

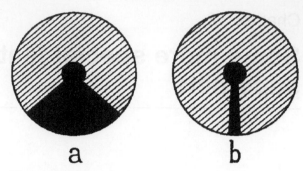

Figure 12.3 The shadow on the screen of a 'magic eye' (a) in the absence of a signal and (b) with a strong station tuned in

signal strength of the received station. This tells us why the grid is connected to the detector load rather than the AVC line, especially when the latter is delayed; we want the grid to receive bias even on the weakest stations and thus indicate the state of tune even if they are not powerful enough to start the AVC working.

A later variation of the magic eye, found only in Philips/Mullard receivers, is the dual sensitivity type having two control electrodes. It contains two triodes of different characteristics, one of which will respond to very weak signals, the other only to powerful stations.

Towards the end of the valve era came three other, miniature, versions, one of which has a fan-shaped display, another with a 'thermometer' display and the third with something that resembles an exclamation mark. This last is tiny – about the thickness of a pencil and two inches long – and has a 1.4 V, 25 mA filament making it suitable for battery receivers, but the writer can recall none which actually used it. It did, however, appear in some mains AM/FM receivers.

All magic eyes suffer, just as the cathode ray tubes in television receivers, from dimming of the display due to loss of emission from the cathode. Replacement is the only answer, but before hope is given up the triode anode feed resistor(s) should be checked for open circuit or very high resistance.

Automatic tuning

In the early 1930s the domestic radio receiver largely had ceased to be a novelty which only an expert (usually the father of the family) was able to handle, and had become an everyday appliance that virtually anyone could operate. To make the task

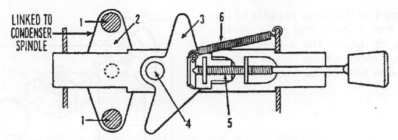

Figure 12.4 Layout of a mechanical pushbutton (edge-on view). 1 and 1 are two rods linking the crank 2 to the other cranks. The dotted ring in the centre of 2 represents a spindle connected to that of the tuning condenser. 3 is a triangular cam, pivoted about 4. When the button is depressed push-rod 5 bears on 3 and moves it towards the rods 1 and 1, pushing the nearest around the spindle until both are touching 3. Thus turning the condenser spindle also to the desired station. Spring 6 returns the pushbutton to the rest position as soon as pressure is released. Turning the button loosens 5, which locks the position of 3.

even simpler, many firms started to introduce some kind of automatic tuning that would find the listener's favorite stations. The most simple method used was mechanical in operation and consisted of fitting a dialling device, like that of the telephones of the period, attached to the spindle of the gang tuning condenser and with the finger holes marked with the names of stations. To receive any of these the listener inserted a digit in the appropriate hole and pulled the dial round, to stop automatically at the correct place on the tuning condenser.

Another mechanical device employed a row of push buttons with adjustable cams bearing on a swivel plate which converted depression of a button into rotation of the tuning condenser. A simple means of adjustment made it possible for the listener to select any desired station along the dial. As in the telephone-dial system, initially no provision was made for selecting wave bands automatically so it was customary to have all the preset stations on the popular MW band. Later, a modified system made it possible to have perhaps two or three of the pushbuttons set for LW stations with automatic change over to and from MW as required.

The great advantage of mechanical systems was that they required no modification to the receiving circuits of the sets and consequently were the cheapest to offer to the public. The only drawback was that, as with any other mechanical device, any wear in the components caused mistuning and occasional resetting of the stations might be required.

A more costly method, but one which involved no moving parts other than in the pushbutton unit itself, had each button operating a small flat switch which brought into play a pair of RF and local oscillator coils pre-tuned to a particular station. It was up to the individual manufacturer as to whether the coils were air cored with small variable condensers, or iron-dust cored with fixed

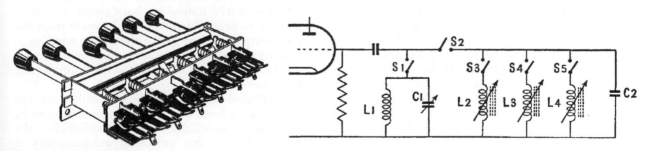

Figure 12.5 Above left: A typical pushbutton switch unit of the late 1930s. Above right: how some of the P/B switches could be used to select (S1.2) manual tuning for the local oscillator or (S3, 4, 5) present iron-cored coils each tuned to a different station

condensers; the latter was the more popular of the two. Bush Radio was a great exponent of the system, advertising it under the slogan 'Bush Button Tuning', but most of the major manufacturers used it at one time or another.

Motor tuning

The undoubted king of automatic systems was the motor tuning introduced in the UK by Ekco in 1938. With this all effort was taken out of the hands of the listener, the work of both tuning and wavechange switching being carried out by two motors. The idea had originated in the USA a year or two earlier but there was a drawback in that

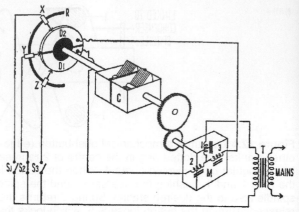

Figure 12.7 Basic layout of the motor tuning system (see text)

Figure 12.6 The small electric motor used in motor tuned receivers

system whereby, no matter how close two stations might be to each other, when the listener switched from one to the other the motor had to traverse the entire tuning range and then reverse back to the new station. Ekco refined this by arranging that the motor would automatically travel in the correct direction to suit any change of station. Figure 12.7 shows how this was done. The motor (M) has three windings: 1 is always used while 2 or 3 is selected as required to make the motor run clockwise or anti-clockwise. It drives the tuning condenser C through reduction gearing, and an extension on the spindle of C drives a selector disc made of insulating material and carrying two semicircular strips of copper, D1 and D2, mounted so that there is a small gap between them. Around the periphery of the disc is a semicircular rail carrying a number of small spring contacts each of which bears on the surface of either D1 or D2

according to its position at any one time, and makes contact with it. These contacts may be moved around the carrying rail to any desired position to correspond with the stations to be tuned in. In practice there would be up to ten, but for clarity only three, X, Y and Z, are shown, connected to S1, S2 and S3 which correspond to three press-button switches.

In the diagram S2 has been closed but its associated spring contact, Y, is pressed against the insulating cap between D1 and D2 and no electrical contact is taking place. Suppose now that S1 is pressed, by which S2 is opened automatically. Contact X is presently pressing on D2 so an electrical circuit is formed from one side of the secondary of the mains transformer T through to winding 3 of the motor and thence back through winding 1 to the other side of the secondary. This causes the motor to run, turning both the tuning condenser and the selector disk until the gap between D1 and D2 arrives beneath X, breaking the circuit and stopping the motor at the point where a desired station has been tuned in. Should S3 now be pressed contact will be made from the secondary via Z and D1 to the motor winding 2, causing the motor to run in the opposite direction until the gap arrives beneath Z at the point of tune for another station.

Additional contacts make it possible for the correct wave band for any particular station to be selected automatically, with provision for the listener to use this switching independently if required. If manual tuning is required it is aided by a couple of small buttons which switch the motor to tune the set in the required direction.

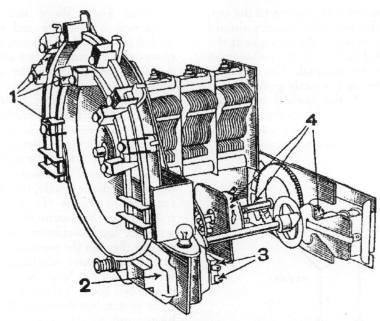

Figure 12.8 Diagrammatic view of the motor tuning system used by Ekco. 1: station selector contacts. 2: The driving motor. 3: muting contacts which short out the loudspeaker when the motor is running to give silent tuning. 4: wavechange switch components. Just above the motor will be seen a pilot lamp, used for setting the station's selector contacts accurately

By 1939 a number of leading manufacturers were offering motor tuning in their most expensive models. Unfortunately changed economic conditions after the war meant that motor tuning all but disappeared, the sole survivor being in a very large and costly 15-valve radio-gramophone made by EMI.

Automatic frequency control

Both non-mechanical press-button tuning and motor tuning were vulnerable to the effects of tuning 'drift' that might make listening to some stations unpleasant or impossible. To avoid this possibility the better manufacturers incorporated automatic frequency control (AFC) which automatically compensated for slight mistuning. It did this by altering slightly the tuning of the local oscillator by taking advantage of something called the Miller Effect. This describes the alteration of the input capacity of a triode when its gain is varied by means of changing the grid bias voltage. By shunting a triode across the local oscillator tuning coil(s) the tuning may be altered over a fairly wide range simply by applying the appropriate bias to its grid. The leading method of

obtaining AFC bias was the use of the Foster-Seeley Discriminator, invented by D.E. Foster and S.W. Seeley in 1936 and the subject of a paper read to the Institute of Radio Engineers in March 1938. It is based on the fact that there is a 90° phase difference between the primary and secondary windings of an HF transformer when it is at its resonant frequency. This phase angle changes if the applied frequency varies. If the secondary is centre tapped each end is 90° out of phase with respect to the tap and if each of the two ends is connected to the anode of a diode the voltages at the cathode

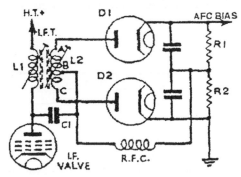

Figure 12.9 The Foster-Seeley Discriminator

will be equal and opposite with respect to the tap, i.e. zero. However, if the applied frequency should vary off the point of resonance the voltages at the secondary will be unbalanced and the voltages on the cathodes also will be unequal. The resultant change of voltage at the tap is exactly proportional to the amount by which the applied frequency varies and if fed to the grid of the control triode it will automatically adjust the internal capacity until the point of zero voltage is attained again.

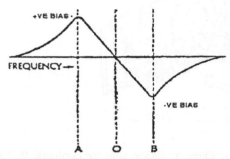

Figure 12.10 The Foster-Seeley Discriminator response curve. O is the point of correct tune for the IF. A and B respectively 5 kc/s above and below the IF, producing positive or negative bias to be applied to the control valve

In practice the discriminator transformer is tuned to the IF of the receiver and fed from the IF amplifier, so that when a station is being received correctly the applied frequency is the IF. Should the set go off tune by, say, 3 kc/s, the IF also changes and the discriminator reacts by producing unequal voltages on the diode anodes and thus a certain amount of AFC bias. This causes the control triode to change its input capacity and to retune the local oscillator until the IF is back to its correct figure. Should the local oscillator drift again the AFC bias produced will again come into play and rectify matters. The usual range of control

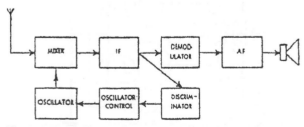

Figure 12.11 Block diagram of the AFC system used in Ekco receivers

is about 5 kc/s above and below the correct frequency of a station and when an AFC-controlled set is fitted with a magic eye it is possible to observe the varying shadow angle as correct tuning is maintained even when the set is deliberately detuned.

The AFC system just described was employed by Ekco in a number of receivers. EMI developed a slightly different system which used a derivative of the Foster-Seeley Discriminator but dispensed with a control valve. The initial AFC bias in this system was used to control the anode current of the double-diode-triode, the cathode of which was returned to chassis via the primary of a small transformer, the secondary being placed in series with the local oscillator tuning coil(s). The DC passing through the primary caused the inductance of the transformer windings to change, which in turn affected the local oscillator tuning. This was probably a good deal cheaper to produce than the Ekco system but the range of control could hardly be as good.

Whatever AFC system was used it was customary to fit an over-ride control to cut it out when manual tuning was being effected.

Silence between stations

To eliminate the babble of sounds produced when a receiver is tuned rapidly along a band from one station to another a control system called Quiet Automatic Volume Control (QAVC) was developed. It soon attracted the handier nickname of 'Squelch'. It works by applying some form of paralysing bias to the frequency changer and/or IF amplifier in the absence of a strong, steady signal. Just how strong the signal must be to overcome the bias may be decided by the listener, and it was common for the control to be marked in divisions between 'strong' and 'All Stations'. It is probable that after the novelty had worn off most listeners left their controls in the latter position.

Amplified AVC

The effectiveness of any AVC system depends on its ability to produce a sufficient range of bias voltage to control fully the pre-detector valves. In conventional AVC designs, whether delayed or not, the amount of bias generated is dependent utterly upon the signal strength of the incoming signal. For complete control, some of the early variable-mu valves needed up to –40 V bias which was outside the scope of the ordinary AVC rectifier. To overcome this problem amplified AVC was developed, in which the control bias is not supplied by the incoming signal but is still controlled by it.

A typical amplified AVC system as used by EMI is shown in Figure 12.12.

The first necessity is a source of voltage that is negative with respect to chassis. This is easily obtained by the use of negative HT smoothing, when the voltage dropped across the smoothing choke or loudspeaker field may amount to –100V or more. As EMI habitually used this type of smoothing in many of its early 1930s receivers the voltage was ready to hand, as it were. The point at which it is tapped off is at the junction of the resistors R24 and R25.

Now look at the cathode of the double-diode-triode: you will see that it is returned to this negative point via load resistors R18 and R24. Not quite so easy to to follow is the way in which the detector diode and the triode grid both return to cathode. the path for the diode is via the secondary of the IF transformer and resistors R10 (the diode load) and R11 (the HF filter), whilst the grid path is via R17, VR3, VR2 and R10.

Ignoring its rather tortuous route, the AF to the triode grid is fed from the detector to the grid of the triode in much the same as in any other first AF amplifier circuir, except that in this instance the negative bias developed at the detector diode load is allowed to reach the grid instead of being blocked by a coupling condenser. It will thus be seen that the grid bias on the triode varies directly with the strength of the incoming signal, which in turn varies the anode current. The triode anode is coupled to the grid of the following output valve in the usual manner by means of a load resistor and coupling condenser, with the value of the resistor chosen carefully to match that of the cathode load resistors.

In the absence of an incoming signal the bias on the triode will be at zero with respect to the cathode. The anode current will be high and there will be a considerable voltage drop across the cathode load resistors. Their values will have been calculated so that at this stage there also will be zero volts with respect to chassis.

If you look at the AVC diode you will see that it is not supplied in the usual way with the IF signal, but

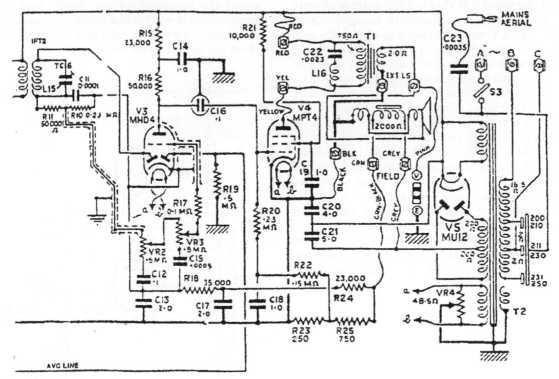

Figure 12.12 Circuit of a typical delayed amplified AVC system (EMI), simplified by the omission of components not associated with the AVC

simply has a load resistor down to chassis and a feed resistor going to the AVC line. As we have seen, for a diode to conduce, its anode must be at a positive potential with regard to its cathode and at present both electrodes are at the same – zero potential.

Now let's see what happens when a signal arrives. According to its strength a certain amount of negative bias will be applied to the grid of the triode. At this stage something very interesting starts to happen. The anode current of the triode begins to fall, reducing the voltage drop across the cathode load resistors, which in turn starts to make the cathode negative with respect to chassis. This effectively makes the anode of the AVC diodepositive with respect: to chassis and current starts to flow through it until the anode attains the same negative value as the cathode. This negative bias is fed to the AVC line as control bias for the preceding valves.As the signal strength increases the bias on the triode grid becomes more negative, reducing further the voltage drop across the cathode load resistors, increasing the negative voltage at the cathode and thus at the AVC diode anode. The really important point is that, the amount of increase in AVC bias depends not just on the bias applied to the triode grid but upon the amplificaton factor of the triode. If this is 20, a voltage of only $-0.1\,V$ on the grid is sufficient to bring about an AVC bias of $-2\,V$, and an increase to $-1\,V$ at the grid will result in an AVC bias of $-20\,V$. It will thus be seen that complete control of the receiver gain may be obtained with signal strengths of only few volts.

EMI's use of amplified AVC in was very effective. In a reasonably good reception area it is possible to attach just a few feet of wire to the aerial terminal of one of these sets and to receive dozens of stations along the MW band all at the same loudspeaker strength. However, it must be emphasised that the equilibrium of the circuit in the absence of a signal depended a great deal on the 'goodness' or otherwise of the double-diode-triode and results could become very variable as the valve aged.

The relative complication of amplified AVC and the coming of more easily controlled variable-mu valves meant that it virtually disappeared by about 1936.

Push-pull output

Mains output pentodes (and tetrodes) are capable of giving much greater power than battery types and thus there was less incentive for manufacturers to employ push-pull output merely for that reason.

Although high output was a consideration in large luxury radio-gramophones, a major factor was the benefits to be had in quality of reproduction. Certain types of harmonic distortion associated with amplifiers is cancelled out in push-pull circuits, while the opposing anode currents passing through the split primary winding of the output transformer reduce the tendency for DC saturation of the core, which consequently may be made smaller in size. Economy of anode current not being a necessity in mains sets, Class 'A' push-pull is normally employed. It is necessary for each of the output valves to be supplied with an AF input of the same magnitude but in opposite phase, obtained by the use of a phase splitter valve following the first AF amplifier. There are a number of circuits in use but the most popular, and the one which gives the clearest picture of what is involved, is known as the 'concertina'; the circuit appears on the left-hand side of Figure 12.13.

It will be seen that there is one resistor in the anode circuit of the triode and two in that of the triode, R1 and R2. R2 has the same value as that of the anode load resistor whilst R1 is the usual bias resistor for the valve. Strictly speaking the value of R2 ought to be adjusted so that it and R1 together equal the resistance of the anode load but in practice this doesn't make much difference; in any case, unless very close tolerance resistors are used there are bound to be imbalances.

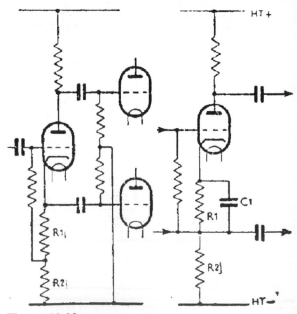

Figure 12.13

The outputs are taken from the triode, one from its anode, the other from its cathode, which are exactly 180° out of phase with each other, by coupling condensers connected to the grids of the push-pull output valves. As no by-pass condenser is fitted from the cathode of the triode to chassis, very heavy negative feedback takes place, to the extent where the gain of the valve is reduced to unity. This is usually of no account because ample drive is available from the first AF amplifier, and the circuit gives excellent fidelity with very low distortion. However, if there is no preceding AF amplifier, as in a short superhet, the phase splitter may be made to give normal gain by feeding the incoming signal between grid and cathode instead of between grid and earth, and using a cathode by-pass condenser, as seen in the right-hand diagram of Figure 12.1. The price is the loss of the special advantages of the original splitter plus a greater tendency for hum to occur. In addition the input to the triode is more difficult to arrange, in particular the use of a gramophone pick-up, due to it being impossible to earth one side; predictably Ultra Ltd, with its predilection for short superhets, was one of the few makers to use the method.

Another type of phase splitter known as the 'paraphase' uses two triodes wired as resistance-capacity amplifiers, of which one (V1 in Figure 12.14) receives its input from the first AF amplifier or directly from the detector as the case may be.

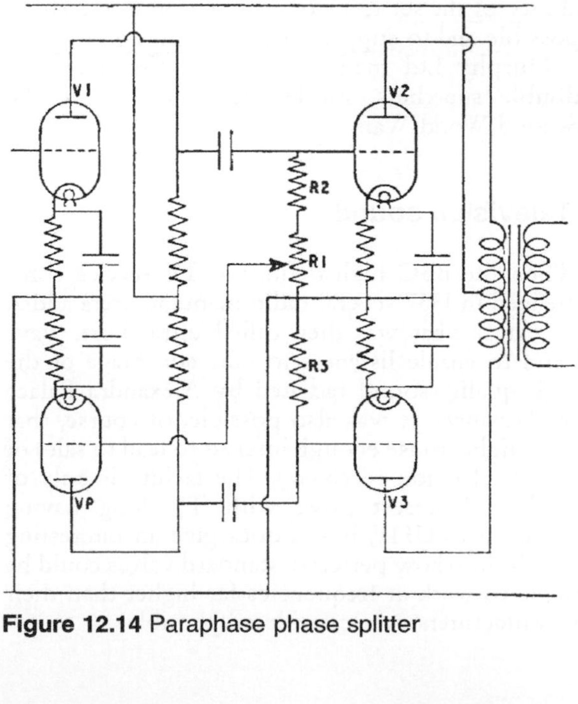

This valve gives the usual amount of gain and drives the grid of one of the output valves (V2). The grid return resistor for this is made up of R2 and R1 in series; the latter is set to give a small percentage of the total signal to the grid of VP, the paraphase valve. Its anode delivers an opposite phase signal to the grid of the other output valve, V3. It is essential that the level of signal reaching V3 is the same as that delivered to V2, so the values of the various resistors have to be chosen with great care, and to maintain their accuracy, to ensure that the overall gain of V2 is unity; close tolerance types are therefore essential. The advantage of the paraphase is that it does give useful gain without the drawbacks of the modified concertina, including near immunity to hum.

Various other types of phase splitter have been used by different manufacturers, most of them based on the paraphase type but using all sorts of devices to enable the total valve count to be reduced. These are too many in number and too complex to be discussed here and the relevant service information is essential for anyone tackling faults on them. Fortunately they did not achieve great popularity, most firms preferring to stick to the tried and true methods. A completely different alternative, found particularly in certain sets made by EMI just after the Second World War, was the use of a 1:1 auto-transformer as a phase-reverser following the first AF amplifier.

Better tone control systems

For most pre-war sets (and a lot of those made after 1945) the term 'tone control' means no more than a way of reducing the treble response of the set. Although the term high fidelity had originated early in the 1930s, listeners in general preferred what came to be called in the trade the 'mellow bellow' of a good wooden cabinet reproducing the lower and middle registers vigorously with not much more than a trace of top notes. This was regrettable since in those days AM broadcasts were capable of giving more extended high note coverage than is the case today. It may not generally be realised that 78 r.p.m. gramophone records were even better, especially when Decca introduced its Full Frequency Range Response (ffrr) records in 1947. Good quality radio-gramophones produced by such firms as EMI, the Radio-Gramophone Development Co. (RGD) and Decca itself incorporated sophisticated tone-control systems worthy of the music that was obtainable, most of

Figure 12.14 Paraphase phase splitter

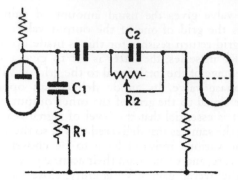

Figure 12.15 Tone control system with both treble (R1) and bass (R2) controls

which were based on the use of negative feedback (but sometimes with an element of positive feedback as well). Instead of the simple 'top-cut' control, such sets gave the listener the facility to vary up and down as required both the bass and treble response. It is sad to reflect that, apart from true music lovers, owners of these sets probably left the bass control right up and the treble control right down.

Band-spread short wave tuning

Starting in the mid-1930s there was a great upsurge in the number of listeners wanting to tune into distant short wave stations, especially to receive American commercial music and comedy programmes sent out on MW and SW simultaneously. Then, as the political situation in Europe worsened, people wanted to learn what America thought about it via their news broadcasts. The trouble was that even with the aid of a 'magic eye' it was tricky to tune in to any particular SW station simply because there were so many of them. The main SW band of 16 m–52 m covers something like 12 mc/s of air space with a potential to carry over 1300 radio channels spaced 9 kc/s apart. In fact the SW band was subdivided into smaller ones each about 1 mc/s wide, known as the 16 m band, 19 m band, 25 m band, 31 m band, 40 m band and 49 m band. Each of these could accommodate over 100 stations within perhaps half an inch of pointer movement on a radio dial; no wonder it was difficult to select the one you wanted. The radio manufacturers got over this by the introduction of band-spread tuning, which means exactly what it says: by using special coils and a small-value tuning condenser each SW band occupies the full length

of the dial, making tuning in stations as easy as on MW or LW. Pye Ltd in particular took up this idea with enthusiasm, first with its 'International' model of 1939 and then with a number of models through the late 1940s and the 1950s. Murphy Ltd also made a notable band-spread set which was good enough to be used at the BBC's official listening station through the Second World War along with sophisticated military 'communication' receivers.

The double superhet

This other approach to SW listening has, as the name suggests, two frequency changer sections, the first of which is broadly tuned and produces an IF of 1600 kc/s or more with a wide bandwidth of several hundred kc/s. When tuned to, say, the 49 m band virtually all its stations appear in the IF. If amplified and detected this would produce a useless jumble of sounds but the secondary winding of the wide-band IFT acts as the grid tuning coil for the second frequency changer, which can select any desired station and pass it on at a standard IF of about 450 kc/s. The IFTs for this frequency have the usual frequency response and are followed by a normal detector, AF and output stage.

The extra valves used in the double superhet give it more gain than an ordinary type and in addition help to reduce the signal to noise ratio, that is, the ability of the set to bring in a station as clearly as possible and to suppress background noise.

Murphy Ltd made one of the first domestic double superhets, Model A36, just before the Second World War.

'Television sound'

When the BBC high definition TV service commenced in 1937 several radio manufacturers added an extra, what was then called ultra-short, wave band to enable listeners to take advantage of the high quality sound radiated by Alexandra Palace on 41.5 mc/s (it was also possible, of course, that this might arouse enough interest to lead to sales of actual television receivers). The facility is only of academic interest now, with TV long having departed to UHF, but it does give an interesting insight into how perfectly standard valves could be made to work at frequencies far higher than their manufacturers ever considered possible.

Chapter 13

Battery and mains/battery portable receivers

The portable receiver is very nearly as old as the domestic set, and as early as 1925 GEC was offering a superhet portable claimed to have a magnificent performance. At its price of 50 guineas, half that of the contemporary small Morris 'Bull-nose' motor car, it ought to have been good. By about 1927 portable design had settled down to a near standard of a three- or four-valve TRF following the lines of domestic sets in every respect except for the 'chassis' – often made of plywood – and the use of a frame aerial. Two volt battery valves were used with the LT supplied by a special type of accumulator with celluloid case and jellified electrolyte that was virtually unbreakable and unspillable. In most cases a standard size 90 V dry battery provided the HT. Two main types of cabinet design were used, the upright, with receiver section at the top and the loudspeaker and batteries at the bottom, with the frame aerial mounted around all, and the 'suitcase', with receiver and batteries in the lower part and the loudspeaker and frame aerial in the lid. Portable sets on these lines were still being produced ten years later, the only main difference being that they tended to be much smaller. Other than a few isolated examples, superhet portables did not become generally popular again until the late 1930s. As with domestic sets, the four-valve arrangement of frequency changer, IF amplifier, detector/AVC/AF amplifier and output was the norm.

Fault finding on the above follows exactly the same lines as for domestic battery sets, except for the fact that the frame aerial(s) have to be considered. These were normally of robust construction and unlikely to suffer from o/cs, but in the 'suitcase' portable with the frame in the opening lid there are flexible interconnections with the receiver chassis which may fray with long usage. Another possible cause of trouble is corrosion due to acid fumes from the accumulator, which so often was in a confined space close to the chassis. Keep an especial eye on switch contacts and valve pins and holders.

'All-dry' portables

In 1939 portable set design was revolutionised by the introduction of a new series of valves specifically intended for superhets and having filaments rated at 1.4 V and intended to be powered by dry batteries, hence their name of 'all-dry' valves. In this country the ranges produced by M-OV and Mullard both comprised a heptode (pentagrid) frequency changer, a variable-mu RF pentode, a diode-triode and an output pentode, but whereas the M-OV types had the international octal base, those from Mullard had the 'P' side contact base. These latter had only a limited production run and soon were replaced by types with octal bases. Another improvement to both the M-OV and Mullard types was the fitting of a tapped filament on the output valve only. With the two halves in parallel the voltage and current rating was the same as that of the previous types, i.e. 1.4 V, 0.1 A, but by running both halves in series they were able to operate at 2.8 V, 0.05 A, making it possible to run all four valves in a series chain. This in turn made possible the introduction of true 'universal' receivers that would work on either batteries or AC/DC mains. These too started to arrive in Britain in

1939, having been pioneered in the USA. Before discussing these, however, we should first look at what became virtually a standard design of four-valve superhet and lasted for at least ten years unchanged.

Coverage was invariably restricted to MW and LW, small frame aerials being provided for both bands. The heptode frequency changer was followed by a pentode IF amplifier with the single diode of the diode-triode used as detector and the source of AVC bias and its triode section as first AF amplifier resistance-capacity coupled to the output pentode. It was usual in battery-only sets for the filaments to be run in parallel with a total consumption of 0.25 A, giving an acceptable number of hours' operation from an LT battery. HT consumption was usually about 10 mA from a 90 V battery but the valves would in fact give a reasonable performance at no more that half that voltage.

The way in which AC/DC mains operation was achieved was to use a conventional valve rectifier such as a 25Z6 to give both the HT and LT voltages via suitable dropping resistors. The rectifier heater had to have its own dropper, usually in the form of a resistive line cord. In the original USA sets all this was easy enough to arrange from a mains voltage of only 110/120 V but in Britain all the resistors had to be of rather inconveniently high values. For instance, the rectifier heater alone at 25 V @ 0.3 A required a line cord capable of dropping 200 V, which at the usual 60 Ω per foot worked out to just over eleven feet in length. A wire-wound resistor both dropped the DC from the rectifier down to approximately 90 V for the HT and provided reasonably effective smoothing. A further resistor dropped another 83 V or so to give a nominal 7.5 V for the valve filaments, operated in series. For this kind of work the design centre rating of each was no more than 1.3 V (2.6 V total for the output valve) to provide a safety margin against the possibility of voltage surges on the mains.

With some sets the changeover from battery to mains operation and vice versa had to be done by a manual switch, but others had automatic change-over by means of a relay. In its normal position, at rest, the batteries were connected as in a normal receiver with only the usual on/off switch to be handled. When the mains plug was inserted, as soon as the rectifier began to give its full output the relay operated, switching in the mains HT and LT supplies and disconnecting the batteries. If the mains should fail or otherwise be disconnected the relay immediately changed the set back to battery operation without any interruption to the programme being received. This was so impressive as far as owners were concerned that it is little wonder that one firm (Pilot) called its mains/battery portable the 'Twin Miracle'. Be warned that in this particular set the very unusual ploy is used of connecting the two strapped *cathodes* of the rectifier to the live mains input with negative HT being drawn from its *anodes*. This means that a lot of the HT and LT dropping and smoothing circuitry looks 'back-to-front' and the greatest care is necessary when faults are being traced. Fortunately this eccentricity was not emulated by other manufacturers who stuck to the conventional way of wiring the rectifier.

A close examination of the circuitry of most AC/DC/battery sets will show that various resistors are connected in parallel with the valve filaments. The purpose of these is to balance out the actual current flowing through each, because that of the HT as well as of the LT passes through them, and this is cumulative along the chain. If not protected by the shunt resistors the filament at the bottom end of the chain would be made to pass about 10 mA more than that at the top – not a great deal but sufficient to affect its long-term life.

When UK production of mains/battery sets resumed around 1946 small selenium rectifiers took over from valve types, obviating the need for mains droppers of resistive line cords. For details of how these rectifiers may be replaced when they wear out, please refer to the main section on power supply repairs.

Miniature valves

Around 1940 a new generation of 'all-dry' valves appeared in the USA. Physically very small – about two inches long by three-quarters in diameter – they were of all-glass construction (later known as B7G in the UK) with the seven base pins also supporting the electrode structure. A number of different types appeared but of those which became best known the frequency changer, the RF pentode and the output pentode were electrically the same as the octal types just discussed. Taking over from the diode-triode was a diode-straight RF pentode which offered more gain when used as an AF amplifier. Had it not been for the Second World War these valves undoubtedly would have appeared in the UK shortly afterwards, but as it was British set makers had to wait until 1945 before they could use them. Meanwhile they began to arrive in this country in small 'personal' portables that had become popular in the USA and were easily carried in the luggage of American servicemen posted over here. The circuitry of 'personals' differed very little from that of the standard size all-dry portables except for the extra resistor and condenser used for the screen grid of the diode-pentode. Many of the components also were miniaturised, such as the tuning condenser, IF transformers, loudspeaker, etc., while special layer type HT and LT batteries reappeared (they had first been made back in the late 1920s) in very compact form. The LT remained at 1.5 V whilst the most popular HT voltage was 67.5 V. Consumption was 250 mA and 8 mA approximately for the LT and HT, respectively.

Octal and B7G types were used side by side in conventional and personal portables until about 1950, when the octal began to disappear, leaving the B7Gs supreme. This state of affairs continued until around 1954, when valve manufacturers introduced a new generation of B7Gs with low consumption filaments that demanded no more than 25 mA for the heptode. RF pentode and diode-pentode, and 50 mA for the output pentode. Although the HT consumption remained the same, the considerable saving in LT current gave much better battery life.

Both the original 50 mA types and the new 25 mA valves were used in mains/battery receivers including revised types suitable only for AC mains operation. In these the valve filaments were operated in parallel and supplied with LT by a small mains transformer and full-wave rectifier. Another winding on the transformer gave around 90 V for the HT via another rectifier. The advantage of the AC-only arrangement was that heat-generating and vulnerable dropping resistors were no longer required, and in addition the chassis could be isolated completely from the mains.

Chapter 14

Automobile receivers

For servicing purposes the receiving circuitry of automobile radios resembles closely that of contemporary mains powered domestic sets, and like the latter the vast majority settled into the four-valve plus rectifier superhet configuration (with, however, some interesting exceptions); therefore the same fault-finding techniques are applicable. It is in the power supply section that the two types differ because in the automobile receiver both LT and HT have to be derived from the car battery. In this respect it has to be remembered that during the vintage radio period both 6 V and 12 V systems were in general use and most manufacturers produced separate models for either voltage.

The LT presented no problem as valves with 6.3 V or 12.6 V heaters could be powered directly from the battery. (In fact some manufacturers preferred to use 6.3 V heater valves in both cases, wiring them in parallel for 6 V systems and in series-parallel for 12 V use. Where necessary, shunt resistors were added across certain heaters to balance the current flow.) The big difficulty for the designers of early automobile sets was how to step up the 6 V or 12 V of the battery to around 200 V for the valve anodes and screen grids, etc. Initially motor generators were tried but these were both noisy and inefficient as regards battery consumption, and were soon abandoned in favour of what became the universal provider of HT, the *vibrator power pack*.

The basic *non-synchronous* vibrator is basically similar to an electric buzzer, powered by the car battery, with the moving reed acting as the centre pole of a two-way switch. It works in conjunction with the centre-tapped primary winding of a step-up transformer, the tap of which is connected to the 6 V or 12 V supply. All the time the reed is moving to and fro, at around 115 times per second, it earths alternately one end or the other of the primary, causing interrupted DC current to flow first one way, then the other, in constant repetition. This induces an HT voltage in the secondary winding, usually of around 200 V to 250 V. Incidentally, the fact that vibrators run at 115 c/s means that the size of the transformer core need be only half that of a type intended for 50 c/s use, giving a useful saving on space.

The HT output from the transformer has, of course, to be rectified, and for many years the favourite valve for the job was the 6X5/6X5GT, a full-wave type with a 6.3 V heater and highly insulated cathode. Another popular rectifier was the 0Z4, a gas-filled valve needing no heater supply; the cathode reached its operating temperature from ionic bombardment when the anode voltage was applied.

No separate rectifier at all is required when a *synchronous* or *self-rectifying* vibrator is used. Extra sets of contacts are arranged on the vibrating reed which switch the HT output at either end of the HT secondary winding in exact synchronism with the LT input. Contrary to what is familiar in most HT rectifier applications, the centre tap of the secondary winding is positive, not negative. As the current flow in a particular half of the primary induces one of the same polarity in the secondary, the appropriate outer end is earthed to complete the circuit.

The 115 c/s output of the vibrator has another advantage for the car radio designer apart from that of transformer size. The output from the rectifier, of whatever type, pulses at 230 c/s and a much simpler HT smoothing arrangement is called for.

The efficiency of the vibrator pack is evidenced by the fact that the estimated HT wattage of a four-valve + rectifier set is about 10 W whilst that of the heaters is about 12 W, a total of 22 W. The actual consumption of such a set from a 6 V or 12 V battery would be about 30 W total.

Negative or positive earthing

One terminal of a car battery is connected to the chassis of the vehicle (called 'earth', but not strictly so), which is used as a return path to avoid the need for two supply cables to every piece of electrical equipment. In the early days of motoring negative earthing was popular until the 1930s when informed scientific opinion declared that positive earthing gave a better sparking effect at the plugs and resulted in less chassis corrosion. It was almost universal until the mid-1960s, when for some reason negative earthing made a comeback. Is it only coincidence that cars now seem to corrode very easily? Be that as it may, the double changeover has caused much trouble for almost everyone concerned with car radios. From a technical point of view an ordinary non-synchronous vibrator pack should be indifferent to the type of earthing, but the synchronous type is definitely polarity-conscious. If one intended for use on a positive earth system were to be connected to a negative earth system the HT voltage too would be reversed in polarity with disastrous results for the smoothing condensers. To guard against this it was usual to make the vibrator reversible to suit either polarity. All that needed to be done was to withdraw the vibrator from its socket, turn it through 180° and plug it back in. To show which polarity was which, + and – signs were printed on the top of the vibrator to be lined up with an arrow pointer.

Constructional aspects

Sets built before the Second World War, especially the archetypal Philco models, used standard sized valves such as the American UX series, which called for considerable ingenuity in packing them into small cases. The convention at that time was to have the main 'works' in a floor-mounted box, with a control head on or near the steering column carrying the tuning, wavechange and volume controls. They all acted in a purely mechanical manner, the two units being interconnected with heavy-duty cable drives. The same system was revived after the war by a number of makers, but using the much smaller GT octal valves. Later two-unit types usually had the complete RF/IF section of the receiver in a dashboard mounting with an amplifier/output/power supply fitted remotely (possibly in the engine compartment) and interconnected with a multi-cored cable. There was sound reasoning (literally!) for installing the power supply, at least, outside the passenger area, since this virtually eliminated the characteristic buzz of the vibrator which could be annoying in an otherwise quiet car.

EMI built an excellent example of the genre for sale under the Radiomobile trade mark in the late 1940s, and this appeared in cars such as the Standard Vanguard, the Austin 16, the Jaguar 3½ litre and the exotic Invicta Black Prince Byfleet Drophead Coupé. This receiver had the loudspeaker mounted in front of the main unit, slightly slanted downwards, whilst a subassembly carried the controls and dial even further forward. Two slim knobs flanked a narrow but attractive dial, above which were two rows of four pushbuttons each. Four were for preset station selection, the rest for wavechange (long/medium) and tone (speech/music). The circuit included an untuned RF amplifier but was otherwise conventional. The valves were of the Marconi/Osram 81 series, having bases similar to the American local. In standard basic form the output stage was contained in the main unit, but the power supply was in a separate case which could either be bolted directly to the other or, more usually, connected via a multi-way cable. There was an alternative output stage, employing two KT8ls in push-pull, which could be added on to the main unit, in which case the original output valve was modified to become a driver stage. A special 24 V model was made for use in coaches, and this had the option of a microphone input for sightseeing commentaries *en route*.

A later Radiomobile had a small dash unit containing an RF amplifier and a frequency changer only, the IF stages being in the main unit. This too could be had in single-ended or push-pull output form. It used the small M-OV '70' range of B7G-based valves.

Permeability tuning

The general adoption of permeability tuning brought about a useful reduction in size of car radios, and in the 1950s it was possible to standardise the size of the hole needed in the car dash for mounting. It was also much easier to include a tuned RF amplifier in the specification, but few makers took advantage of this. One which did was Ekco, with the CR152 model. This has an ingenious manual/preset tuning system using three small combination knobs and dials connected to individual sets of RF and oscillator coils (the frequency-changer grid circuit was not tuned). One set covered long waves, 1150 to 1850 m, the other medium waves in two sections, 190 to 340 m and 330 to 570 m. The dials could be set on certain stations, selected by turning the wavechange switch or rotated at will. Incidentally, a loud-speaker muting switch, ganged with the wave-change, was fitted to this model, and should be checked if the set appears to be 'dead'.

Some unusual sets and features

Both Ekco and Pye produced car radios having a number of band-spread short-wave ranges. A separate dial was provided for each band, changed in unison with the wavechange switch.

The continental firm of Becker made some extraordinary radios featuring automatic, self-seeking tuning with AFC. When used manually, the permeability tuning was operated conventionally by a knob, but on 'self-seek' it was power driven by a clockwork mechanism. The spring for this was tensioned by a solenoid which was switched on automatically as the tuning reached the high frequency end of the dial. This threw the tuning instantly back to the LF end, ready for another traverse of the dial. Included in the clockwork was a nylon-bladed paddle wheel, similar to that used in a musical box, which revolved at high speed during tuning, and which could be braked by a catch moved into position by the action of a small relay. Because the paddle made so many turns relative to a small movement of the pointer, extremely accurate braking was assured. The relay was connected into the anode circuit of a double triode valve controlled by the detector; as a station came on tune the relay closed, so quickly that the mechanism stopped at virtually the precise point. To overcome

any slight error the AFC came into action in the same manner as described elsewhere in the text, by means of a reactance triode. A switch gave the choice of three levels of sensitivity, so that only very powerful, or those plus fairly strong, or all, stations would stop the tuning. The automatic tuning was brought into play by slight pressure on a hinged bar above the dial, but some models had provision for remote control when the set was installed in a chauffeur-driven car. The output and power supply sections were in a separate unit for engine compart-ment mounting. If the normal MW and LW coverage was not sufficient for the owner an optional multi-band band-spread SW adaptor was available. These sets were strictly the province of the wealthy car driver!

AM/FM automobile receivers

One or two AM/FM sets appeared from time to time, such as the Philips model X61V. It covered MW, LW and VHF, had pushbutton tuning for three AM and two FM stations, would work on either 6 V or 12 V systems of either polarity and incorporated a facility to enable the owner to use a mains-type Philishave electric razor from the HT line. What it didn't have, alas, was any great appeal to the public. The truth is that FM reception in cars met with even less success in the UK than the domestic variety, for, with the best will in the world, the performance on FM in BBC-only days was not consistent enough under varying driving conditions to make it really attractive even if the programmes available had given any choice from those available on AM. Nevertheless, a set of this type is worth restoring now that there are far more local stations with transmitters sited closer to the areas where many people have to take their cars; and the adoption of slant polarisation has made the traditional car aerial more suitable for FM recep-tion. The only problem, as with all UK FM receivers in the valve era, is that the VHF coverage was restricted to between 88 mc/s and 100 mc/s, rather frustrating for anyone wishing to receive Classic FM!

Chapter 15

Frequency modulation

The principle of frequency modulation (FM) is that the carrier wave of a transmitter receives a constant level of modulation, but that its frequency deviates by a predetermined amount above and below the nominal figure. The standard deviation has been for many years 75 kc/s. The applied modulation decides the number of deviations that occur per second, e.g. an applied modulation of 400 c/s produces 400 deviations every second, and so on. The system is capable of carrying the full AF range with ease. This method of modulation had been tested as long ago as just before the First World War, when it seemed likely to prove easier to achieve with the primitive equipment then available, but improvements to valves and other technical advances reversed this situation and FM fell into long disuse.

In 1922 a mathematician called J.R Carson 'proved' to his own satisfaction that FM would never be a practical possibility (despite the 1913 tests) since, he averred, the bandwidth requirements would be colossal, or even 'infinite'. Shortly after this E.H. Armstrong commenced experimental FM transmissions. He had concluded that the manmade interference and atmospherics which plagued radio communications were largely in effect amplitude modulated, and he saw in FM the means of combating them. Influenced by Carson's gloomy predictions Armstrong at first tried very narrow deviation bandwidths, and found that there was no discernible improvement in signal to noise ratio over AM. Undeterred, he tried wider bandwidths with very encouraging results. He was soon able to show that an FM transmitter on VHF uses less bandwidth, expressed as a percentage of the carrier frequency, than does an AM station on MW!

Armstrong's work was for some years carried out in Marcellus Hart Research Laboratory at New York's Columbia University, followed by a short period in 1934 when he operated a 44 Mc/s, 2 kW transmitter situated on the top of the Empire State Building, then the world's tallest structure. In 1937 he set about building a high power transmitter at Alpine, New Jersey. Transmissions from this location proved conclusively that FM was able to offer substantial advantages over AM. Since the effective power of the transmitter is not affected by the depth of the modulation, nor its frequency, it is able to operate continuously at or near its maximum. Not only are the signals to a large extent immune from interference as mentioned above, but there is considerable resistance to the other kind of interference due to another station working on the same frequency. This would have to have a power of over 50% of the desired station at the receiver before it could be heard. By comparison, even a very weak AM station can spoil the reception of another in certain conditions. This co-channel interference, as it is called, is particularly troublesome on MW during the hours of darkness, when the effective range of transmitters is greatly increased. Thus FM is able to provide excellent sound quality unimpaired by fading or the several forms of interference.

By the 1940s FM had been adopted generally in the US and magnificent receivers were being manufactured to take advantage of its potentialities. Given that the war and its aftermath were bound to delay the introduction of FM to the UK to a certain extent, progress here was still pitifully slow. The BBC installed a 25 kW FM experimental transmitter at Wrotham alongside another for a system

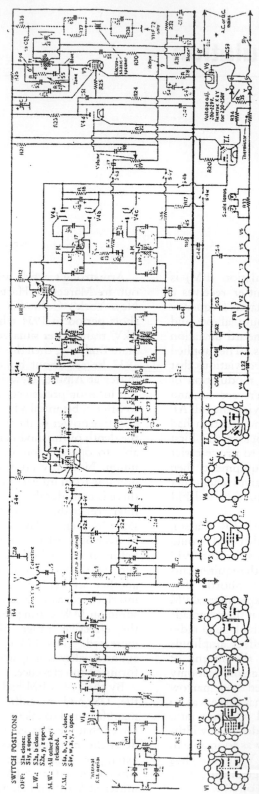

Figure 15.1 The Bush Models VHF71 and 72 provide good examples of typical AM/FM receivers in 1958. Points to note are the use of a UCC85 double triode in the VHF tuner unit, the AM frequency changer acting as an IF amplifier on VHF, the ratio detector employing a UBC80 triple-diode-triode and the electrostatic 'tweeter' (high frequency loudspeaker) in the VHF72

called HIFAM (HIgh Fidelity AM Broadcasting) which seems to have been rather a superfluous exercise since HIFAM had already been tried and rejected in the USA. In the event a full service was not established by the BBC until 1955, and it started with the crippling disadvantage that it merely duplicated the existing Home, Light and Third programmes, thus providing little or no incentive for the public to buy FM receivers. Listeners on the whole remained indifferent to the advantages that FM offered in the way of reduced interference and better sound quality since this could be fully realised only if an expensive new aerial system was installed – and why do that to get the same programmes? It is fair to say that the absence of any competition in FM broadcasting held back its popularity for twenty years or more.

Some aspects of FM receiver design

International standards having been set down many years ago, receiver design must to a great extent follow a pattern. The most favoured band for FM broadcasting (although others are in use) is between about 87 and 108 Mc/s. Note that until comparatively recently this band was occupied above 100 Mc/s in the UK by mobile radio services for the police and fire services, and it was not until these had been banished to much higher frequencies that commercial broadcasting on FM could take place. Channel spacing is nominally 220 kc/s, which is ample to allow for the standard deviation of 75 kc/s (i.e. a total of 150 kc/s per channel). In practice, however, some stations are slightly offset from the frequencies allocated to them, and the actual spacing may differ. The standard IF for FM receivers is 10.7 Mc/s, and for once this seems to be strictly adhered to. The bandwidth of the IF transformers is likely to be rather more than 150 kc/s, to allow for 'drift' in the RF tuning during the warming-up period, the usual figure being 200 kc/s.

A different type of detector has to be used in FM receivers which will respond to its different form of modulation, and one which enjoyed early popularity in the US was that same Foster-Seeley that had been developed originally for use in AFC systems. However, this detector also would respond to any stray AM that may have found its way into a receiver and it had to be preceded by a 'limiter' valve which passed FM but not AM. A later development, called the ratio detector, is largely self-limiting and was used in all UK FM receivers.

De-emphasis

It was discovered in the US that the signal to noise ratio could be further improved if the higher audio frequencies were artificially boosted prior to transmission, this process being called pre-emphasis. At the receiver de-emphasis is applied by a simple resistance-capacity network in the AF circuitry following the detector.

Design aspects of British FM receivers

The lack of competition in broadcasting already mentioned dictated to a large extent how FM receivers should be designed. For instance, at first a number of firms tried the experiment of making small FM only sets, but their restriction only to BBC transmissions prevented them from becoming really popular. Combined AM/FM receivers then appeared, the usual coverage being MW, LW and VHF, short waves being dumped. As regards AM the usual four-valve plus rectifier circuitry was favoured, with a separate tuner for FM. Dual frequency IF transformers were used capable of working at either the usual 465 kc/s or thereabouts for AM and 10.7 Mc/s for FM. To boost the sensitivity on FM the AM frequency changer was switched to act as an FM IF amplifier.

By the time these AM/FM sets appeared the small B9G all-glass valves were almost universally used in UK receivers and the vast majority of AC-only models had a line-up consisting of an ECH81 triode-heptode frequency changer, an EF89 RF pentode as IF amplifier, an EABC80 or –81 triple-diode-triode as AM detector, FM ratio detector and AF amplifier, and an EL84 output pentode. The rectifier would be an EZ80 or perhaps a metal contact cooled type. Some early VHF tuner units used a pair of EF80 high slope RF pentodes as RF amplifier and self-oscillating frequency changer but most makers preferred a special double triode, the ECC85, which did both jobs. In AC/DC receivers the valves used were, respectively, the UCH81, UF89, UABC80/81, UL84, UY85 and UCC85.

Part 2

Chapter 16
Tools for servicing radio receivers

As regard small tools, anyone who is already something of a handyman already will have a selection of screwdrivers with different blades, various pliers, both blunt and snipe-nosed, side-cutters, a hand drill and a good selection of bits, various files and so on. It is very useful, too, to have at hand a selection of nutdrivers from 8BA up to 0BA and open-ended spanners to cover the same range. You will need at least one good electric soldering iron rated at about 25 W with a fine bit for getting into awkward places; if possible, acquire a larger iron of 60 W or more rating for the odd times when you will have to solder direct to a steel receiver chassis.

Solder 'guns'

Ordinary soldering irons have heating elements for mains supplies, wound with resistance wire and insulated in mica and steel cases. They take a few minutes to warm up, then have to be left switched on in order to be ready for instant use in the workshop. When the trigger switch of a solder 'gun' is depressed its element reaches working temperature in a few seconds, so it consumes current only when in use. This is achieved by replacing the usual heating element by a mains transformer with a low voltage secondary rated typically at about 1 V, 80 A. This is made up of a turn or two of thick copper strip around the primary, to which are attached a pair or heavy-duty terminals. The bit, which is fashioned from thick copper alloy wire, is attached to these terminals and when the trigger is pulled a heavy current flows through it, causing it to heat up. At the business end the element is flattened over about a quarter of an inch at which point the greatest heat occurs. The better makes have a tapped primary with two-position trigger, so that full power (full depression) is used to gain operating temperature and reduced power (part depression) to maintain it. Be warned that guns without this feature may carry on getting hotter and hotter all the time the trigger is depressed, so one has to make a joint, etc., quickly before damage is done to heat-vulnerable components. In this context, take care with any type of imported soldering iron, especially from the Far East because their elements may overheat on UK mains voltage. Irons from the USA and intended for 110/120 V supplies will be marked so; they are usually very well made and worth acquiring to be used with a suitable step-down transformer.

Irons with 12 V elements are obtainable for use on automobile batteries or low voltage transformers. Some have very small tips that are useful for working in confined spaces.

Solder

To go with the irons you will need some good quality resin cored solder, preferably about 18–20 gauge. Steer clear of thick industrial types and very thin types meant for printed board connections.

What about instruments?

Most professional service engineers will be able to tell you that in the final analysis a very high proportion of repair jobs may be completed successfully with no more than a good multimeter, which in vintage radio terms is virtually synonymous with the name AVO, the firm which has been making the best on the market since the mid-1920s. The AVOmeter has all the attributes that make up a first-class tool for radio servicing, including a good selection of voltage, current and resistance ranges, a large clear dial and a smooth needle movement of the type known as 'dead beat', which means that it comes immediately to rest on the dial reading to be indicated without any preliminary swinging about.

What ranges are required?

To be able to check anything from a miniature battery portable set up to a massive luxury mains radio-gramophone you will need a multimeter that will read up to about 1000 V both AC and DC – although in theory this ought to be possible with a single 0–1000 V scale for either AC or DC in practice it would be impossible to obtain accurate readings at the low end of the markings. Thus the complete coverage needs to be broken down into four or more sections, say 0–10 V, 0–100 V, 0–500 V and 0–1000 V. For resistance checks 0–1000 Ω, 0–100 kΩ and 0–10 MΩ (note: the unlinear nature of the dial markings on resistance ranges means that roughly the lower half of any range actually spreads over about four-fifths of the dial with the rest compressed at the one end; therefore a resistor of say 750 Ω is much more easily tested on a 0–100 kΩ range than on a 0–1000 Ω range).

Fortunately, multimeters made in the vintage radio days were designed specifically to cater for these kinds of requirements, so in this respect a good old AVOmeter may be a far better practical proposition than a brand new one designed for far different purposes.

The ohms-per-volt game

The sensitivity of a meter is expressed as an 'ohms-per-volt' figure, and the greater the number of ohms the better, at least in theory. What it actually means is this: the heart of any multimeter is a meter movement that measures small amounts of current, typically with a *full-scale deflection* (f.s.d.) of 1 mA. When you make a voltage reading with the multimeter the movement is responding to current

flowing through it, the dial of the meter being calibrated to indicate the amount of voltage that corresponds to that current.

To give an example, Ohm's law tells us that a voltage of 1 V applied across 1000 Ω will cause a current of 1 mA to flow. Thus a multimeter with a basic movement of 1 mA would have a sensitivity of 1000 Ω per volt. The 1000 Ω is made up of the internal resistance of the basic movement, commonly 50 Ω, plus a series resistor which would be of 950 Ω. For a 10 V range the overall resistance would need to be 10 000 Ω, so the series resistance would be 9950 Ω. By this time the meter resistance is becoming such a small percentage of the overall that it may be ignored, so for a 100 V range the series resistance would be 100 kΩ, and so on. Wire-wound resistors are used for the lower ranges and very high stability carbon types for the higher ranges.

What does ohms-per-volt mean in practice?

Suppose you wish to read the anode voltage of a valve which draws 3 mA anode current through a load resistance of 47 kΩ from an HT line of 200 V. The resistor will drop 141 V at the rated anode current so the voltage should be a shade under 60 V. The AVO 'Seven' has a basic movement of 1 mA but it is shunted to make it draw 2 mA at f.s.d., so the sensitivity is reduced to 500 Ω per volt. Thus on its 100 V range the overall resistance is 50 000 Ω which, when applied to the valve anode, will draw around 1 mA extra current through the load resistor and the anode voltage will drop accordingly, giving a meter reading of perhaps no more than 20 V even though there is nothing wrong with the set.

Now let's look at the AVO 'Eight'. The basic movement of this has an f.s.d. of only 50 μA, and to make it register 1 V the overall resistance has to be 20 000 Ω. The resulting sensitivity of 20 000 Ω per volt makes a great deal of difference to the reading than would be obtained using the same example as before. The tiny current of about 10 μA drawn by the meter would drop the anode voltage by less than half a volt, so a much more accurate reading would be obtained.

It would be easy to be misled by the above into believing that the old AVO 'Seven' has no place nowadays in radio servicing. Nothing could be further from the truth, because the 'Seven' was the industry standard through much of the vintage years until about 1955 and nearly all the radio manufacturers used it to produce the voltage check lists in their service manuals. Therefore the readings produced on a 'Seven', even if they are not totally accurate in *actual* voltage, do indicate what is right and what is wrong. Reading given by a meter of high sensitivity, although much more accurate, may in practice be misleading!

For the rest of the valve years the 'Eight' took over and most service data voltage readings were based on its use. What needs to be remembered is that when an 'Eight' is used to check voltages originally taken on a 'Seven' the readings usually will be higher, and this should be taken into account (the reverse is true when a 'Seven' is used to check figures taken with an 'Eight').

There is little doubt that the 'Eight' is now the most popular meter for vintage radio servicing, assisted by its ready availability on the Government surplus market at a fraction of the cost when new; thus does the taxpayer aid the vintage radio enthusiast. There are, of course, many other makes and types of multimeter, not all with conventional dials of the type now referred to as analogue but also with digital readouts and even simulated voice readouts. These last two impose virtually no load on a circuit and thus might be thought to be the best for servicing, but in fact they lack the ability of the analogue type to give the smooth indication of rising and falling voltage or current that is necessary for certain tests on valve radio receivers.

The right way round for meters

I hope that experienced readers will forgive the inclusion of what they may consider to be an obvious point. However, at one of the Radiophile Workshops no fewer than eight participants arrived with beautiful new ex-Government AVO 'Eights' yet had no clear idea of how they should be used. For the benefit of the complete beginners, then . . .

For most DC voltage tests the negative terminal of the meter goes to chassis and the positive to the HT line, valve electrodes, and anything else carying HT voltage. Always start with the meter switched to a range higher than the maximum to be expected, e.g. if the HT voltage on a valve should be around 100 V, don't start with the 100 V range but with the 400 V range on the AVO 'Seven' and the 250 V or 300 V range on the 'Eight'. When negative voltage has to be checked (as, say, the voltage drop across a choke used for negative smoothing) connect the positive terminal of the meter to chassis and the negative to the choke.

For DC current tests always connect the meter in series with the device that is drawing current so that the positive goes to the positive side of the supply.

For most resistance tests it is immaterial how the leads are connected, except when checking metal rectifiers, which should be tried twice with the leads reversed the second time. There normally should be a much higher resistance one way than the other and similar results, either high or low, indicate trouble.

AC measurements may be made with the leads either way round but when voltage checks are being made with respect to chassis it is best to connect the negative terminal to the latter in readiness for further DC tests.

The signal generator

The one other instrument that is essential for repairing superhet receivers is the signal generator or service oscillator (strictly speaking the former is a highly accurate device more for the laboratory than the workshop, whilst the latter is a less expensive general purpose tool). It is used to align, i.e. to bring into correct tune, first the IFTs, then the RF and local oscillator circuits and needs to cover from about 100 kc/s up to 100 mc/s. Anyone reading certain articles in old radio magazines may well come across an assertion that it is possible for an amateur to align IFTs without a signal generator. Forget it; not even the most experienced engineer can do this properly, the best that can be hoped for is that if just one of the IFTs has been mis-adjusted it may be possible simply to peak up the other on a station.

Try to get hold of the best signal generator that you can, those made by Advance being particularly good. The writer has an Advance Model E2 bought new in 1952 which is still in regular use, so reliability is guaranteed. E2s appear regularly at vintage radio events and auctions, seldom costing more than a few pounds.

The most sophisticated and accurate signal generators were those built (erected might be a better term!) by Marconi's Wireless Telegraph Co. through the 1940s, 1950s and 1960s, with model numbers commencing with the letters TF. Some of these are enormously large and heavy and cost huge sums only affordable by large laboratories and the military (i.e. the taxpayer). For instance, the mighty TF86 measures about two feet six inches square and eighteen inches front to back, weighs over a hundredweight and cost new more than a contemporary family car. In recent years

TFs have been appearing at radio events for only a few pounds, their sheer size daunting most potential buyers. However, if you do have the room to house one, you will have the satisfaction of owning one of the best generators available anywhere.

The resistance and capacity bridge

This enables the values of resistors and condensers to be checked accurately, added to which bridges also incorporate a means of checking condensers for good internal insulation by applying suitable voltages in ranges from about 25 V up to 300 V. This feature may also usefully be employed to 're-form' electrolytic condensers. It is certainly an instrument that will come in handy now and again but don't go out your way to obtain one – one is sure to come along sooner or later at a bargain price.

What about an oscilloscope?

What indeed? You may read reams and reams of articles in old radio magazines about how wonderful this instrument is, but in practice it probably will prove to be a fairly expensive dust-gatherer. Place it low on your shopping list unless you are able also to obtain a wobbulator, which is the common name for a frequency-modulated oscillator. When used to align IFTs the wobbulator supplies the 465 kc/s or whatever, rapidly and constantly swept up and down for 25 kc/s or more on either side. This produces at the detector a varying voltage which starts low at say 440 kc/s, then climbs gradually to maximum at 465 kc/s, only to start dropping again until 490 kc/s is reached, whereupon the whole thing starts over again. If the probe of an oscilloscope is connected to the detector the varying voltage will be shown on the screen as a replica of the response curve of the IFTs, which may be adjusted until the correct shape is obtained. A control voltage produced in the wobbulator is applied to the 'scope to keep it running in perfect step. For anyone engaged professionally in repairing high quality receivers or communication receivers the wobbulator/'scope set-up is a good investment but the hobbyist can afford to ignore it unless it comes at a non-refusable price.

The same remarks apply to the next two items, the *AF signal generator* and the *'Q' meter*. The first, with usual coverage from about 10 c/s up to

30 kc/s is useful if one needs to be able to plot the audio response curve of a receiver or amplifier. The job of the second, strictly speaking, is to test the 'Q' or goodness of tuning coils, which it does by applying a signal at the appropriate frequency and measuring the gain by means of special kind of detector connected to a milliameter. It also enables one to check the frequency range of a coil placed in parallel with a variable condenser, the setting of which can be altered from about 10 pfd to 750 µfd. This is only likely to be a useful facility for anyone engaged in rewinding tuning coils on a fairly regular basis.

The output meter

This is no more than an AC voltmeter which can be connected across the secondary of the speech coil in a receiver. With the steady 400 kc/s AF modulation produced by the average signal generator applied to the set, the output meter registers a voltage dependent on the output from the receiver. It is thus an excellent indicator of when the IFTS and the RF and local oscillator coils are brought into correct alignment. If you have a multi meter you already have an output meter, since the same job may be done by a suitable AC voltage range connected across the speech coil.

Valve testers

Oddly enough, these are more likely to appeal to the person who does only the occasional radio repair than to someone with a busy workshop and a good spares section. Certainly in the radio trade, although every shop had its valve tester, few engineers bothered to use them because it was quicker and more positive simply to try a new valve in a set. There is also the argument that in a sense a set is itself a valve tester since, if all the supply requirements are in order, a few voltage tests will indicate the condition of its valves with fair accuracy. However, for anyone lacking a large valve cupboard and wishing for an independent means of checking suspect examples, a tester can be very useful. The best-known types are the various models made by AVO since about 1935, and by the Taylor Instrument Co. from about 1938–1958. The early AVOs and the Taylors worked by measuring the mutual conductance of triodes and other multi-electrode valves, and the anode current of diodes and rectifiers at a given voltage. The AVOs were of twin-panel construction, one carrying all the voltage selector switches, the supply transformer and the actual meter movement, the other housing a wide selection of valve holders and a multiway thumb switch enabling many combinations of connections to be made to the holders to suit almost any known valve. The Taylors used a single large panel carrying the meter, the power supply, the valve holders and three switches which worked in various combinations to give the correct base connections. Either type of tester is likely to be available at vintage radio events at seldom over about £50 and maybe a lot less.

The later AVO Valve Characteristic Meters are far more sophisticated devices enabling very comprehensive and accurate checks to be made on various factors that affect a valve's performance. The military used them in large quantities and it is those versions which are most likely to appear at radio events, although the civilian original does pop up now and again. Be prepared to pay up to £200 for a really good example but look out for the occasional bargain at a fraction of this.

Warning: whichever of the above valve testers you may buy, make sure that it has the essential operating book with it, for if this is missing you may have to pay up to £30 to obtain a replacement (they are not infrequently more valuable than the testers themselves!).

The Mullard 'high speed' valve tester

This offers the most picturesque way of testing valves with its cathode ray tube display and 'one armed bandit' action. It is basically a large metal box with a few knobs on the front below a 3-inch c.r.t., and a large metal lever on the right-hand side. At the top rear is a slot into which is inserted a punched card made of paxolin, individual examples being supplied with the machine for all the popular valve types. Pulling the lever causes contact springs to bear on the card, completing circuits according to the positions of the holes and thereby setting up all the valve heater and electrode voltages automatically. A spot on the c.r.t. moves to show, with the aid of a scale printed alongside, the state of health of the valve. The complete outfit consists of the tester itself and two large steel boxes containing the cards, all supposed to be mounted on a strong cradle which may or may not be present.

These testers were fabulously expensive back in the 1950s but can now be obtained at auction for around £50 complete with cards – but make sure that the popular valve types are still represented and that the essential mains adjustment card also is there. Another tip: watch out for ex-Government versions for which all the cards were for military valves with CV numbers. It is possible to cross-reference a lot of them but it is a tedious job.

Chapter 17

A few words about safety precautions

Before we start on radio repairs a few words have to be said about safety, because mains powered receivers have the capability of delivering very nasty, if not fatal, electric shocks. Fortunately danger may be avoided by taking some fairly simple precautions which soon become second nature.

As mentioned earlier, one side of the electricity supply mains (the neutral) is connected to earth and the other (the live) is at 230 V above earth. The amount of electric shock sustained by anyone coming into contact with the live main will depend to a large extent on whether he is standing on a surface which provides a good or bad conductive path to earth. It is quite possible for someone working in an upstairs room on a dry wooden floor to touch a live object and feel no more than a slight tingle, yet if he were standing on a garden path he would most certainly receive a bad shock. Thus the first precaution is to make sure that the floor of a workshop, if not of wood, is insulated in some way such as laying down boards and rubber or plastic mats.

The next thing to remember is that electric shocks are normally far more serious when they travel through the body across the heart, such as would happen if someone were to touch a live object with one hand whilst the other was in contact with an earthed object. The golden rule here is never to use both hands at a time when working on a radio set: always keep one hand firmly in a pocket out of the way of danger.

The danger points

Most UK AC/DC receivers and a high proportion of part AC-only and AC/DC/battery sets have one side of the mains connected to the chassis, so it is absolutely essential that this should be the neutral main. Don't assume that if the mains lead is fitted with a three-pin plug all will be well if the neutral wire apparently is connected to the neutral pin, for if someone has replaced the on/off switch in the set at some time it is conceivable that the wires may have been crossed and that the neutral lead is now live and vice versa. In any case, don't trust the colour marking of wires because they may be non-standard, the standard in this case being the old one in use during the valve era of red = live and black = neutral. You may find two leads of the same colour, or one red and green, or almost any other combination, so to be sure check the resistance between the chassis and each lead with the on/off switch in the on position; there should be a very low resistance in one case and a fairly high one, 500 Ω or more, in the other. It is the first one that should go the neutral pin and the second that should go to the live.

Resistive mains cords

A resistive mains cord is unmistakable with its one lead made of flexible asbestos and covered with 'wool' of the same material. This lead invariably should be connected to the live pin of the plug.

More than two leads

Some resistive cords have two ordinary conductors plus the resistive one; in such cases the latter should be connected to the live pin along with the lead having the higher resistance of the other two (or no apparent connection) to the chassis. This lead will be the one used to feed AC to the anode of the rectifier.

Spare conductors

In a very few cases you may find a third plain conductor in the mains lead which does not appear to have been connected previously. Leave it alone because it may be (a) an unused spare conductor associated with a resistive mains cord or (b) an 'aerial', especially in the case of a midget receiver.

Nothing certain yet

Even with the neutral mains lead definitely connected to the neutral pin on the mains plug do not assume that all must be well, especially if you are not working on home ground. Be particularly careful if extension or amateur wiring is involved. It is by no means unknown for the wiring to a power socket to be reversed, making the neutral pin live and vice versa. The most dangerous aspect of this is that any appliance remains live even though ostensibly switched off. The present writer has encountered a number of such cases including one in which all the wiring in the house was 'back to front' due to the electricity company having connected up its meter incorrectly.

The safety routine

The safest way to proceed is first to check the mains plug on the set as above, then, with the back off the set and its switch in the 'on' position, plug into the mains and switch on at the socket. Now use a neon screwdriver to check if the chassis is neutral: the lamp should not glow. Since these devices do not 'fail safe', then make sure that it *will* glow on a live surface by touching it on some part of the set that ought to be carrying AC, such as the mains dropper or a fuse if fitted. Only after all this is it safe to assume that the chassis is indeed safe to touch. In fact, the checks take far less time to carry out than to be described and experienced engineers do them by habit.

Don't relax your guard

What you must remember now is that although the chassis itself is dead with respect to earth various items on it, including those just mentioned and the on/off switch, still carry live AC and must be avoided.

Other points in the set, such as the smoothing condensers, the output transformer and sometimes again the mains dropper, will carry the HT + voltage from the rectifier and these will be up to 250 V DC with respect both to the chassis and to earth. So long as you touch these only with the insulated test prod of a multimeter you will be safe.

Genuinely AC-only receivers

These have double-wound mains transformers that isolate the mains from the set completely, and their chassis will be safe to touch whichever way round the mains leads are connected. That is the theory and in the great majority of cases it is realised in practice but things can go wrong and **precautions should be taken before working on any unfamiliar set.** It is not unknown for electrical leakage to occur between the primary of a mains transformer and the core, nor for either wear and tear on the mains lead or careless handling to bring about a break in the insulation where it passes through the chassis. Be suspicious of any anti-modulation hum condensers connected from primary to chassis and also of 'mains aerial' condensers. Always measure the resistance of each conductor in the mains cord with respect to chassis; it should be close to infinity unless, of course, a three-core lead with one of them as an earth connection is being used.

As in AC/DC sets, all the various AC and DC voltages before and after the rectifier will be live to chassis and contact must be avoided except with a meter prod.

The isolating transformer danger

Yes, this is a deliberately provocative heading designed to draw attention to the fact that safety devices don't always guarantee immunity from shock. The writer has seen it asserted in some circles that an isolation transformer affords complete protection against electric shock to anyone engaged on repairing mains radio receivers but this, unfortunately, is nonsense.

An isolation transformer is a double-wound device with a 1:1 ratio, so that when 230 V AC is fed to the primary from the mains the same voltage appears on the secondary. The latter is completely isolated from earth so it is possible to touch either output terminal separately without receiving a shock. When an AC/DC radio is powered from such a transformer its chassis will not be live with respect to earth whichever way round the plug may be inserted.

What the isolation transformer *doesn't* do is to alter in any way the position regarding all the points on the set that are live to chassis when the set is powered directly from the mains. In other words, exactly the same precautions have to be taken as before.

Another point: if two or more AC/DC sets are connected to an isolating transformer, unless great care is taken it is quite possible that their individual chassis might be at mains voltage with respect to each other.

To sum up: by all means use an isolating transformer for its ability to make one chassis at a time safe to touch, so long as you don't trust it to do anything else.

Other hazards

Electric shock is not the only potential hazard for anyone engaged in radio repairs. If you have lived long enough to read this book you presumably will already have learned to treat ordinary tools with respect, such as always keeping sharp ones pointing away from yourself, but you may not previously have done much soldering. Always try to arrange what you are doing so that the iron and the joint are above the surface of the bench and not above your person. Molten solder dripped onto trousers or socks soaks irretrievably into the material and can causes very painful burns. Even worse is any action that results in hot solder being flicked into your face: you have only one pair of eyes (see next paragraph). Incidentally, despite the near automatic reaction of trying to catch any falling object, never try this when a hot soldering iron falls off the bench.

Fumes from solder

The flux agents used in some cored solders produce very unpleasant fumes when they are being melted by the soldering iron. By definition, most of the time you will be working close to a soldering job to make sure of a good joint, thus placing yourself in a good position to breathe in all those fumes. Whether or not they actually are harmful the writer does not know, but what is certain is that in the confines of a small workshop they can cause sore eyes and throats (this is likely to be far worse for non-smokers whose lungs are unused to being assaulted by fumes). If you are vulnerable a full-face respirator (not a dust mask) will be a good investment – and it also keeps hot solder out of your eyes!

Chemical warfare

Some of the multifarious cleaning fluids and aerosols also can be unpleasant, if not actively hazardous in a small workshop. It is not always the more arcane substances that are the worst in this respect: many years ago the writer left an employee working on a television set in a fairly large workshop. This chap had decided to clean the screen of the set with ordinary methylated spirit and by the time the writer returned the fumes had overcome him and he was lying slumped over the bench. Fortunately he soon revived when brought out into fresh air, which brings us to another very important point . . .

On your own

It is more than likely that when you are working on radio receivers you will be alone in a workshop which may be in an outbuilding well away from other people. Can you be certain that anyone would hear you call for help should an accident occur? This is meant to be a sobering thought which hopefully will bring home to you the necessity of taking care.

Chapter 18

A logical approach to fault finding

It cannot be stressed too highly that only a systematic approach to fault finding can guarantee a speedy and efficient repair. Haphazard testing is a waste of time, for there is a certain order in which checks must be made, which is from the power supply stage back to the aerial socket. The following chapters of this book dealing with the various sections of receivers have been arranged in the same order, to facilitate servicing. If you study the chart (Figure 18.1) you will find that it commences with 'set dead', and recommends actions to be taken, starting with the power supply. This, then, is the subject of the first detailed chapter, followed by AF amplifiers and output stages, detector/AVC/AF amplifiers, IF amplifiers, frequency changers, and so on. The approach, in all cases, is first to deal with the basic, conventional circuitry found in the majority of sets, followed by a discussion of 'unusual features', found in perhaps only certain makes, or at fairly well-defined eras in the development of receivers.

Obviously, it is possible to join the chart at a lower point, e.g. 'hiss but no stations' but after remedial action has been taken it is still sound policy to run through the check list again from the top. A radio which has not been used for some time, due to the fault given as an example, may well develop others after being put back into service. Something as simple as replacing a leaky condenser in the output stage could forestall an HT overload leading to an expensive or impossible mains transformer replacement job.

Initial tests

When you are faced with a radio set about which nothing is known, it is advisable to carry out some simple checks before plugging it into the mains. The presence of a mains transformer or dropper will indicate whether the set is for AC only or AC/DC operation, unless it is so old as to be one of the comparatively rare DC-only models. If there are any suspicions be guided by the valve numbers and/or the absence of a rectifier. Assuming, however, that the set is for AC or AC/DC operation, test across the mains lead with the ohmmeter, with the radio switched 'on'. AC-only sets should show a reading of $30\,\Omega$ to $70\,\Omega$ approximately, whilst AC/DC sets will be much higher, usually from $500\,\Omega$ upwards, depending on the current rating of the valve heaters. Should a dead short be registered, check for perished insulation on the lead, particularly where it enters the set.

Many sets had a condenser wired from live input to chassis, or in AC/DC sets, rectifier anode to earth. A value of 0.01 μfd to 0.1 μfd is common, its purpose being to reduce 'modulation hum' – a 50 c/s buzz derived from the mains and apparently tuned in with a strong station – and it is not uncommon for them to go dead short. To replace, use one rated at a minimum of 1000 V DC or 300 V AC.

In AC-only sets the presence of a large value condenser from mains to chassis can render the latter slightly live even though the component is in first class condition. A shock from this source can be unpleasant, because although not dangerous in itself, it can frequently lead to a hand being grazed as it is snatched hurriedly past a sharp object!

If all seems to be in order on the mains input, turn to the HT side, and check the resistance of the main smoothing condenser to chassis, using an ohms range of about $100\,k\Omega$ maximum. The needle should flick right over for a second or two, then drop back fairly rapidly to above $10\,k\Omega$

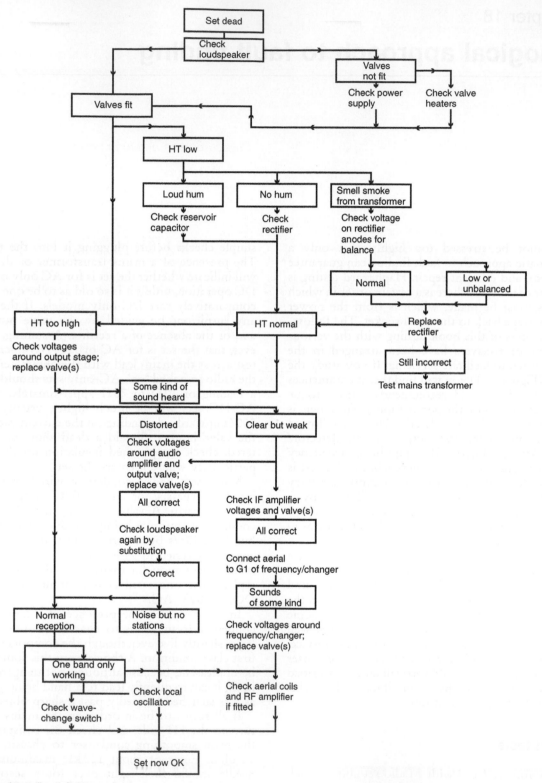

Figure 18.1 Fault-finding chart

at minimum, indicating that the condenser is charging from the meter battery, and thus reasonably efficient.

The standing resistance to chassis may be due to a slight short somewhere in the set, or to the presence of 'bleeder' resistors used to drop the voltage for certain valves. If this test is satisfactory, plug the set into the mains, bearing in mind that with AC/DC sets the neutral side must be connected to chassis for safe working. Switch the testmeter to the 500 V DC range and connect to the previous HT point and chassis. Within about three minutes maximum, readings should be obtained, and the set should produce some kind of noise.

You are now in a position to employ the fault-finding chart and to refer to the appropriate section of this book. Obviously one cannot claim to be 100% comprehensive, but the faults and cures listed are those which have been experienced most often by the writer in 50 years of practical experience in the radio trade.

Please note the reference on the chart to 'check loudspeaker'. Many sets were fitted with switches, or plug-and-socket connections to enable the internal speaker to be silenced while an extension unit was in use. Always give these devices a close look, and if necessary connect a test speaker to make sure that complete silence is not due to a simple fault in this part of the set. In fact, always look for the simple things before commencing high powered fault finding. Fuses can be open circuit even if they appear to be in perfect condition, on/off switches do stick in the off position, and 'dry'

soldered joints can mysteriously make themselves a nuisance after 50 years of working well enough. You could say that Murphy's Law – if anything can go wrong it will go wrong – and its many subclauses might have been propounded with vintage radio in mind but this is not really fair, because so many valve receivers have lasted for what would have seemed totally incredible lengths of time to their makers. The last valve receivers were made in the early 1960s, which is long enough ago in all conscience, but what worker, assembling a set in 1929, could possibly have imagined it still being in working order 70 years later? In fact, what other domestic appliance even approaches the radio set in large-scale longevity? On this optimistic note let us commence practical servicing work.

Look for previous work!

Experience – bitter experience – has taught the writer always, on principle, to mistrust any previous repair work done on a set. All too often an enthusiastic but careless repairer will have changed components but put in replacements either of incorrect value or of a completely unsuitable type. This may sound both pessimistic and cynical but over the years the writer has encountered the problem so many times as to make it almost an occupational hazard. Always look out for suspiciously new-looking components and check them out before doing anything else to a set.

Chapter 19

Repairing power supply stages

AC-only receivers

If there should be no reading across the mains lead when checked with an ohmmeter, connect the test prods directly across the set side of the on/off switch, as either it or the lead itself may have gone open circuit. The remedies here are obvious. It is rare for the mains transformer to give trouble unless subjected to gross overloading, so investigate all other possibilities before condemning it. It is worth remarking that a keen sense of smell is as good as a testmeter where transformers are concerned – the odour of a burned-out winding is quite unmistakeable.

Check any fuses with the ohmmeter – don't trust to appearances. In addition, it is not unknown for a fuse holder in ostensibly perfect order to have developed high resistance contacts, breaking the input circuit even though the fuse itself is all right.

If there is no obvious fuse to test, have a close look at the mains tapping panel, because it often incorporated a small two-pin plug with a fuse in its body. Even where there was no fuse it has been known for these plugs to go open circuit and effectively disconnect the input.

Should all these tests prove negative, and the meter reveal that the transformer primary is indeed open, there is a problem, since replacements are rather expensive if bought new. It may prove to be more practicable to wait until a good second-hand component (perhaps from a similar set) can be

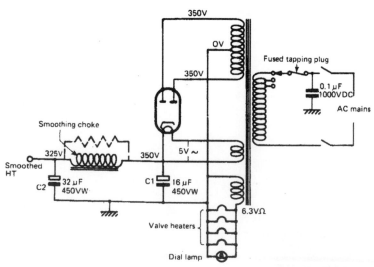

Figure 19.1 Typical AC-only power supply unit

obtained. When the work is carried out, first make a careful diagram of the original connections to the transformer, including any colour coding for cables, showing the leads to the primary, HT secondary, rectifier heaters, LT secondary, etc. Don't rely on memory – something unforeseen may delay the job by which time confusion has had time to set in.

If the valves light up but no HT voltage appears the first step is to replace the rectifier valve. At this stage always test the AC voltage on the anodes of the rectifier valve. It may be anything between 200 V and 500 V, but the important thing is that it should be the same on both anode pins. Any discrepancy other than a volt or two inevitably points to a faulty mains transformer, which will overheat and destroy itself in time.

If the rectifier shows signs of overheating with reduced output, test for a short on the HT line, initially across C1. Where there is a very low resistance of a few ohms this condenser is itself suspect, but a reading of up to 1500 Ω indicates that the fault lies on the other side of the smoothing choke or resistor, possibly due to C2 having shorted. The test meter should be transferred to this component, but the procedure is likely to be a little more complicated for this reason. Very often several HT connections will be found on C2. They should be disconnected one by one until the reason for the short is found. This procedure holds good right through the set. If need be, note the connection then unsolder or cut the wire, and test. The most elusive short may be traced in this way, with the exercise of patience – but do be careful to keep those notes!

Low HT volts, with a loud hum from the speaker, points to C1 having gone open circuit or low in capacity. If C2 goes faulty in this way the HT voltage will be normal and the hum not quite so bad.

Use the correct replacements

Proper electrolytic condensers made especially for the job must be used in HT smoothing circuits, especially in the reservoir position. As has been explained elsewhere, the DC output from a rectifier carries an AC ripple which the reservoir condenser must be capable of handling. If the HT current drawn by the receiver is, say, 75 mA, the AC ripple current rating of the reservoir must be at least 150 mA. Although the ripple current is not always marked on condensers, those sold for smoothing purposes will have been designed to handle an ample amount for domestic receivers. When two or more condensers of the same capacity are housed in a single can, for instance 8 μfd + 8 μfd the section intended to be used in the reservoir position usually will be colour coded red and the section for smoothing coded yellow. It can normally be assumed that when two different values are housed in a can, such as 8 μfd + 16 μfd, the smaller value will be intended for the reservoir position.

Miniature electrolytic condensers are not suitable for use in HT smoothing circuits and should be avoided.

Negative HT smoothing

This is more likely to be found in receivers made in the middle 1930s. Its presence is detected by the fact that the centre tap of the rectifier anode winding is not taken to chassis directly, but via the smoothing choke or speaker field. The reservoir condenser is connected between the rectifier cathode or filament and the centre tap, making its negative side anything from about 50 V to 150 V negative with respect to chassis. The point to bear in mind when smoothing condensers have to be replaced is that the new one may have to be insulated with a few turns of tape to prevent its case (which is usually connected to negative) being earthed by a fixing clip. If this is not done the smoothing choke or/speaker field will be shorted out. If a choke is employed the result will be a powerful hum from the loudspeaker, but if a field is used there will be no sound at all because the loudspeaker will not be energised!

Low voltage negative bias

Some sets have the centre tap of the HT winding returned to chassis via a fairly low resistor, perhaps about 50–100 Ω. This is not negative smoothing but simply a means of obtaining a few negative volts for bias purposes, typically in connection with delayed AVC. The reservoir negative terminal is usually taken to chassis as usual but there may be a fairly large value, low voltage electrolytic connected across the bias resistor. Check this if there is an HT hum not traceable to the main smoothing condensers.

RF by-pass and other HT decoupling condensers

Some firms, notably EMI, made a practice of wiring a paper condenser across the HT line to by-pass any RF energy that might appear on it and cause instability. These have been known to go 'leaky' or even dead short, taking the HT line down to chassis and causing severe overloading of the power supply stage in general. The value employed was generally from 0.05 μfd to 0.5 μfd, and when fitting a replacement use a standard size non-inductive type meant for valve radio use.

AC/DC receivers

As with AC sets, the switch should be checked first along with the mains lead, followed by the mains dropping resistor. Since the valves and dial lamps (if any) are in series, the failure of any one will extinguish all the rest. Dial lamps are more vulnerable than valves and should be checked first. Two alternative means of wiring them into the circuit are shown in Figure 19.2; (B) is to be preferred. R2 is an optional refinement which allows the set to continue working even if the lamps should fail.

The thermistor also is optional. It is a special resistor having quite a large resistance when cold (commonly 200 Ω) which falls to around 50 Ω when heated by the passage of current through it. It protects the valve heaters from harmful surges when the set is switched on from cold. Thermistors are robust devices, but they do sometimes crack, or have a connection fail.

If the valves are lit but there is no HT, try the rectifier. AC/DC types are far more likely to pack up altogether than are AC-only valves due to a separate, highly insulated cathode which can be damaged by a short on the HT line. If R1 has gone open circuit and removed the anode volts from the rectifier, check C3, the anti-modulation hum condenser.

The HT smoothing arrangements are very similar to those of the AC set, but as explained earlier, the voltage is normally equal to, or less than, that of the mains so arrangements are made to get as much of the available voltage as possible in to the anode of the output valve. Very often the HT end of the output transformer is taken directly to the cathode of the rectifier, with a fairly high value resistor (typically 1 kΩ–5 kΩ) used to 'smooth' the HT to the rest of the set. Check this resistor for damage from overheating. In the odd cases when the output transformer has an overwind for HT

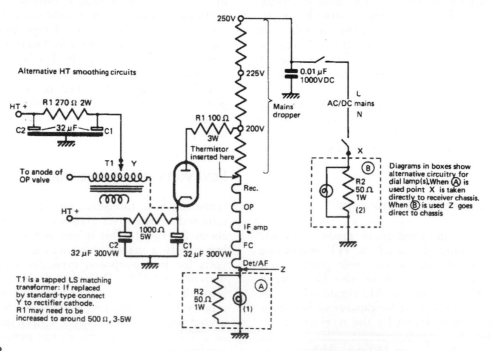

Figure 19.2

smoothing and this has gone o/c a conventional transformer may be used as a replacement, ignoring the overwind and taking the HT end of the primary to rectifier cathode.

Tracing HT shorts should be carried out as described for AC-only receivers.

When the mains dropper fails

Open circuits on mains droppers may be treated in two ways, according to the extent of the damage. When only one section of a multi-tapped resistor has gone it is worth checking to see if a slight rearrangement of the connections will put things right. For instance, with UK mains now standardised at 230 V any windings fitted exclusively for 240 V and 250 V may be adapted for other purposes. Otherwise, it is often possible to wire a replacement resistor across an o/c section to take its place. Connections here must be good and firm: use high melting point solder or nut-and-bolt joints.

Should the entire mains dropper be open circuit, all is not lost, even if the resistance value is unknown. Once the total voltage required by the valves and dial lamps is established, the correct resistance can be calculated by Ohm's law.

A typical valve line-up would be UCH42, UF41, UBC41, UL41 and UY41. The voltages of the heaters are, in order, 14 V, 12.6 V, 14 V, 45 V and 31 V, a total of 116.6 V. Two dial lamps of 6.3 V each would bring this up to a shade under 130 V. Thus the dropper would have to make up the difference between 130 V and the 230 V mains. Referring again to Figure 19.2, it will be noted that not only the heater current, but also the HT current, flows through the top part of the dropper. This must be taken into account when calculating. According to the data, the current under average working conditions is in round figures 70 mA. Added to the heater current of 100 mA, the total flow through R1 and R2 in Figure 19.2 will be 0.17 A. Each resistor will be required to drop 20 V at the given current so Ohm's law gives us just under 118 Ω. The nearest commercial value of 120 Ω ohms should be used. The wattage is given by 20 V × 0.17 A = 3.4 W, so a 5 W type will be well within its rating.

Moving on to R3, we find that this has to drop the remaining 70 V (200 V minus 130 V) at the heater current of 0.1 A. This works out to a convenient 700 Ω ohms at 7 W, and a suitable resistor should easily be obtainable. R4 is called a limiting resistor, and its purpose is to damp any surge of current through the rectifier when the set is first switched on from cold. A 100 Ω 3 W resistor makes a very good replacement.

Part AC-only sets

Generally speaking the power supply sections of these sets follow the same lines as those of AC/DC receivers apart from the use of an auto-transformer in place of a mains dropper and the same fault-finding and repair techniques apply.

Experience has shown that the auto-transformers used in these sets are extremely robust and unlikely to fail in normal use. However, if this does happen consideration should be given to converting the set to full AC/DC operation by replacing the transformer with a mains dropper.

Resistive mains cords

These were used in many small and midget AC/DC receivers, both UK made and imported (especially from the USA) from about 1933 to 1953 as an alternative to a normal mains dropper. Unfortunately replacement cords have not been made for very many years so the chances of finding spare lengths are very slim. Some old stocks may have survived in long-established radio shops but even then they may not be in usable condition. When resistive cords are left rolled up for years on end there is a tendency for o/cs to occur at short intervals along their length, rendering the whole thing useless, so if you do find a 'new' coil of line cord check it carefully. You *might* find some odd straight lengths that may happen to be of the correct resistance value but as this would fall into the category of a near miracle some more down-to-earth possibilities need to be examined.

Your course of action will depend to a large extent on the exact job done by the cord in a particular model. Most UK-built receivers were made to work with HT lines of around 200 V, the least voltage that might be expected from the 200/250 V mains, so the rectifier anode was supplied almost directly from the mains and the resistive cord had only to handle the heater current. In this case it may be possible to route a new conventional mains lead into the set via a voltage dropping device for the heaters, such as an auto-transformer, a mains dropper in a metal safety cage or simply a mains electric light bulb of suitable wattage. Note: this will always be higher than would at first appear likely to take account of the lower current the lamp will pass when in series with the heater chain. For instance, a 230 V 100 W lamp will pass about 0.43 A in normal service but will be very suitable for a 0.3 A chain. For 0.2 A chains use a 60 W lamp, and this size probably will be suitable also for 0.16 A and 0.15 A chains because these are likely to add up to a much higher voltage than 0.2 A types. Finally, for 0.1 A chains try a 40 W lamp.

Extra precautions necessary with AC/DC and part AC sets

It was common practice to use only a single-pole mains switch to break the side of the mains lead that went to the HT-line and the bottom of the heater chain, which in almost every known British set was the chassis of the receiver. The interesting situation arises that if the mains lead is plugged in to make the chassis neutral while the set is working, it will become live when switched off and vice versa, due to the path from the top of the heater chain back down to chassis. It is therefore essential to replace most carefully all the various devices such as covers for chassis and grub screws employed to prevent owners from accidentally coming into contact with live metalwork. It is also worth considering replacing the existing mains switch with a double-pole type which will break both sides of the mains input.

Note: it is not possible to switch all the conductors in a three-core resistive line cord, so always either switch off at the mains socket used to supply such sets or withdraw the plug.

Barretters

In place of a mains dropper, certain AC/DC and DC-only sets had a device known as a barretter. This resembled an old-fashioned light bulb, and used a principle which had been discovered many years ago, namely that a resistance wire made of iron enclosed in a hydrogen atmosphere tends to pass the same amount of current within quite wide variations of applied voltage. This is, of course, ideal for regulating the current through a series heater chain, especially where the mains voltage is subject to severe fluctuation. Barretters were made at one time or another to suit virtually any range of valves drawing between 0.1 A and 0.3 A. The type

of base varied widely according to type. Some had an Edison screw as used for US light bulbs, others had valve-type bases matching those employed in the receiver.

In the event of a barretter failure it is worth while making enquiries of the various vintage valve specialists, because there is a very good chance of obtaining an exact or near-equivalent replacement. Otherwise, give consideration to the use of an ordinary electric lamp, as discussed earlier.

American radios often had special barretters or 'ballast' tubes incorporating shunt resistors for pilot lamps. Either glass or metal/glass envelopes were fitted, and the bases were either UX 4-pin or octal.

Comprehensive lists of both British and American barretters will be found in Appendix 2.

'Watt-less' droppers

Ever since the 1940s radio magazines have been talking about using large value paper condensers in place of a conventional droppers in series-heater receivers, on the grounds that the condenser consumes no power, produces no heat, cannot possibly go wrong and generally is just about the best thing since sliced bread. The fact remains that the idea was never adopted commercially except in one particularly horrible little television set which gave so much trouble that it had to be recalled for modifaction to a conventional dropper. However, for anyone keen to try the method, as an example, a dropper for a typical 0.15 A range of valves operating from 230 V mains would be about 800 Ω. A 4 μfd condenser has a reactance of 796 Ω at 50 c/s and could thus be used as an alternative to a dropper or barretter or even an auto-transformer. Remember though, that all these are fail-safe – if they stop working it is because they go open circuit, preventing any voltage from reaching the heater chain. On the contrary, when a condenser working under these conditions fails it is almost certain to go dead short and expose the chain to a catastrophic overload in which the valves are bound to suffer damage – *cf.* Murphy's Law.

Replacing 'metal' HT rectifiers

This term includes every type from the original old finned copper-oxide types to the later contact-cooled models. They all shared a common characteristic of having a fairly high forward resistance, which means that they had an inherent current-limiting action so that when a set is first switched on there is a certain restriction on the output voltage and current. They were also all prey to the effect of falling output with age, and it is all too common to find that the HT in a set using a metal rectifier has dropped to half the normal figure. In such cases it is tempting simply to fit as a replacement a miniature silicon rectifier which produces no heat, cannot possibly go wrong and – well, you know the rest.

In fact, silicon rectifiers, good as they admittedly are, are not suited technically for use in vintage radio sets without special precautions being taken. They have an extremely low forward resistance which means that (1) heavy peak currents flow at initial switch-on and (2) the DC output is very likely to be equivalent to the peak value of the AC input. With (1), when a silicon diode is fed from a mains transformer the initial surge is capable of causing overload damage to the latter, while with (2) the voltage on the HT line may rise so much above the design rating as to cause component or even valve failure.

To examine a typical case, consider an AC/DC receiver in which the original metal rectifier was fed with AC directly from the mains input. Its output would have been expected to be the equivalent DC voltage, say 230 V, and the ratings of the smoothing and other condensers will have been designed to suit this voltage. A silicon rectifier used as a replacement might well produce a DC output of towards 100 V more than that of the AC input so the potential danger of overloading is quite evident. This is of particular importance in mains/battery sets such as the Ultra 'Twin' series in which a small metal rectifier produced both the HT voltage and the LT for the valve filaments; an ill-judged replacement with a silicon rectifier could easily cause one or more of those filaments to burn out expensively.

The answer is to fit a suitable limiting resistor between the AC input and the anode of the silicon rectifier. A certain amount of experimentation may be needed to find the exact value required, starting with something likely to be on the high side for safety. In the Ultra 'Twin' the output from the original rectifier was measured for the service data as 223 V DC with an AC input of 230 V. The HT current consumption was 10 mA and the filaments drew 47 mA. Since a silicon rectifier may be expected to deliver well over 300 V DC, the limiter should initially be capable of dropping 100 V @ 57 mA (i.e. the combined HT and LT current). For

this a 2.2 kΩ wire-wound rated at 5 W would be suitable. Check the DC output under working conditions and if necessary reduce the value of the limiting resistor a little at a time until the correct output voltage and current are measured. Note: in mains/battery receivers the usual practice was to run 1.4 V rated filaments at 1.3 V for longevity and to give a margin of protection against voltage surges. On no account should this voltage be exceeded in service.

A few very early AC-only mains receivers used metal rectifiers in voltage-doubling circuits which also must be treated with care when silicon diodes are to be used as replacements. It would be better to use a separate limiting resistor for each rectifier rather than a common one which might run too hot for comfort. Again work on the basis of starting with large values and reducing them if necessary until the correct output voltage is obtained. The same remarks apply to the replacement of bridge rectifiers such as the contact-cooled types used in receivers of the late 1950s.

Dealing with energised loudspeakers

The general design of these has been touched on in an earlier chapter. When the field winding is used in place of, or to augment, a smoothing choke for the HT supply of the receiver a failure of the winding obviously will stop the set dead. If a replacement speaker is not available consider substituting an LF choke or even a large resistor for the field and replacing the loudspeaker itself by a permanent magnet type. The DC resistance of the smoothing choke or resistor should match that of the old field winding so that the HT voltage remains correct. When a smoothing resistor is employed it should be mounted on an insulated tag strip in a place where it will receive some ventilation.

Conversely, if an energised speaker has suffered irreparable cone damage but the field winding is still in good order, it may be possible to remove the latter (complete, not just the winding) and to mount it, say, on the baffle board with wood screws where it can continue to do its job of smoothing the HT. A permanent magnet speaker may then be used without further modification being required.

A few sets, notably some American 'midgets', had energised speakers whose field windings were connected between HT and earth, being used as pure electromagnets. These speakers may be replaced by permanent magnet types without any bother, except that the HT should be measured to ensure that with a reduced load it does not become too high. In most midgets 90 V is quite sufficient, and above this the output valves tend to dissipate more power than is good for them. An extra series resistor in the rectifier anode circuit is the best way to deal with this situation. Incidentally, it was not

unknown for the speakers in these sets to be of the high impedance moving-iron type, and this should be checked if an output transformer is not immediately visible.

Note: whenever faults are found in a power supply unit that are the result of overloading, always check the coupling condenser feeding the grid of the output valve – see next chapter.

Chapter 20

Finding faults on output stages

Since there is little difference in AC and AC/DC techniques other than in the power supplies, in this chapter and those following the notes will apply equally to either type. The only point to remember is that AC sets can be expected to have higher HT voltages, perhaps even twice as much, than AC/DC models and replacement components should be rated accordingly.

Figure 20.1 shows a typical output stage having the triode section of a double-diode-triode driving a pentode, with conventional matching transformer to a permanent-magnet loudspeaker. Condenser C1

across the transformer primary is included to reduce the high frequency response. If it should happen to become a dead short it would effectively silence the set. The testmeter should show 200 Ω to 500 Ω across the primary, and a low reading indicates a fault, hopefully in the condenser, because it is a lot cheaper to replace than the transformer itself!

It is rare, but by no means unknown, for the primary winding to short internally with a drastic reduction in volume. If, on the other hand, C1 should go open circuit it can sometimes result in

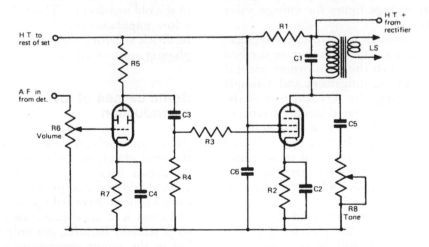

Basic AF amplifier and single-ended output stage. Typical component values:

R1 1000 Ω 5 W	*R6 1 MΩ pot.*	*C3 0.02 µF 500 V*
R2 180 Ω 1 W	*R7 2.2 kΩ ½ w*	*C4 50 µF 25 V*
R3 1000 Ω ¼ W	*R8 50 kΩ pot.*	*C5 0.05 µF 1000 V*
R4 370 kΩ ¼ W	*C1 0.01 µF 1000 V*	*C6 32 µF 350 V*
R5 220 kΩ ¼ W	*C2 50 µF 25 V*	

Figure 20.1

high pitched whistles in the loudspeaker, giving the impression of RF or IF instability. If a replacement has to be fitted, use one rated at 1000 V DC, because very high peak voltages are built up on the transformer. This is particularly so if the set is run with the speaker disconnected for any reason. Sets have been seen in which the voltage has been so great as to break down the insulation of the output valve holder and to burn part of it away.

Replacing output transformers

The new transformer (whether genuinely new or second hand) must match (have the correct primary to secondary ratio) the output valve, or valves in the case of push-pull stages and it must also be capable of carrying the same amount of anode current. The service data for the set usually will give the anode current for the output valve but otherwise it will be found in the Valve Data Appendix (2) to this book.

The ratio is determined by taking the square root of the figure obtained by dividing the optimum load for the valve by the impedance of the loudspeaker voice coil. Optimum ratios for all the common output valves will be found in Appendix 2 to this book. The loudspeaker impedance may be quoted in service data for a particular set, but if not an average figure for vintage valve receivers was 3 Ω.

If you are lucky enough to have a new replacement the specifications may be shown on the box or on a label fastened to the transformer itself; if not you will have to use judgement and a simple test. Generally speaking, the physical size of the transformer will reflect the rated anode current, so as long as the new transformer is at least as big as the old, all should be well. In fact, you will find that there was a surprising amount of standardisation of size and fixing centres for fastening bolts which often helps a good deal when replacements are made.

As regards ratios, it is no good trying to compare the resistances of the primary and secondary windings as they will be of completely different gauges of wire. What you do is to connect the secondary to a source of low voltage AC, say about 2 V, and measure the voltage that appears on the primary. Divide the latter by the former and you have the ratio. Don't take too long about this in case of possible overheating of the output transformer.

Watch out for negative feedback

Many sets had tone control or tone compensation systems using negative feedback derived from the secondary of the output transformer. A replacement transformer must be correctly 'phased' to give negative feedback and you will soon be made aware of a mistake in this respect. When the secondary is connected the 'wrong way round' the result will be positive feedback which will provoke instability and maybe ear-splitting oscillation. The remedy is, of course, to reverse the connections to either primary or secondary, whichever is the more convenient.

Transformers with three windings

Philips and Mullard receivers in particular used negative feedback derived from a tertiary winding on the output transformer. In the absence of an exact replacement the following expedient usually will give complete satisfaction.

Experience shows that it is always the primary winding of these transformers which fails by going open circuit. Fit a replacement transformer as detailed above but leave the old one *in situ*, with the leads to the tertiary winding undisturbed. Connect the secondary of the new transformer to the loudspeaker as usual, and also to the secondary of the old transformer. This will then act simply as a low impedance transformer delivering negative feedback as before; the same precautions regarding phasing must, of course, be observed.

Some causes of low or distorted reproduction

An all too common cause of distortion, but with the level of volume little affected, is the coupling condenser between the triode anode and the control grid of the output valve. If it starts to 'leak' even a little positive voltage will get onto the grid and cause it to draw excess anode current. This is so potentially harmful, not only to the valve itself but to the output transformer and HT supply as well, that a check on the coupling condenser should be regarded as top priority in any set. The tests are a little different for sets with conventional cathode bias and older ones having negative bias.

With cathode bias, the control grid voltage should be zero. If a positive reading is obtained on the meter, it can usually be put down to a faulty coupling condenser, with an internal short in the valve as an outside chance. There is a very simple test to determine which actually is the culprit. With the meter still connected to the grid, short the anode of the triode to chassis with the blade of an insulated-handle screwdriver (it is perfectly safe). This removes HT from C3, and if the meter reading drops sharply the condenser clearly has been leaking. If no difference is recorded, the valve itself probably is at fault, the most common failing being a leakage between grid and cathode.

With negative bias, there should be an actual negative reading on the control grid; it may not be large, even when a sensitive meter such as the AVO '8' is used, because the voltage probably will come via one or more high value decoupling resistors. Carry out the same test of shorting down the triode anode and see if the negative voltage increases. Alternatively measure the anode voltages of the output valve with and without the triode anode shorted down. If the voltage rises when the short is applied you again need to change the coupling condenser. If no rise is recorded, don't immediately assume that all is well as another possible fault must now be investigated. It has just been stated that the negative bias arrives via high value resistors. Associated with these are various decoupling condensers which, if leaky, can reduce severely the negative voltage applied to the grid. The only true test here is to remove one end of each condenser from circuit and measure across it with the ohmmeter. Even a very slight leakage means that a replacement condenser must be fitted.

Replace faulty coupling or decoupling condensers with good quality types of adequate voltage rating. Experience again shows that a very high proportion of receivers coming in for repair suffer from this type of failure and unfortunately, before it has been discovered and put right, damage may have been done to other components, plus the output valve. For instance, the excessively high anode current due to the positive grid bias may have caused the bias resistor R2 to overheat and change its value. In extreme cases it is known for the voltage to rise above the working value for the by-pass condenser C2, breaking it down completely. If it should not actually read dead short, try the effect of bridging another across it, since an open circuit condenser can reduce the output considerably in sets not too well endowed with gain. Note, however, that certain models were deliberately deprived of this component in order to introduce negative feedback and thus improve the bass response of the output stage.

The most serious result of a persistently over-run output stage is, apart from damage to the valve itself, an overheated and ultimately shorted output transformer, plus, of course, consequent harm to the HT supply components. It is certainly not rare for a simple coupling condenser with a 'leak' to cause a mains transformer to burn out, so never neglect to apply the simple tests outlined above.

A less common cause of low volume is failure of the triode anode load resistor, R5 in the diagram. When this goes open circuit the effect is usually not to silence the set completely, as might be expected, but rather to reduce the volume to a whisper. Here's a curious fact: the writer has found over fifty years of repairing radio sets that for some reason or other the value of anode resistor most likely to fail is $220\,k\Omega$. Why this should be is inexplicable, but it has happened time and time again. Whatever its value, though, the triode anode load resistor should always be a prime suspect in cases of low or distorted sound.

What about the output valve?

It was stated elsewhere in this book that a radio set could act as its own valve tester to a certain extent, and here is a good example. If all the coupling and bias components around the output valve, and its HT supplies, are in good order but there is still low or distorted output, you need to know whether or not the valve is drawing the correct anode current. You could, of course, disconnect the HT feed to the output transformer and insert a milliammeter, but a much easier and effective method is to measure its cathode voltage. The correct voltage will usually be given in the service data for the set; otherwise consult the valve tables in Appendix 2. A low reading, indicating low anode current, suggests that the valve has lost emission, a likely sequel to persistent overloading. A high reading, indicating excess anode current, suggests an internal short in the valve. In either case replacement is going to be the only answer.

In receivers using negative grid bias there will be no handy cathode voltage to check, as it will be connected directly to chassis. In this case, discover the DC resistance of the primary of the output transformer, either from the service data or by direct measurement with the ohmmeter, then determine how much voltage is being dropped across it. Simple application of Ohm's law will then tell you how much anode current is flowing through the winding, with the same interpretation being put upon the readings as for cathode voltage.

Heater-to-cathode leaks

It is worth mentioning at this point that even a slight breakdown in the insulation between the heater and the cathode of a valve can put AC on the latter and bring about a 50 c/s hum in the loudspeaker. If a hum of this type is experienced that does not appear to be due to an HT smoothing problem, suspect an h/k leak. Note: this fault is largely confined to AC/DC valves with high heater voltages.

Push-pull output

This feature is found mainly in 'luxury' table models and large radiograms. Decca, HMV/Marconiphone and RGD were probably the leading exponents both before and after the Second World War, all producing large (sometimes immense!) record players and radiograms with tremendous outputs, measured it must be stated in good old British watts (RMS). Push-pull was by no means confined to the expensive end of the market; however, one of the cheapest examples being provided by the Barker '88', a mail-order set from the eponymous firm which boasted eight valves and eight watts push-pull output for a price of only eight guineas.

Fault finding on push-pull output stages follows the same lines as for conventional (what are called 'single-ended') stages. Be warned, though, that generally speaking the larger the set, the higher the HT voltage, and in the real monsters it is frequently around 500 V, so they need to be treated with respect.

Quite often the output valves used in the very large receivers were directly heated power triodes, exemplified by the Marconi-Osram types PX4 and PX25. Their low internal impedance facilitated good matching to loudspeakers and extremely high quality sound reproduction could be realised (the term 'hi-fi' dates from the 1930s). Even today the PX4 and PX25 are in great demand and fetch large sums of money in specialist radio auctions such as those run by *Radiophile* magazine. For this reason it is sensible to make quite sure before assuming that one of these valves has ceased to work.

Incidentally, don't trust entirely to a resistance reading to assume that the filament is in good order, for only part of it may be continuous. There is a good visual test which may be applied. Their multi-section filaments are supported and kept taut by four tiny springs at the top of the electrode assembly. If any of these springs should appear to be loose or even missing it is a sure sign that part of the filament has collapsed.

Various types of bias are employed with pairs of directly heated valves. The least expensive method, with a single LT winding to supply both, is to use the filaments as a virtual cathode by wiring a 'humdinger' across each with the sliders connected down to chassis via bias resistors. A dearer alternative is to employ two separate LT windings, each centre tapped with the taps returned to chassis via bias resistors. In both these cases the actual values of the resistors should be checked carefully to ensure that they are the same within a few ohms, especially after the sort of overloading brought about by leaky coupling condensers.

A third method is to use grid bias, usually supplied by negative HT smoothing. The same remarks apply regarding possible leakage via decoupling condensers as for single-ended output stages.

The other favourite output valve for high quality push-pull stages was and is the Marconi-Osram KT66 beam tetrode. This valve is almost identical electrically to the American 6L6G but such is the power of cachet that it costs five to ten times as much to buy. It is fairly common for the valve to be triode connected, with the screen grid strapped to the anode. Either cathode or grid bias may be used. A little further down the ladder came most of the popular beam tetrodes and pentodes, which could offer very good quality but did not have the charisma of the types just mentioned. In all cases, when cathode bias is employed, there may be a single common resistor or separate ones for each valve. Negative bias is not quite so likely to be found. Carry out the same checks as already described above.

For a push-pull stage to work properly it is, of course, essential for both valves to be 'matched' as regards emission. This may be checked in the set itself by using the same techniques as described above for single-ended stages. Don't forget when you measure the resistance of the output transformer primary, or check the voltage drop across it, to apply the meter from the center tap to each anode in turn. The resistance check also will reveal any serious imbalance in the windings such as might be due to shorted turns on one half of the primary.

Note that push-pull stages driven by phase-splitting valves usually will provide a reasonable output with one of the output valves removed.

This makes it possible to compare the performance of the two valves by removing each in turn. It will not work, however, with some of the rather cheap and nasty push-pull stages used in certain 'high quality' sets of the late 1950s, in which one of the output valves acted as a phase reverser to drive the other. Beware of this feature particularly in sets using EL41 output pentodes.

It is also necessary for both output valves to receive the same amount of AF drive. This is one of the few times when an oscilloscope comes in really handy, especially the 'double beam' type with which both grid inputs may be viewed and compared simultaneously. Otherwise it is a matter of ensuring that all the resistors in the phase splitter, whichever type this may be, are of the specified values as shown in the service data. It is particularly essential for the anode and cathode load resistors in a 'concertina' splitter to be matched accurately.

Hot valves

Experience has shown that the later all-glass valves such as the EL84, ECL82 and ECL86, which all run very hot in service, are all subject to internal leakage rather more than the old-fashioned large valves, which had wider electrode spacing and dissipated heat better. When these small valves are used in push-pull amplifiers it is essential to keep an extra close eye on the bias resistors and condensers for signs of overloading.

'Crackly' tone controls

When a simple variable top cut type of tone control consisting of a fixed condenser and variable resistor in series is wired between the anode of the output valve and cathode or chassis, a leakage on the condenser will permit DC voltage to flow through the track of the resistor. This will cause a 'crackling' sound as the control is turned, the intensity of which will vary with the amount of voltage leaking through the condenser: if this should be very serious the track of the resistor could burn out.

It may be possible to clean up a control that has suffered from only a minor DC leak. Take off the metal cover by bending out the three or four small tabs which hold it to the moulded part of the assembly, which will give access to the track. Gently rub a soft-lead pencil (2B or more) over the track to deposit graphite on it. Then smear it with silicon grease or Vaseline and reassemble.

With any luck this treatment will have done the trick if not there is one more old service engineers' trick worth trying. Gently pull the spindle of the control outwards as you turn it to see if this removes or reduces the crackling. If it does, with the spindle pulled outwards wrap a turn of fine wire around it between the retaining circlip and the threaded bush. Twist the ends of the wire together to hold it in place and then secure it with a blob of solder.

Note: the same methods are also applicable to crackly volume controls.

Chapter 21

Faults on detector/AVC/AF amplifier stages

When a receiver uses a double-diode-triode with top cap grid connection in this stage a very simple and convenient test for the AF amplifier section is to apply a finger to the cap, whereupon a loud hum should come from the loudspeaker (assuming, of course, that the output stage has been passed as satisfactory). In the case of single-ended DDTs such as the 6SQ7 try touching the blade of a small screwdriver on the centre contact of the volume control, with this fully advanced. If no hum results measure the anode voltage on the valve, which normally will be between about 40 V and 100 V. Very low or no voltage suggests that the anode load resistor has gone high or open, whilst a high reading points to the triode section of the valve itself, for which the easiest test is substitution.

A very occasional fault encountered by the writer has been the fitting by a previous repairer of the wrong value volume control such as a 5 KΩ type being fitted in place of a 500 kΩ. It is well worth measuring the resistance of the centre contact to chassis with the control at maximum to make sure all is well here.

When the AF amplifier is working well but no signals are heard the detector must be investigated. The diode sections of DDTs are usually extremely reliable and seldom give trouble. A remote possibility is hum due to heater/cathode leakage which may be cured by swapping over the detector and AVC diodes. In cases where the diodes are strapped try disconnecting each in turn. If the detector is working correctly, touching the meter probe (on a suitable voltage range) onto the anode of the preceding IF valve should cause crackling from the loudspeaker, but compare the remarks regarding oscillation in the section dealing with IF amplifiers. With this in mind measure the voltages on the detector diode anode, which should be zero with respect to chassis. As mentioned elsewhere in the text, if there is negative voltage on the anode the result as far as the listener is concerned will be 'ploppy' reception with the odd few stations appearing suddenly along the dial with wide dead spaces between them. In extreme cases the detector may be cut off altogether, with no stations at all receivable.

First of all, check to see if you are working on one of those sets fitted with 'squelch' or some other means of obtaining 'silent tuning'. It ought to be possible to switch off the feature, but if operating the switch appears to make no difference the first thing to do is to check the switch itself with an ohmmeter. Usually the switch closes to override the squelch and if it fails to 'make' a dose of switch-cleaning fluid may put things right. If not it may be possible to fit a replacement unless the switch is a special type made to work with an unusual operating system. Given that there is very little call for 'silent tuning' nowadays it probably wouldn't hurt simply to short out the switch permanently.

When, as in the majority of sets, 'silent tuning' is not featured, proceed as follows.

If you have a circuit diagram for the set being repaired, check it to see if the cathode of the detector diode (usually that of a double-diode-triode) returns to chassis via a bias resistor. In certain cases no cathode bias is used on the DDT, which will affect the test procedure. In the absence of a circuit a quick check is to measure the resistance between the cathode and chassis; if a bias resistor is in fact used you may expect it to have a value of between about 500 Ω and 5 kΩ.

Assuming that there is a bias resistor, the anode of the detector diode should return to it via the diode load, which may be anything up to 2 MΩ. If there is no path the diode load is either open circuit or someone has rewired it incorrectly.

Now disconnect the lower end of the bias resistor from chassis and test for any resistance between the diode anode and chassis. Theoretically it should be infinity, any reading at all showing that a leakage is taking place, most likely via a coupling or decoupling condenser. Try disconnecting one end of each of these in turn until the resistance returns to infinity.

In cases where a bias resistor is not fitted and the cathode is taken directly to chassis, the resistance of the diode load should appear between the anode and chassis. When the bias resistor is omitted 'grid current' biasing is used for the triode section of the DDT, effected by returning the grid to chassis via a very high valve resistor, typically 10 mΩ. It is essential that the negative voltage which is developed at the grid should not reach the detector diode anode, so a DC blocking condenser is inserted between the two. To test this, check to see if there is any resistance between the grid of the DDT and the detector diode anode – if there is it should not be less than 5 MΩ and a more usual figure would be well in excess of 10 MΩ.

As with any circuits where high value resistors are employed in conjunction with coupling and decoupling condensers, the latter must be completely free from 'leaks'. This is particularly important with DC blocking condensers used to prevent the detector diode anode from returning to chassis and thus receiving negative bias. It is here necessary to repeat the warning about keeping an eye out for previous work done on a set, in case someone has done some incorrect wiring around the volume control, especially if this has been replaced.

General AVC problems

Whichever type of AVC is employed check for leaky decoupling condensers along the bias line. Note that in some cases negative delay bias is applied to the AVC diode from the negative HT line, this voltage also being used as minimum operating bias for the IF amplifier and frequency changer valves (and RF amplifier, if used).

When a 'clamp' diode is used to delay the AVC check the resistor used to provide the positive clamp bias. This is likely to be of anything up to 22 MΩ and experience shows that high value resistors are all too likely to go higher and higher until they become the next thing to an open circuit. When this happens and the clamp bias disappears the performance of the set may well be restricted quite badly.

This latter remark applies equally to the failure of any means used to delay AVC. As always, look out for previous and faulty repair work.

Amplified delayed AVC, being complex in design, naturally stands a greater risk of going wrong than conventional systems. This will usually result in the AVC action being restricted and causing severe overloading on strong stations.

The chief thing to remember is that no AVC bias can be developed without the negative voltage source to which the cathode of the DDT is returned. Measure this point to see that it is at least 50 V negative with respect to chassis. If this appears to be in order, check the resistors used to connect it to the DDT cathode.

If much lower than normal, or even no negative voltage at all is found the fault is almost certain to lie in the HT smoothing circuitry. Check the resistance between the centre tap on the mains transformer HT winding and chassis to see if it corresponds to that given in the service data for the set. If not, check the resistance of the smoothing choke or field coil in case it is suffering from shorted turns, then the values of any series resistances used. As always, be on the look out for incorrect replacements fitted by an earlier repairer.

Because the AVC action depends so much on the amplification factor of the triode section of the DDT, any drop in emission in the latter will have a marked effect on the efficiency of the system. The best bet here is to check the valve by substitution.

Detection and AVC in 'short' superheats

A few early sets of this type employed a separate double-diode valve, but this arrangement was soon superseded by the double-diode-output pentodes such as the Pen4DD and EBL1/31. In nearly all cases to make up for the lack of an AF amplifier between detector and output stages the AVC was heavily delayed by the use of higher-than-normal cathode bias resistors. It was common to employ two in series with the AVC diode returned to chassis to receive the maximum bias, the pentode grid returned to the junction to receive its normal bias and the detector diode returned to the cathode to receive zero bias. Obviously there is a lot of scope here for things to go wrong, especially if a previous repairer has been careless. As for DDTs, check the various bias voltages and look for leaky condensers or high/open resistors.

Indirectly heated double-diodes in battery receivers

Points to watch: first, check that the voltages on the anodes of the diodes are within limits. Manufacturers were shy about giving figures, but the following tests should be quite adequate. Connect the meter negative lead to chassis and measure the cathode voltage, which should be approximately 6 V. Then check the signal diode to ensure that it has nearly the same potential. If it is much less the diode would be unable to respond to anything but very powerful signals and it would suggest that the resistor returning it to the cathode had gone high value or open circuit. The AGC diode, however, should be at zero or even slightly negative for the delay to be effective. A low reading on the cathode in a receiver such as the Cossor, where a voltage divider across the HT line was used to provide voltage, suggests that one or other of the resistors has altered in value. In the type of circuit used by Decca, in which the cathode is connected directly to a bias battery, it theoretically would appear to be impossible for the cathode to be below par but wires have been known to break before now.

Secondly, note that the AF appearing at the detector finds its way to the volume control, in most designs via a condenser of around 0.1 µfd. Alternatively the latter would be in series with the slider of the control and the following valve's control grid. The purpose of this was to obviate a DC path which would result in either the diode or triode receiving some unwelcome bias of the wrong polarity. There are some high value resistors around this part of the circuit, and as always condensers must be absolutely free of leakage to ensure correct working conditions.

All makes of indirectly heated 2 V double-diodes ceased to be produced a number of years earlier than the majority of 2 V types and replacements will be harder to come by as a result. In most cases, with ordinary directly heated battery valves, so long as the filament and bulb are intact there should be some kind of performance available but it is possible for a cathode to lose its emissive properties

and a simple ohmmeter check cannot guarantee that the 2D2, 220DD, etc., will work. It is worth asking the specialist valve dealers if they have a replacement available, but otherwise Mullard published an official modification to enable a TDD2A double-diode-triode to be used in its place. The valve holder does not need to be changed and only three connections have to be altered:

Pin 1 on 2D2 – disconnect and take to pin 5 on TDD2A.
Pin 2 on 2D2 – leave as found
Pin 3 on 2D2 – leave as found
Pin 4 on 2D2 – leave as found
Pin 5 on 2D2 – Disconnect and insulate.

In addition, connect the lower end of the detector diode load resistor to LT+, making sure that in doing so you don't accidentally short out the GB supply.

This should enable the set to work as before but the volume will be reduced to a certain extent due to the lack of AVC delay.

Grid leak and anode bend detectors

These are found mostly in TRF receivers although a few early superhets did use them. Prior to about 1935 triodes were used almost exclusively, but after that date 'straight' RF pentodes became popular because of their greater gain. Little can go wrong with a triode grid leak detector other than the valve itself or the grid condenser and resistor or the anode resistor. Experience shows that the most likely candidate for failure is the grid leak, because as with all high value resistors it can go very high indeed or even o/c. When this happens the negative voltage built up on the grid in the process of detection cannot escape back to chassis, and it continues to rise until the valve is 'cut off' altogether. Note that in battery receivers the grid leak should return to the positive side of the filament.

Because the grid leak detector generates its own grid bias the triode cathode normally is returned directly to chassis unless a gramophone pick-up facility is incorporated in the set which makes use of the triode as its first AF amplifier. In this case the act of switching the set to the 'gram' position, or in some cases by merely inserting the connecting lead from the pick-up, a cathode resistor is brought into circuit. Always check that the cathode does indeed return directly to chassis when the pick-up is not in use.

An opposite arrangement may be used in the case of an anode bend detector. Its grid has to receive a considerable amount of negative bias for the detection process, usually obtained by means of a cathode resistor of up to 20 kΩ. Check that there is a resistance of this order between cathode and chassis if the detector does not seem to work properly. On the other hand, for gramophone pick-up use the triode needs far less bias and there is usually some means of shorting out all or part of the cathode resistor. Check that whatever switching method is used that the cathode returns via the correct amount of resistance when either 'radio' or 'gram' is selected.

When an RF pentode is used for detection the principles are the same as for triodes in both grid leak and anode bend modes, the only difference being that the screen grid has to be supplied with voltage – but not very much. When a feed resistor from the HT line is used it is likely to be at least 500 kΩ and maybe up to 2.2 MΩ. When readings were taken with an AVO 'Seven' the service manuals reported the screen voltage as 'very low' or 'not measurable'. An AVO 'Eight' will give a reasonable idea of what voltage is present but to be on the safe side check the resistor on the ohmmeter. Clearly any leakage on the associated decoupling condenser will have a considerable effect on the voltage (*cf.* diode-pentodes used in 'all-dry' receivers).

In some receivers, especially of USA origin, the screen-grid voltage is obtained from the cathode of the output valve, this being quite sufficient for a pentode used as a detector. The anode load resistor also will be high and it too should be checked on the ohmmeter.

'Westector' diodes

These were small copper-oxide rectifiers made by the Westinghouse Brake and Saxby Signal Company from about 1932. Several types were produced in two main groups, the 'W' series and the 'WX' series. They were suitable for use as detectors, the first up to frequencies of about 200 kc/s, the second up to 1500 kc/s. Thus 'W' types could be used in the early superhets with low IFs whilst the second could function in LW/MW TRF receivers, although being stretched to the limit at the higher end of the MW band. They could also be used to provide AVC and in 'battery economy' circuits.

Although cheaper than equivalent thermionic diodes Westectors never really achieved great popularity. They appeared in a few of the commercial superhets of the early 1930s, before double-diode-triodes had been introduced, but seldom afterwards. Apart from the fact that the DDT could do three jobs at once, at a time when the larger the stance number of valves in a set the better, as a selling point a tiny component, hidden away under a chassis, simply wasn't so attractive as a valve in full view. In fact, the only real instance of Westectors being used in a mass produced set was in the Wartime Civilian Receiver of 1944, when valves were in short supply.

Experience shows that Westectors give very little trouble. The writer does not recall ever having to replace one in fifty years of radio servicing but there can be a first time for almost anything. The method of testing would be to disconnect one end of the device and compare its 'forward' and 'backward' resistances on the ohmmeter. The forward resistance (the direction in which conduction takes place) should be a fraction of the backward resistance. If the two figures should happen to be much the same the Westector almost certainly would be faulty. If an exact replacement cannot be obtained it may be possible to press a small germanium diode into service.

Chapter 22

Finding faults on IF amplifiers

The valves used in the IF amplifiers of the very early superhets were 'straight' screen grids, but these were soon replaced by variable-mu types and then by RF pentodes. In all cases complete failure of the amplifier can be brought about by a screen-grid HT supply resistor going open circuit. In some cases a single resistor was used to supply either two IF amplifiers or a single IF amplifier and the frequency changer, the current passed being high enough to warrant the use of a wire-wound type. Always check these for signs of overheating. Some firms, notably EMI, slipped sleeving over such resistors, in which case look for signs of its having become brittle with heat. Any overloading may be due to leakage on an associated decoupling condenser.

A common practice in the early 1930s was to seal most of the paper condensers used in a set within a pitch-filled metal box sprouting a dozen or more soldering tags for connection purposes. Unfortunately, as so often happens to inaccessible components, these condensers go 'leaky' and draw heavy current through the feed resistors. The easiest cure is to snip off the leads from the sealed box and to fit a replacement modern condenser close to the valve holder. Whilst this works perfectly well, some purists are moved to take out the boxes, melt out the pitch and fit new condensers in place of the old, subsequently re-pitching the box and fitting it back as before. The writer leaves it to the individual to decide whether or not to go to this length, but if you do for goodness' sake make a very detailed diagram of the connections before you start!

Cathode bias resistors seldom give trouble, nor for that matter do their decoupling condensers. Although the latter have as much chance of going leaky as any other elderly condenser, it really doesn't matter very much as long as the capacity is unaffected. A leak of 50 kΩ on a screen grid decoupler could cause problems, but on a cathode decoupler connected in parallel with a resistor of only about 300 Ω maximum it is immaterial.

Deathly silence

Occasionally you may encounter self-oscillation in the IF stage. It is difficult to describe the effect of this except as a most unnatural silence which can almost be 'heard'. This is due to two things: first the grid of the IF valve is driven heavily negative by the osicllation and secondly the latter causes a large AVC bias to be developed. A strange concomitant is that the voltage as measured on the anode of the IF valve will be very considerably higher than that of the HT line itself! The most likely cause is an o/c screen decoupling condenser.

Another source of self-oscillation

The majority of British screen grids and RF pentodes were made with 'metallised' envelopes, a thin layer of metal being sprayed onto the glass and connected down to cathode or a separate earth pin on the base. The metal screen thus provided protected the valve from mutual interaction with its nearby neighbours or other components. Some makes of valves, especially MOV, were prone to losing chunks of the metallising over the years, to the detriment of the screening effect, resulting in IF instability. It may be possible to effect a cure by wrapping a layer of metal cooking foil tightly around the valve and earthing same to the original metallising connection. This took the form of a thin wire sprouting up from the valve base and glued onto the metallising with the aid of a thin washer. With some other makes of valves the metallising tended to stay intact but the earth connector came loose. The old engineers' trick is to wind a number of turns of thin fuse wire around the metallising near the top of the valve base and then solder them to the connecting wire. This simple trick can effect a wonderfully efficient repair and enhance an engineer's reputation out of all proportion to the deed.

Beware

Many of the octal-based RF pentodes and 'kinkless tetrodes' made by M-OV were made without metallising and metal screening were fitted with cans by the set-makers using them, chiefly EMI and GEC. This left a 'spare' tag on the valve holder which very often was used as an HT anchoring point. An electrically equivalent metallised valve should never be substituted for the original in one of these sets as the metallising will become connected to HT+ and therefore 200 V DC or more above chassis.

Repairing faulty or damaged IF transformers

IF transformers fall into two main groups, those having fixed inductance coils trimmed by small variable condensers, and those having adjustable iron-dust cored coils with parallel fixed condensers. It has been known for the latter to change capacity with age, de-tuning the transformers and making it impossible to bring them back to correct alignment. In this respect be especially vigilant for small disc-like condensers about the size and shape of a pajama coat button; these had silvered-mica elements which tended to crack. Replacements should be of the close-tolerance type. Sometimes the originals have odd values and it may be necessary to use two new ones in parallel to obtain these. If you can't get the exact value, a pfd or two either way shouldn't matter since the adjustment of the core should make up for this.

Trimmer condensers were usually fitted with thin slices of mica to act as a dielectric. It is a brittle material which can and will break up under rough treatment, so take care if you need to adjust them (but see the later remarks on realignment before touching them!).

It is inevitable that at some time an IF transformer will be encountered which has suffered damage at the hands of an unskilled person. The most common victims are those with iron-dust cores, for these are easily damaged if an attempt has been made to adjust them with an ordinary screwdriver, a tool quite unsuited to the job. The result is all too often a core which is split and jammed in the coil former. There is a remedy, but it calls for a steady hand and much patience. The broken core must be drilled out with a fine drill, used in a hand brace, until only a shell remains. Extreme care must be used, as any deviation from a straight line will be likely to ruin the former and probably the winding as well. The remains of the core are then broken out little by little with the aid of a slim screwdriver.

When every last scrap has been removed the thread may be cleaned up by running a suitable size of bolt through its length. Replacement cores may be obtained as kits, or salvaged from scrap radio or TV sets. Smear them with silicone grease before fitting, and run them through the formers, end to end, using only light pressure and reversing the direction as required to pass any tight places. It is imperative that the tool used to turn cores fits the slot perfectly. It must not be a loose fit, or the slight to and fro movement will soon wear a slot into a useless hollow. Incidentally, many cores are double-ended, and one which has worn away at one end may still be usable if it can be eased out and reversed.

The time-honoured method of making trimming tools among radio engineers is to obtain a hard (but non-metallic) knitting needle of the appropriate thickness and to shape it to fit the slot with a file or sandpaper. If the needle is brittle, so much the better, for the chances are that *it* will break before a core if the latter sticks. Later cores were of smaller dimensions than the original pre-war types, and some will be found to have hollow centres, either still slot shaped, or of hexagonal section. Tools to fit these may be obtained from component firms who advertise in such journals as *Practical Wireless* and *The Radiophile*.

It was common practice for manufacturers to seal cores after factory alignment to prevent movement in transit. Some used wax, others quick drying paint. Another method was to insert a piece of wool or thin rubber in the former to exert a gentle pressure on the core. Wax may be softened with a soldering iron, which should be applied for just enough time for the wax to start to melt, but no longer, or it could run down the threads and set again quickly. The core should be unscrewed whilst the wax is still soft, then both it and the former can be cleaned up properly. The kind of paint usually employed was a cellulose-based varnish, which will respond readily to a few drops of nail varnish removing fluid. As to how you obtain this, let your conscience be your guide. Follow the treatment with a spot of thin oil.

There is another type of core in which the iron-dust material is moulded onto a brass former carrying a thin threaded rod for adjustment purposes. It passes through a fixed nut at the one end of the coil assembly which gives in-and-out movement of the core when the rod is rotated. A second nut may be fitted onto the rod to lock it when alignment is completed. There are two alternative means of turning the rod; it may have a thin slot cut into the end, to take a screwdriver, or it may be filed off into a semi-rectangular shape. For this you will need to make a simple trimming tool consisting of a short length of fine-bore copper tube, with about ½ inch of one end flattened by gentle hammering to grip the rod. Sometimes the slotted variety will be found to be broken, leaving only a forlorn stump of brass as memorial to a too-vigorously applied screwdriver. If you have a very fine hacksaw it may be possible to renew the slot, otherwise solder a suitable brass nut to the rod and adjust it thereafter with a nutdriver.

With either of the above rodded cores, if turning produces no noticeable effect on the tuning, inspect them to see if the iron-dust portion is still attached to the rod. You may well discover that they have parted company, in which case you will probably have to dismantle the transformer to get at the two sections. An interesting manifestation of this has been known to puzzle severely amateur repairers. A receiver chassis standing upright on a bench may demonstrate a marked lack of IF gain, but when it is turned over for inspection the gain returns to near normal, due to the cores sliding up and down within the IFTs.

It may be possible to re-glue the cores onto their brass rods but the writer wouldn't bank on it. As usual, service engineers have methods to deal with this situation. One of these is to pass a thin compression spring up the middle of the core between the two cores so that it tends to keep them apart. The rods can then be used simply to shove the cores inwards against the spring until the correct point of alignment is obtained; if you have to back off an adjustment the spring moves the core back for you. Some purists have been known to criticise this method on the grounds that they worry about the effect of the metal spring on the inductance of windings. The way round this is to wrap thin paper or tape around each core to make it a tight but movable fit inside the former and then to push it into the alignment point with a thin trimming tool. This works but you don't get a second chance if you push a core too far.

Obviously, if any of the repairs detailed above have had to be carried out, the transformer concerned will need to be realigned. If only one or two cores out of four or more have been disturbed it is permissible in dire necessity to adjust them for maximum signal strength on a fairly weak station, but this is the only exception to the golden rule of never attempting a full alignment job on IF transformers unless you have an accurate signal generator. *Anything else can lead only to inferior performance and disappointment.*

Realignment with a signal generator

The output leads from this should be connected between the chassis and the control grid of the frequency changer, which usually can be obtained via the one section of the tuning condenser. Stop the local oscillator working by shorting out the other section of the condenser. Connect the output meter or the AVO on a low AC voltage range across the secondary of the output transformer.

Inject the appropriate IF (a comprehensive list of these, dating back to the very early superhets, is given in Appendix 1) using as small an output as will provide a usable reading on the output meter, in order to prevent the AVC from coming into action and masking the points of maximum response. The exact order in which the trimmers or cores should be adjusted is a matter of opinion: some engineers favouring reverse order, starting with the last (that nearest the detector) and working back to the frequency-changer anode coil; others prefer the opposite way. A few years ago the writer checked on the recommendations of various radio manufacturers and found that one third said one way, one third said the other and the other third made no recommendation at all. For what it's worth he tends to use the reverse order method.

Regularly reduce the output from the generator as the gain of the set improves during the alignment process. Repeat all adjustments at least once to account for interaction between windings, then seal the trimmers or cores with paint.

'Staggered' IFTs

In order to obtain a good response curve from the IFTs, and thus good quality reproduction, some firms employed a system of tuning the IFTs not to a specific frequency but to 'stagger' two, three or more different ones across the required bandwidth. These may be identified in the IF data in Appendix 1 by groups of figures separated by dashes, e.g. 128.8–122.8–126.5–122.5 (the actual set is the Marconiphone Model 535). In this case, the first IF winding, that in the anode circuit of the frequency changer, is peaked at 128.8 kc/s, the second winding, in the grid of the IF amplifier, is peaked at 122.8 kc/s and so on. It is even more essential that 'staggered' IFTs should be adjusted and readjusted extremely carefully than with conventional IFTs.

Do not forget to remove the short on the local oscillator after the IF alignment is completed.

Aligning with a wobbulator and oscilloscope

This method has already been outlined in the chapter on instruments. In practice the output from the wobbulator is connected in the same way as a conventional signal generator and the input to the oscilloscope is connected between chassis and the diode load resistor near to the last IF transformer. The terminal on the wobbulator marked 'X' should be connected to that marked 'X Plate' or 'X Plate Direct' on the 'scope. This synchronises the two instruments and ensures a steady response curve on the latter.

The tuning control on the wobbulator is set for the nominal IF and the 'sweep bandwidth' control to about 25 kc/s. The resulting response curve on the screen of the 'scope should be something like that shown in Figure 22.1 – but don't be surprised it is anything but to start with. Adjust the trimmers or cores in sequence, a little at a time until the curve begins to come into shape and eventually approaches the optimum. It may not be possible in practice to attain an exactly symmetrical shape, but this is not so important as the avoidance of steps in

the sides, or nearby spurious peaks, which would bring about co-channel interference when the set was tuned to one of a group of closely spaced stations. Note that the top of the curve should be flattened off to give a sensibly even response over about 2 kc/s above and below the nominal IF, and that it then falls away fairly steeply. This is a good compromise shape for a domestic receiver where reasonable selectivity has to be balanced against adequate bandwidth, to preserve the higher AFs.

Alternative IFs

When consulting the list of IFs in Appendix 1 you may find that occasionally more than one figure is quoted, such as '128 or 134' (Cossor Models 634 and 635). The reason for this is to do with our old friend the image frequency, with its relationship between the IF and certain transmitter wavelengths. It was found that a slight change of IF was beneficial in certain areas of the country, the particular firm's dealers in those parts being instructed to realign the sets as required. After sixty or more years the original need for the particular IF probably will have gone, but it is as well to keep to the one chosen for the set, because if this is altered all the RF and local oscillator alignment will have to be readjusted as well.

Beware of HT!

A final word of warning regarding transformers fitted with trimmer condensers. These may need to be adjusted with a metal tool due to stiffness and if this proves necessary, take care, for trimmers across the primary windings will be at HT, normally 200 V and more in mains sets. Insulate all but the tip of the tool to prevent shocks or shorts.

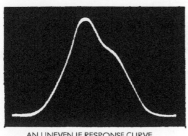

AN UNEVEN IF RESPONSE CURVE
(TYPICAL RESULT OF "TUNING BY EAR")

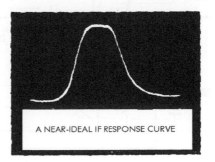

A NEAR-IDEAL IF RESPONSE CURVE

Figure 22.1

Chapter 23

Faults on frequency-changer circuits

In the average circuit diagram the frequency-changer (FC) stage appears to be the most complicated part of the radio receiver. This is due mainly to the waveband switching arrangements, which are only too often drawn in a spidery fashion with lines crisscrossing each other all over the place. It is much more helpful initially to regard the FC as a single waveband device, as in Figure 23.1, with the items in the broken line boxes being those which are concerned purely with tuning. Note that the two coils in the local oscillator circuit (extreme right) are shown as having variable iron-dust cores, and that the padder, C11, is shown as a semivariable condenser. This is to illustrate both possible arrangements; an actual receiver would employ

only one or the other (fixed padder with variable core, variable padder with fixed core).

In this circuit the popular triode-hexode type of frequency-changer valve is shown, but it is applicable to most other types. When a heptode (pentagrid) or octode is employed the first grid corresponds to the triode grid and the second grid to the triode anode.

Each additional waveband would have its own pair of boxes connected into circuit by the wavechange switch. The latter may develop poor contacts over fifty, sixty or even more years of use, resulting in crackling or perhaps the complete loss of one or more bands. With any luck a squirt of cleaning fluid may cure these problems. Certainly,

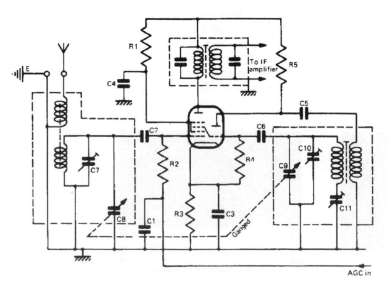

Typical component values:

C1 0.1 μF	C7/10 0-100 pF
C2 100 pF	C11 0-500 pF
C3 0.1 μF	R1 47 kΩ
C4 0.1 μF	R2 1 MΩ
C5 100 pF	R3 330 Ω
C6 100 pF	R4 47 kΩ
C8/9 500 pF swing	R5 33 kΩ

Figure 23.1 A simplified frequency-changer stage

this should be the first thing to be tried if a set refuses to work on one or all bands.

Let's suppose you are confronted with a receiver that will receive no stations on any band, but is 'lively' from the IF amplifier onwards, as demonstrated by injecting an IF signal into the grid of the IF amplifier. Now transfer the signal generator output lead to the control grid of the FC valve. Does it pass through the set at increased volume? If so, at least the mixer section of the valve is all right. If there is a much reduced sound or none at all, check the voltages on the anode and screen grid of the mixer. If low or absent, look at the associated HT feed resistors and decoupling condensers.

If the voltages appear to be on the high side, check the cathode voltage (if cathode bias is used). Little or none suggests that the valve itself is at fault and it should be tried by substitution. A very high reading, say 100 V or more, suggests that the cathode resistor may be o/c. Note that certain Ekco sets of the mid-1930s used large value wire-wound cathode resistors in connection with 'silent tuning' systems, and these are known to be vulnerable.

Let's suppose now that an IF signal will indeed pass through the mixer at good strength, yet no stations can be heard. With a good aerial connected tune along each band in turn, listening carefully. Does a 'rushing' noise come in at the low frequency end of the MW band or the high frequency end of the LW band? If so, this is an indication that the LO (local oscillator) is not working. Measure the voltage on the triode grid (or G1 of a heptode or octode), expecting to find it several volts negative with respect to chassis. If not, check the triode anode or G2 voltage. A zero reading points to an o/c feed resistor or shorted decoupler, but a low reading does not necessarily mean the same. When the LO fails to oscillate it draws a much higher than usual HT current and thus the voltage will fall considerably.

Test the resistance of each winding of the LO coils for the non-working band(s). You may find that one or more has gone o/c, in which case it must be removed for close inspection. If you're lucky the break may be visible and repairable without too much trouble, but all too often a subsection of Murphy's Law will cause the break to be in the middle or at the inner end of the winding. At this stage, working on the 'why me?' principle, you may be wondering what makes a coil that has worked satisfactorily for all those years suddenly go o/c. In most cases it is due to a phenomenon called 'green spot', the development of a small speck of corrosion due to the wire having been handled by damp hands back in the dim and distant past.

You are now faced with the job of rewinding the entire coil. Try to remove the old wire carefully and take it up on a suitable spool. You may be lucky enough to find a single break and to repair it without too much difficulty, but multiple breaks may necessitate all the wire being replaced. In this case saving as much as possible of the old will give you a good idea as to how much you need for the rewinding job. Large diameter coils are fairly easy to rewind tidily, but you are unlikely to be able to do the same with small coils unless you are lucky enough to have a winding machine. Don't worry about this unduly as untidy coils seem to work pretty well, all things considered.

The 'Q' meter mentioned in the chapter dealing with test instruments comes into its own for getting the inductance of a coil right to tune over a particular range of frequencies. If you are not lucky enough to own one, take comfort from the fact that coils with iron-dust cores have a reasonable amount of latitude in this respect and a certain amount of trial and error should eventually win the day.

If you have to remove any metal screens to gain access to coils for repairs, always replace them before you commence realignment.

A case in point

Whilst this book was being prepared the writer was asked to look at a Murphy receiver which had baffled several previous experienced but non-professional repairers. A wartime model, it had only SW and MW bands, of which neither worked. The presence of the 'rushing' noise at the low frequency end of the MW band indicated that the LO was not working, and indeed several components had been replaced in this area, including the triode anode and grid resistors and the small coupling condensers. Since this work demonstrably had been fruitless, attention was turned to the local oscillator coil windings. There should have been resistance readings back to chassis on the grid windings and both appeared to be o/c when measurements were taken from the wiper of the wavechange switch. Now, to paraphrase Oscar Wilde, to lose one winding may be attributed to accident but to lose two points to carelessness; in any case, the SW winding was of such a robust nature as to make 'green spot' highly unlikely. Although the wavechange switch contacts looked perfectly good, just in case readings were taken directly across the coils, which then proved to be in order. Gentle pressure on the switch contacts with a thin bladed screwdriver restored the connections and normal reception was obtained on both bands. It was impossible, however, to 'set up' the contacts to make them work reliably and in normal circumstances it would have been necessary to replace the switch. However, when the set was made wartime shortages must have caused Murphy to use any switches that were to hand, and this example had a complete set of spare unused contacts right alongside the faulty set, and to which all the connections could be transferred without difficulty. The owner of the set was suitably impressed, but the fact is that a little thought on the part of the previous would-be repairers might have suggested to them the same answer, and the moral is: don't make assumptions from visual examinations. The wavechange switch, as mentioned, looked so good as to be above suspicion, but as Sherlock Holmes famously remarked, when all the possible solutions have been investigated without result the answer must lie with the impossible.

Faults on the aerial coils

Now let's consider how to proceed if the LO stage is working correctly but only the very strongest stations can be heard, probably with considerable background noise or whistling. See what happens when the aerial is taken directly to the control grid of the mixer. Do the signals become louder or remain unchanged? These reactions point to either the primary or secondary of the HF transformers in the aerial circuit being o/c. Measure each as for LO coils and if necessary carry out repairs using the same methods.

To return briefly to the wavechange switch, occasionally you may find a case of persistent noisiness, although all bands are present, which does not respond to cleaning. This may be due to HT tracking across the wafers. The RF circuits are nearly always switched on the 'dead' side of coupling condensers, but in sets having a 'gram' position the HT to the FC and IF stages may be switched out to silence them. It is frequently found that the section of the switch used for this job breaks down, causing deafening results. The cure is to remove the HT leads from the switch and join

them permanently. Not having the 'gram' position is no great loss as in practice it is seldom used.

When realignment is necessary

The most probable source of poor performance is interference with the adjustments. Owners of radios have been known to take them in for repair after 'tightening all the loose screws', as they put it! Sometimes the trimmers and cores are sealed with paint, which not only reveals any subsequent adjustment, but also indicates the original settings. You could try returning to these in the hope that this will restore the performance of the set, but all too often a complete realignment job will be necessary.

Before starting ensure that the IF alignment is correct. Now make sure that the dial pointer is at minimum frequency when the tuning condenser is fully closed, adjusting its position if necessary by sliding it along the drive cord or runner, etc.

It is always best to follow the set manufacturer's instructions as given in service manuals, but if these are not available the standard procedure to be described should give satisfactory results.

With the simplest medium and long wave receivers, and in the absence of specific instructions, it is advisable to start on the medium wave band, because some cheaper sets do not have a separate long wave oscillator coil, but load the MW one with an extra condenser.

Connect the output leads of the signal generator to the receiver via a 'dummy aerial'. Many generators are furnished with one of these when new, but there is always the chance of its having been lost over the years. Figure 23.2 shows a suitable substitute that may be made up quite simply.

Connect an output meter or the AVO on a low AC voltage range across the secondary of the output transformer.

Tune to 550 m (545 kc/s) and adjust the signal generator to this frequency, noting whether or not it comes in at the correct position on the dial. Adjust gently the core of the local oscillator coil

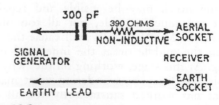

Figure 23.2

(or the padder as the case may be) to obtain this, then adjust the core of the aerial coil (if fitted) for maximum reading on the meter, keeping the output of the generator as low as possible to prevent the AVC from coming into operation.

Now retune the set and the generator to 200 m (1.5 mc/s) and adjust the oscillator trimmer for correct pointer position and the aerial trimmer for maximum signal. In all probability this will affect the tuning at the other end of the band so repeat the entire operation several times until overall accuracy is obtained.

Switch to long waves and tune the set and generator to 2000 m (150 kc/s). Adjust the the local oscillator coil or padder to set the pointer, then the core of the aerial coil (if fitted) for maximum output. Retune to 1000 m (300 kc/s) and adjust the oscillator and aerial trimmers for pointer position and maximum output. As for MW, repeat several times for overall accuracy.

Short waves present a bit more of a problem. The reaction to adjustment of cores or trimmers is much sharper, particularly at the higher frequencies and very slight movements are called for. Non-metallic trimming tools are essential if hand capacity effects are to be avoided.

The standard short wave coverage for receivers was approximately 16 m to 52 m. Suitable alignment points are 50 m (5 mc/s) at the low end of the dial and 20 m (15 mc/s) at the high end.

For band-spread SW ranges choose alignment points towards the top and bottom of each. As an example, the '41 m Band' in the Murphy A126 covers from 40 m 50 m (7.5 mc/s 6 mc/s), so use 7 mc/s and 5.5 mc/s.

Seal cores and trimmers with wax (as from an old condenser) or quick-drying paint when the alignment process is completed.

Optical bandspread

An alternative to the purely electrical band-spread described above was to retain continuous coverage of, say, 16 m to 50 m, but to expand the dial artificially so that it had enough effective length to make tuning simple. Ferranti pioneered the idea in 1936 with their Magnascope dial. It consisted of a separate 180 inch scale mounted inside the set, with a small optical system to project the markings on to a screen above the main tuning dial. By this method the scale was made to be equivalent to one 6 ft in length. Murphy produced an updated version some twelve years later and quite wrongly claimed it to be

an original idea, which provoked a very angry reaction from Ferranti. In both systems maintenance is confined to keeping the lamp, lenses, mirror, screen, etc., clean, but don't scrub the scale!

As the actual frequency coverage is still the same as with ordinary receivers, use the 5 mc/s and 15 mc/s alignment points.

RF amplifiers

Sometimes known as signal frequency amplifiers, there are a number of advantages to having a tuned RF amplifier preceding the frequency-changer stage. The extra gain makes the set more sensitive and improves the signal to noise ratio, but there is another useful consideration. Conventional superhets rely largely on their IF stages to provide selectivity, and there is often only one tuned circuit prior to the FC, which in itself may be relatively inefficient. A good RF amplifier immediately at least doubles the number of pre-FC tuned circuits, and receivers intended for serious short wave listening may have two or more RF amplifiers. The stage gain diminishes fairly rapidly as the frequency increases, but this may not be so serious a consideration as the maintenance of good selectivity.

Sometimes the RF amplifier was called a 'preselector', especially when it took the form of a separate add-on unit.

As far as fault finding is concerned an RF stage resembles an IF amplifier, frequently employing the same valve type and similar component values. There has to be an extra section or two on the wavechange switch to cope with the various tuning coils, and there will also be another set of trimming condensers plus, of course, the required additions to the gang condenser. Alignment again follows the usual rules of cores at low frequencies, trimmers at high.

Rejectors

We spoke earlier of aerial rejector circuits ('wave traps') intended to counteract swamping by powerful stations. With the various changes in the wavelengths of stations over the years it could well happen that the wavetrap might be reducing the strength of a weak but desired station, and slight readjustment may be necessary. Usually the adjustment core or trimmer needs to be altered only very slightly to achieve the desired result.

Another type of rejector is designed to reject signals at IF. A signal at this frequency should be injected into the aerial socket, strong enough to break through the tuned circuits and be heard in the loudspeaker. Tune the rejector for minimum sound.

On completion of adjustment, rejector cores or trimmers should be sealed.

Image rejectors

The problem of the 'image' is largely confined to superhets with low IFs. The local oscillator of a superhet may be designed to operate either above or below the signal frequency, but in practice it is usually higher, i.e. the sum of the RF and IF. Suppose that a set with an IF of 125 kc/s is tuned to a station on 1000 kc/s. The oscillator has to run at 1125 kc/s, and should there be a strong station on 1250 kc/s, capable of breaking through the aerial tuning, it too will beat with the local oscillator to produce 125 kc/s. The result will be either crosstalk between the two stations or an intrusive whistle of varying pitch. This is not all. If the receiver is tuned towards the low frequency end of the band, when the pointer reaches the 750 kc/s mark the local oscillator will be running at 875 kc/s, and thus capable of combining with the 1000 kc/s signal once again to give the IF. In this case the station will be heard again, although at lower volume, and the false point of reception is called an 'image'. Many early superhets had image rejectors fitted in the aerial input circuits and adjusting them has to be carried out in accordance to the specific instructions of the set manufacturer. If these are not to hand, it is better to leave the rejector well and truly alone, particularly some of those used in EMI receivers which involved the physical movement of coils with relation to each other.

An image is always twice the IF away from the genuine station, and so raising the former to 450/470 kc/s alleviated the problem greatly, at least as far as medium waves are concerned, by spacing the wanted and unwanted signals more widely apart.

Adjusting ferrite aerials

Receivers manufactured from the mid-1950s onwards were often fitted with ferrite rod aerials. The windings are not always adjustable, but where they are the rule is the same as for any tuning coil: the core to be set at the low frequency end of the band. In fact, the accepted method is to slide the appropriate coil along the rod to the point of maximum volume and to seal it into position with wax. Broken rods need not be replaced if the pieces can be easily glued together, because magnetically, joints don't matter.

Faults on the tuning condenser

Crackling or intermittent tuning can be due to shorting of the main tuning condenser, or occasionally bad earthing of the moving vanes. The latter is sometimes caused by dirty or broken spring contacts on the moving shaft, resulting in earthing taking place through the bearings instead of directly. Even when the earthing arrangements are in order, lack of lubrication on the bearings can result in slight crackling as the set is tuned, particularly at the higher frequencies. Shorting of the vanes might be caused by conductive dust getting between them. It has been known for very enthusiastic amateur repairers to remove the condenser from the set and to clean it in a dishwashing machine! I will refrain from describing the professional's most effective method since it is a little hazardous and simply suggests a high pressure air blast.

Another explanation for shorting condenser plates is that some enthusiastic but incautious person has adjusted them, either directly by bending them, or via the spacing screw which is to be found at the rear of the condenser. Make sure that this is set properly, so that the fixed and moving vanes are equally spaced. There should be a locknut to ensure that the screw stays put. If it is found that reception ceases at certain parts of the tuning range, find the exact spot with the aid of an ohmmeter, as follows.

Switch to long waves, where coils connected across the condenser will have the least effect on the meter readings, and place the test prods between earth and each set of moving vanes in turn. Rotate the tuning knob slowly until a sharp drop in resistance is shown on the meter. Knowing now which set of vanes is at fault, and where, it should be possible to clear the short circuit by gentle bending of the moving, not the fixed, vanes. The former, incidentally, were usually slotted on the rounded edge so that precise adjustments to alignment could be made at the factory by slight bending. Don't confuse this intentional warping with out and out mishandling.

Tackling broken dial drive systems

The dial drive probably has been the one item in radio sets responsible for the greatest amount of sweat, tears and profanity on the part of service engineers, so be warned.

The simplest sets have but a knob fitted directly on the tuning condenser shaft. The next up the scale have a small epicyclic reduction gear, and then come the complicated ones!

Drive cord setups can be divided into two broad classes: (a) Philips, (b) The Rest. There must have been someone at Philips from the 1930s to the 1950s who was a cross between a mechanical genius and a sadist. Not content with intricate cords for the condenser, he also used fiendishly clever Bowden cable and steel wire drives for the tuning dial. To attempt to describe one of these mechanisms is only slightly more difficult than expounding nuclear theory on the back of a postcard, and to restring one calls for all the patience attributed to the prophet Job.

Approach a Philips dial drive warily. Unless the cord or the wire is actually broken, leave well alone, and merely lubricate the condenser bearings and the runner which carries the pointer, to relieve the load on the drive as much as possible. If you cannot obtain steel wire you might like to explore the chances of using fine brass wire as used for picture hanging.

Whatever you do, before you start to restring a Philips (or any other) dial, make quite sure that you know which way the pointer should travel with respect to the dial markings. There is nothing more likely to make a strong man weep than to spend hours installing a cord drive only to find the pointer moves in the wrong direction.

It is to be hoped that the entire tuning drive system is mounted on the chassis of the receiver so that it may be attended to with the latter out of the cabinet. Unfortunately, all too often Philips made this impossible by mounting the dial itself and all the pulleys and rods, etc., inside the case which makes matters very difficult indeed. The very first drive cord ever tackled by the writer, at the age of fourteen, was just such a one as this and it took several evenings of solid hard work eventually to figure it out. However, it did give the satisfactory feeling that nothing ever could be as bad again.

Some common types of dial drive

It is impossible to enumerate all the types of dial drives, but at least there is a basic pattern from which to work, shown in Figure 23.3. Some more expensive models featured a flywheel on the inner end of the tuning spindle to facilitate tuning from one end of the dial to the other.

Should the tuner knob turn without effect, and yet the cord drive appear to be intact, check the tightness of the grub screws in the centre of the drum, and for the presence of oil on the narrow part of the spindle where the cord runs. The spring itself may have lost tension over the years so either replace it or stretch it further to a convenient point of the drum. Incidentally, when withdrawing the chassis of a radio from the cabinet for any reason, have a look to see if the drive cord has to be disconnected from the dial pointer first. This simple check can save a broken cord and much hard work!

If the cord has broken but is still in position around the drum and various guides, do not disturb it until you have made a sketch of how it is lying and of the number and direction of turns round spindles, etc. If the worst has happened, and a tangle of ancient drive cord is found lying in the bottom of the cabinet, try to find all the broken pieces, both to reclaim the spring(s) and in order to get an idea of the length of new cord required, thus avoiding waste. Incidentally, good flax stranded fishing line is just right for restringing cords. It was, in fact, officially recommended by EMI back in the 1930s.

Before starting to restring the drive, ascertain the direction in which the pointer has to move. When the condenser is fully closed the pointer needs to be at the low frequency end of the dial and this varies from set to set – it may be on the left or right according to the whim of the designer. Knot one end of the cord to the spring after threading it through the hole or slot in the drum.There may be more than one hole, so use the one which allows the greatest amount of cord to go round the periphery of the drum.

Where the cord passes over small pulleys, ensure that these are free to turn, oiling sparingly if necessary. Do not be tempted to put more than two turns around the tuning spindle as it will not increase the available grip, but merely cause piling up and binding of the cord. The loose end of the cord should be passed around the drum in the direction opposite to the initial wrapping, so that as it is drawn off in one direction, it is taken up in the other. Finally pass the end back through the slot and knot it to the spring as well. The latter should be stretched to hook into one of the holes usually provided.

The drive should now operate correctly, but note that the initial stretching of the cord will probably have to be taken up by altering the spring tension.

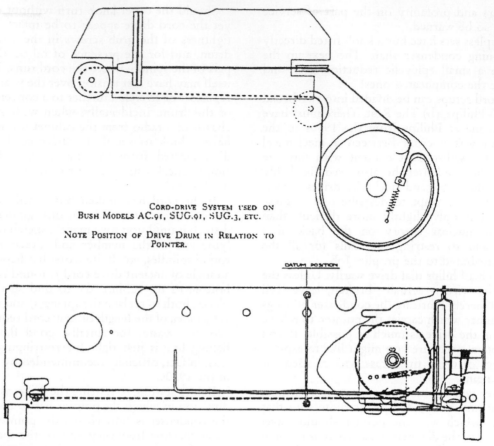

CORD-DRIVE SYSTEM USED ON
BUSH MODELS AC.91, SUG.91, SUG.3, ETC.

NOTE POSITION OF DRIVE DRUM IN RELATION TO
POINTER.

Figure 23.3 Typical arrangements for cord drives

Permeability tuning

This was discussed earlier. In its practical form the tuning coil is made long and thin, with a movable core mounted on drive cord passing through it. Permeability tuning was used in a few table radios just after the war, but is more likely to be found in FM receivers and car radios. The only fault normally encountered is restriction of the tuning range caused by the jamming of one or more of the cores in its coil. A smear of silicone grease will usually effect a cure.

Dial pointers

Pointers are of two basic types, those that are actually carried by the cord, and those which ride on some kind of rail and are merely pulled along. In the latter case make sure that it can travel freely by applying a little light oil to the rail.

To set up the pointer correctly, examine the dial glass or the backing plate for a datum mark and slide the pointer to coincide with it. If such a position is not marked anywhere, set the pointer so that it travels equally just beyond the upper and lower limits of the dial markings. If its previous position has been incorrect, it may be necessary to slightly realign the set to obtain the correct readings for wavelengths.

A word of warning regarding the cleaning of glass dials. In some sets the paint used to mark them was such poor quality that even water would remove it, so it is best to experiment first with something of no consequences, e.g. the maker's name, which nearly always appeared in an unobtrusive place. If this resists a good hard rub it should be all right to use a mild detergent on the rest, but be very careful. Far too many dials end up as plain glass thanks to overvigorous cleaning.

Chapter 24

Repairing American 'midget' receivers

American midgets are engaging little sets in two respects. One, they are masterpieces of the *multum in parvo* school of construction, and two, they often have very attractive cabinets, especially those made of the plastic material called Catalin, which enabled almost any shape in any colour to be produced with ease. Although midgets were inexpensive they were seldom cheap and nasty, as is evidenced by the large numbers of 50- and 60-year-old $5 bargains that are still in good working order.

Servicing these sets is an art all of its own, requiring as it does a good knowledge and appreciation of US design techniques, which differed in many respects from those employed by UK manufacturers. For a start we have to bear in mind that midgets, which were invariably of the AC/DC type with series heaters, were designed to work with HT voltages of no more than 90 V, due to the US mains voltages which were between 110 V and 120 V, either AC or DC. Most, up to about 1939, used heater chains rated at 0.3 A and employed resistive line cords to drop the mains voltage to suit. After this date valves with higher voltage heaters rated at 0.15 A began to take over, and in many cases the chain voltage was close to that of the mains and no dropping resistors were necessary.

The first thing to do when you have a midget on the bench for repair is to ascertain whether or not a resistive line cord should be fitted, and if so if it has the correct resistance. This is a necessary task because all too often previous owners or repairers will have either shortened the cord or even removed it altogether, and this must be put right before the set is plugged into the mains. Incidentally, it is worth repeating the advice given elsewhere in this book that it is far better to use a step-down auto-transformer to power midgets than to attempt to modify them for the UK 230 V mains voltage.

Look at the valve line-up to see if it consists of 0.3 A or 0.15 A valves. If you are not familiar with the valve types find their heater ratings in the data section in Appendix 2. Add the voltages up to find the total chain voltage, and subtract the answer from 117.5, this being the accepted mean voltage for US mains. If, with a 0.3 A chain, the difference is more than about 30 V expect to find a resistive line cord. 0.15 A chains will usually be within 10 V or 20 V of 117.5 V and seldom require one.

You can work out the required resistance of a line cord using Ohm's law, all that is necessary is to divide the difference between the chain voltage and 117.5 by the chain current. For instance, the average TRF midget with 0.3 A heaters has a total chain voltage, including that of a pilot lamp, of about 65 V. Subtracted from 117.5 V this gives 52.5 V to be dropped, and dividing this by 0.3 gives the answer 175 Ω. The usual standard was 60 Ω per foot for 0.3 A line cord so its length will be about a yard. A 0.3 A superhet with one more valve needs about four inches less. Check the actual length of cord on the set, bearing in mind that there will be about four or five inches under the chassis, and if the lengths are more than a few inches less than those given don't plug in the set but measure the exact resistance with a meter.

Tackling faults on resistive line cords

What can you do if the line cord is too short or, worse still, open circuit? As has been stated earlier, resistive cords worked perfectly well when left to themselves and gave no trouble. When problems occurred they were usually traceable to owners having wound them into neat coils which overheated them, or to snipping odd lengths off them. Remember, though, that most American midgets were converted to work on UK mains by the addition of an extra length of line cord, and although this might now be too short for 230 V it would be more than adequate for 110/120 V operation from an auto-transformer. For this reason it is a good idea to collect and store any odd lengths of line cord that you might come across, because they might well be of use in the future.

Should you find that the cord is only a few inches short you may be able to get away with fitting an extra dropping resistor on or under the chassis. Six inches of 0.3 A cord equals 30 Ω, which could be replaced by a 30 Ω wire-wound resistor. The dissipation would be 2.7 W, which should not be troublesome even in the smallest of midgets. Likewise, a foot of cord could be replaced by a 60 Ω resistor; as the dissipation would be 5.7 W a 10 W type should be used. The heat produced is probably about as much as could be tolerated in a midget so don't go beyond this.

If the cord appears to be o/c, don't give up hope until you have tried chopping off about an inch and a half from the mains plug end, where constant handling may have caused the resistance wire to break. Many a cord can be brought back to useful life in this way.

If all else fails you may like to consider using an external mains dropper in a metal safety cage. Remember that in sets with 0.3 A chains the cord feeds only the valve heaters, with the anode of the rectifier going either directly to the live mains input or via a small stopper resistor of around 30 Ω. Another possibility is to replace a 0.3 A chain with 0.15 A equivalents, this bringing up the heater voltage to the point where only a small dropping resistor, if any, is required.

Extra precautions necessary with midget sets

It was common practice to use only a single-pole mains switch to break the side of the mains lead that went to the HT– line and the bottom of the heater chain, which often was the chassis of the receiver. The interesting situation arises that if the mains lead is plugged in to make the chassis neutral while the set is working, it will become live when switched off and vice versa due to the path from the top of the heater chain back down to chassis. It is therefore essential to replace most carefully all the various devices such as covers for chassis bolts and grub screws employed to prevent owners from accidentally coming into contact with live metalwork.

To their credit, many US manufacturers were aware of the danger of shock and eliminated it by isolating their chassis altogether. Instead of using it for the hearer and HT– returns a busbar was fitted within the set to do the job. As this was unfamiliar to British eyes it was all too easy for an ignorant engineer to frustrate the good design by shorting down the HT– line to chassis, typically when replacing smoothing condensers. Always examine carefully the wiring of American midgets around the on/off switch. If one side does not go to chassis but to various insulated points these are almost certainly the HT– busbar in an isolated-chassis set. Check the resistance between the busbar and chassis – it should not be less than about 250 kΩ. If a low resistance is registered the cause must be investigated and put right or the safety of the isolated chassis will be lost.

Types of midgets

Most of the midgets you are likely to meet will be either TRFs using an RF amplifier, a detector, an output valve and a rectifier, or superhets using a frequency changer, an IF amplifier, a detector/AF amplifier, an output valve and a rectifier. Thus most TRFs have four valve stages and most superhets five, but in practice the use of combination types means that the actual valve count may be less. For instance, a triode pentode such as the 12B8GT or 25B8GT may carry out the work of both the RF amplifier and the detector in a TRF, while rectifier-pentodes such as the 25A7GT or 70L7GT may provide both the output and the HT supply. By using, say, a 25B8GT and a 70L7GT together the equivalent of a four-valve may be built with only two. However, the circuitry will be much the same as for all TRF midgets.

Rectifiers and HT smoothing

Although, as with all AC/DC receivers, half-wave rectification was used in midgets, in 0.3 A sets you will quite often find that a full-wave type was used with the sections strapped. If the set also has an energised loudspeaker it is quite possible that one half of the rectifier may be used simply to supply the field winding, with the latter being connected between one of the cathodes and HT−. A condenser of about 8 μfd to 16 μfd will be shunted across the field to smooth the HT passing through it. Incidentally, don't be surprised to find that the speaker is of the moving-iron type.

As regards the main HT smoothing, it was standard practice in all midgets to connect the top end of the output transformer primary directly to the rectifier cathode. From the same point a resistor of up to 10 kΩ both smoothed and dropped the HT for the screen grid of the output valve and the rest of the valve electrodes. It was usual to employ a twin electrolytic condenser (typically 20 μfd + 20 μfd) for smoothing. The working voltage rating normally was no more than 150 V, so a UK type with a 250 VW rating will be more than adequate − but watch the ripple current rating and how the negative is connected. When an isolated chassis is used a replacement condenser should have its metal case insulated by a plastic sheath or by several turns of tape.

No decoupling required

Because the smoothed HT in a midget is no more than 90 V it is unnecessary to employ dropper resistors and decoupling condensers for the screen grids of RF or IF amplifiers, or frequency changers, these being connected directly to HT+. It is also possible in many cases to dispense with cathode bias resistors and condensers for IF amplifiers, so the underside of a midget can looked surprisingly uncluttered.

Output stages

The same basic circuitry is used for both TRFs and superhets. The valve will be either a pentode or beam tetrode capable of giving more than ample output and it is common for the by-pass condenser across the cathode resistor to be omitted, thus giving a measure of negative feedback to improve the tone. As with all types of radio set, always check the coupling condenser that feeds the grid for leakage.

Detectors in TRF midgets

Either grid-leak or anode bend detectors were used. Note that when the former type is employed the grid condenser and resistor may not necessarily be connected between the top end of the tuning coil and the grid of the valve, but may be inserted between the bottom of the coil and chassis/HT−.

The necessary grid bias for an anode-bend detector is obtained by using a high value cathode resistor, typically between 10 kΩ and 22 kΩ. This will have to be decoupled, but the condenser seldom fails.

Reaction was very seldom used in midgets; on the few occasions when it was a preset reaction condenser was fitted.

The most popular type of valve for the detector position was a straight RF pentode although a few midgets used a high-mu triode. When a pentode is used the very low screen-grid voltage necessary may be derived from the cathode of the output valve.

RF amplifiers in TRF midgets

A variable-mu pentode is always employed, with volume control effected by a potentiometer (usual value 25 kΩ–50 kΩ) in the cathode circuit. Its wiper is connected to chassis/HT−, with one end going to the cathode of the valve and the other to the top of the aerial coil. As the volume is advanced the cathode voltage is reduced and the amount of resistance across the aerial coil is increased, but when the volume is turned down the cathode voltage is increased and the amount of resistance across the aerial coil is reduced. It is usual to employ some means of preventing the cathode from being taken directly to chassis/HT− at full volume, sometimes by a fixed external resistor of about 100 Ω–300 Ω, sometimes by an extra length of track on the control after the stop for the wiper at maximum.

A point to watch: Americans tended to use M as a symbol for 1000 Ω on resistors so a control marked 25 M will have a resistance of 25 kΩ. Meg was used to indicate 1,000,000 Ω.

The Americans were good at producing highly efficient tuning coils. A feature of them not usually found over here was the use of a coupling loop, a single turn of fairly thick wire, to improve the coupling between primary and secondary. This was employed in both the aerial and RF coupling coils, which were very similar in design. The standard coverage was from 550 kc/s to 1700 kc/s, with the last digit ignored on the dials, which were marked from 55 to 170. A typical feature of the dial was to have the word 'Broadcast' inscribed near the 55 end and 'Police' near the 170 end, a reminder of when many US police forces had radio communications working on the low end of the MW band.

Before the Second World War a certain amount of American midgets were exported to the UK, and these were adapted to cover long waves by the fitting of extra coils and a wavechange switch. As mentioned elsewhere in the text, during the war some 100 000 American midgets were imported into this country to make up for the shortfall in domestic radio production. At that time the LW band was not being used for domestic broadcasting so there was no need for the sets to be modified.

Detector/AF amplifiers in superhet midgets

Almost without exception a double-diode-triode was used with its two diodes strapped. Minimal IF filtering was used, possibly just one condenser of about 100 pfd to chassis from the top of the diode load, which itself might well be the volume control. Grid current biasing was usually employed for the triode section of the valve, with the cathode taken directly to chassis/HT–. AVC bias was drawn from the detector load with seldom more than one decoupling resistor and condenser to feed the IF amplifier and frequency changer.

IF amplifiers

These were simple in design and there was very little to go wrong, with no feed or bias resistors and decoupling condensers to worry about. Note that the second IFT might be unscreened and fitted beneath the chassis, while in very small midgets it was sometimes omitted and the detector diode fed from the IF amplifier anode by inductance capacity, a simple tuned coil being fitted in the anode circuit. Very occasionally two IF amplifiers were employed, but not always exploited fully since one of the IFTs might be untuned.

Frequency changers

Heptodes generally were preferred to triode-hexodes. In many cases the local oscillator coils were wired into the cathode circuit of the valve. When a feed resistor is used to supply G2 + G4 it has to carry a fairly heavy current and should be checked if oscillation ceases.

The usual IF was 455 kc/s, and the frequency often was printed on the cans of the IFTs. 'Tracked' tuning condensers were popular in midgets as they enable conventional padding condensers to be omitted.

RF amplifiers in superhets

These were sometimes employed, possibly more as a sales feature rather than for their actual performance. There was seldom room or the inclination to fit a three-gang tuning condenser so the RF stage usually was aperiodic, with resistance capacity coupling being used between it and the frequency changer. Curiously enough, the RF amplifier valve might well be a triode.

Aerials

TRF midgets nearly always had a 'throw out' aerial consisting of about eight to ten feet of thin flex which could be draped around a room more or less out of sight behind furniture, etc. Usually small clips were provided on the cabinet back or underside onto which the aerial could be wound when out of use or for transportation.

Superhet midgets also often had throw out aerials but latterly frame aerials became popular, usually glorified by names such as 'wave master'. Experience shows that they sometimes leave much to be desired in the way of sensitivity, so if a midget with a frame aerial gives a disappointing performance try the effect of bringing your hand close to the frame to see if its pick up is improved. If so it might be worth considering adding your own throw out aerial coupled into the top of the frame via a 0.001 μfd, 300 VW isolating condenser.

Chapter 25

Repairing faults on automobile radios

As mentioned earlier, the receiving sections of car radio receivers function in the same way as those of domestic sets so the same fault-finding procedures should be used in the output, detector/AVC/AF amplifier, IF amplifier, frequency changer and RF amplifier (if fitted). Only the power supply stages are different and we shall address ourselves to them.

LT problems are infrequent as in general the heaters of valves used in car radios are robust and reliable. In a set for 6 V operation all the heaters will be in parallel, and in the case of glass type an o/c valve will be given away at once by the lack of heater glow. Metal and metal-clad valves will, of course, need to have their heaters tested with an ohmmeter, but before removing them from the set, first narrow down the field by establishing, in the usual manner, in which part of the set a fault lies.

Twelve volt receivers using valves with 12.6 V heaters in parallel may be treated just as for 6 V sets, but as mentioned earlier there was a tendency for manufacturers to use 6.3 V heater valves in series parallel. For instance, the frequency changer and IF amplifier valves might have heaters rated at 6.3 V, 0.3 A each connected in simple series form. The heater of the popular 6X5/6X5GT rectifier is rated at 6.3 V, 0.6 A. Typically, this would be placed in series with that of a 6V6/6V6GT output tetrode (6.3 V, 0.45 A) which itself is in parallel with the heater of a 6Q7/6Q7GT double-diode-triode (6.3 V, 0.3 A). As the total current of the last two is 0.75 A, a 42 W resistor would be wired in parallel with the 6X5 heater to pass the difference of 0.15 A. It will be seen that if any of the three valves should go

o/c the others are going to be affected. Should the 6X5 fail, for instance, the 6V6 and 6Q7 will receive no LT and also fail to light. On the other hand, should either the 6V6 or 6Q7 fail, the survivor will be over-run considerably due to excess current passing through its heater. Fortunately, as mentioned above, heater failures are rare and experience shows that when they do happen it is the 6X5 that is more likely to fall by the wayside. Nevertheless, it is necessary for the repairer to be aware of what *could* happen.

When the HT supply is absent the first thing to be checked should be the vibrator. In fact, this is in effect self-checking since it should reveal itself as in working order by emitting a characteristic buzz as soon as the set is switched on. If no buzz is heard you need look no further, but don't assume straight away that it must be the vibrator itself that is at fault. Between it and the battery will be a number of components that could give trouble, such as a fuse, one or two RF chokes and the receiver on/off switch itself. The fuses in automobile receivers pass continuously much heavier currents than their counterparts in mains sets, and they can just wear out. The RF chokes, fitted to suppress interference from the car's electrical system, are wound with thick wire and are unlikely to go o/c, but beware of bad soldered joints. The on/off switch is also called on to pass a heavier current than in a mains set and its contacts may simply burn away after a number of years. Multiway interconnection cables used with two-unit receivers are a possible source of trouble due to fraying of individual wires.

If after all these items have been passed as satisfactory the vibrator still doesn't buzz, use the voltmeter to trace the 6 V or 12 V, as the case may be, from the battery input to the pins of the vibrator. If no voltage is found look for the point at which it has disappeared, but if it is present things look bad for the vibrator. A test by substitution is now the only real way forward. It is possible for an experienced service engineer to dismantle and repair a faulty vibrator but it is a difficult job that may not prove to be very reliable, so regard this as a last resort. To open a vibrator the seal around the base has to be bent back with the aid of side cutters and no man is capable of restoring this to its original condition. If you find the vibrator base looking the worse for wear you may be pretty certain that someone previously has had a go at it, which more or less indicates that you might as well give up and look for a replacement.

The primary winding of the step-up transformer is robust and unlikely to give trouble. The secondary, however, like any other that delivers HT, is vulnerable. A feature of vibrator transformers not found in domestic receivers is that both primary and secondary windings are normally shunted by condensers, which should also be inspected closely. A bad leak on one of those across the secondary could cause fatal damage to the winding. These are called 'buffer' condensers and must be replaced by types having an AC rating of 800 V, due to the high peak inverse voltages involved. *Ordinary coupling types will not do.*

As mentioned earlier, the 6X5 (and other rectifiers used in automobile receivers) has a highly insulated cathode, between which and chassis the rectified HT appears. An insulation breakdown here can have serious results, since it shorts the HT down to chassis and imposes a heavy load on the transformer which, if left too long, could burn it out. Always watch a glass valve rectifier for signs of internal sparking when trying a set for the first time, even though its heater/cathode insulation may seem satisfactory when checked on a meter.

The 0Z4 heater-less rectifier needed a minimum of 300 V (peak) to make it start when new, and this requirement increased somewhat with age. It can become erratic in time, and if a new one is not to hand it may be replaced by a 6X5 with heater supply derived from the battery, via a 10 Ω, 5 W resistor in the case of 12 V sets.

A certain number of sets in the late 1950s had small contact-cooled rectifiers, and in common with those found in domestic receivers, their output drops off sharply with age. If the HT voltage is found to be very low this is the component to suspect. Because of its small size this type of rectifier could be hidden away very effectively, and it may need a very sharp eye to spot it! If a similar unit is used as a replacement, remember that it is most important to bolt it down effectively, since the heat generated in a car radio has far less chance of escaping than in a table model. Do not replace with silicon rectifiers before reading the list of essential precautions regarding these devices given in Appendix 2.

A failure of the reservoir condenser would drop the HT, but might not give as much audible warning as it would on a 50 c/s supply. As always, ensure that the working voltage and ripple current are sufficient for replacements.

Realigning automobile receivers

The usual procedure as used for domestic sets, and detailed elsewhere in this book, should be followed but take this special precaution when adjusting the RF and local oscillator trimmers.

Aerials used in automobiles are connected via screened cables which have significant internal capacities. The set designers are aware of this and adjust the inductance of the aerial coils accordingly. They also provide an aerial trimmer which is adjusted for best results in the particular vehicle in which the set is installed. To ensure correct alignment, use a special dummy aerial to terminate the signal generator output; a suitable design is shown in Figure 25.1. This may also be used when

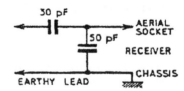

Figure 25.1

'aerial testing' a car radio, in conjunction with a short length of wire hung vertically above the bench. Don't use a long wire aerial because this will give a false impression of the set's sensitivity. It is both disappointing and frustrating when one goes to a great deal of trouble to reinstall an automobile radio which sounds fine on the bench, only to find that it has a very poor performance in the vehicle!

Can the polarity be changed?

Non-sync vibrators will work 'either way round' without any adjustment being required, but you may have to modify the battery input wiring. It was usual to have the RF anti-interference chokes by-passed by one, two or even three fairly large value electrolytic condensers, typically 50 μfd, 25 VW. Sometimes special reversible types were used which will work 'either way round' and don't need to be touched, but these were probably in a minority. You are far more likely to find that the ordinary polarised types are fitted which will have to be unsoldered and reversed for a change of battery polarity. For obvious reasons, mark the case of the receiver very clearly to indicate positive or negative earth as the case may be.

Sets with synchronous vibrators always need to be adjusted for different polarity. In many cases it is necessary to withdraw the vibrator from its socket, turn it through 180° and reinsert it. Normally arrows and + and − signs are provided on the vibrator and surrounding metalwork to indicate its working position. Check for the electrolytic by-pass condensers as for non-sync vibrators.

Predictably, Philips evolved systems of their own to make their receivers capable of working on either polarity and often on either 6 V or 12 V as well. The adjustments consisted variously of reversible plugs, terminals with sets of connecting links, and flying leads secured either by screws or by soldering. All are so complicated that no ordinary person can make sense of them, and it is absolutely essential to follow the maker's instructions. Sometimes these will be found printed on a label within the case of the set, but if not you should be able to obtain servicing data from suppliers such as *The Radiophile*.

Occasionally you may encounter a set which bears a prominent marking 'for positive (or negative) earth only'. Normally this means that although it is not totally impossible to change the polarity it is so difficult as to make it not worth while.

Note that many automobile receiver manufacturers adopted the American style of referring to the car chassis as 'ground', hence 'negative ground' or 'positive ground' receivers.

Chapter 26

Repairing battery operated receivers

Conventional domestic and portable types

The first thing the reader may well ask is how does one go about obtaining the necessary HT, GB and LT voltages now that the batteries used have long since ceased to be produced?

Taking HT batteries first, there are several ways around this, one being to make up your own from the small 'PP3' type 9 V units connected in series. Obviously ten PP3s are needed for 90 V while fourteen will make 126 V, suitable for use in sets using the nominal 120 V HT battery. Intermediate voltage tappings in multiples of 9 V may be made as and when required. With intermittent use PP3s will last a considerable time, but if it is anticipated that a set will be used for long periods at a time it would be better to employ an HT 'eliminator'. These were made from the late 1920s up until the 1950s and are to be found at most vintage radio auctions and other events for reasonable sums.

Beware: make sure that you obtain one intended for AC mains operation, as they were made for DC mains as well. If nothing else the weight should give the game away as AC types contain a mains transformer, rectifier and largish smoothing condensers, whilst the DC versions have only a few resistors and some moderately sized condensers. Most eliminators gave just three or four different HT voltages but the better types included an LT source for 'trickle charging' an accumulator overnight when the set was out of use. Note that these charging devices are *not* suitable for running valve filaments (but see later!).

It may not be widely realised that an HT battery in good condition has *low* internal resistance

whilst an HT eliminator in good condition has a *high* internal resistance. The low internal resistance of the battery serves as effective RF and AF decoupling and when an eliminator is used in its place this is lost and instability may result. The effect will vary from receiver to receiver and from eliminator to eliminator but if it should occur it will be worth while trying a 2 μfd or 4 μfd condenser from HT+ to chassis and/or a 25 μfd from HT– to chassis

GB batteries, which almost always means 9 V tapped at every 1.5 V, may be made up by wiring six 1.5 V cells in series. An alternative method is to use another PP3 in conjunction with a chain of six 15 kΩ resistors wired across it to provide the intermediate tappings. The drain on the battery will be only about 100 μA, which means it will last for a very long time.

On the LT side it is not really practical to consider using dry batteries as the current consumption of even a small receiver would run them down quite quickly. It is far better to use the 2 V lead-acid cells that are still available under the generic name of 'Cyclon' cells, in various sizes that should suit almost any receiver, large or small.

In the late 1940s and early 1950s the then well-known firm of Amplion made some special eliminators under the name 'Convette' which delivered both HT and LT. To cater for different LT loadings a variable output control was provided along with a small voltmeter to check that it was correct. These Convettes turn up now and again at auctions and other events and are well worth buying. Do not confuse them with another type of Convette which may appear at first glance to deliver LT but which in fact is an HT vibrator pack which needs to be supplied with 2 V from an accumulator. These units can be useful for supplying HT from an LT source but be warned that the consumption from the accumulator is fairly heavy and that they emit a very audible hum whilst working.

The LT voltage must be 2 V within 0.2 V. On no account should the upper limit of 2.2 V (that of a fully charged 2 V accumulator) be exceeded, even for short periods. Once battery valve filaments have been subjected to persistent over-running their efficiency and life is reduced sharply. On the other hand, they will not work properly below 1.8 V or perhaps a little higher if they have been abused in the past.

Fault finding

In most respects the same techniques apply as for mains operated receivers and the procedures given for these may be used save, of course, that the section on power supplies may be ignored. Servicing data is available for commercially made sets from about 1930 onwards, but for early receivers, especially home-built types, you will need to work things out for yourself. Fortunately most examples were fairly simple TRFs which departed only in detail from the basic receivers shown earlier in this book. It is good practice to sit down and trace out the circuitry of old sets and thus see the technical features discussed in the book 'brought to life'. As regards components, it is surprising how many 'new' 70- and 75-year-old items are still in circulation and may be obtained at vintage radio events or from specialist dealers. The same applies to valves, and it is only the really old 'pip top' types that are likely to be very expensive, although the £50 that one may cost nowadays is still cheaper in real terms than it was new at 18/6d back in 1923 for someone earning £2 per week!

'All-dry' and mains-battery sets

Generally speaking the maximum HT used in the sets was 90 V in the standard size receivers and 67.5 V in the 'personal' sets, although some American sets did work with as little at 45 V. As before, PP3 batteries in series may be used to provide replacements, ten for 90 V while eight will give 72 V, quite suitable to replace a 67.5 V battery. The filaments of 'all-dry' valves were, of course, intended to be powered by batteries and a couple of highpower 'C' cells wired in parallel will be effective. Grid bias batteries were not used in these receivers. Amplion produced 'Convettes' especially for 'all-dry' portables and personals, in sizes that matched the original batteries. Please note, though, that the types intended to give an LT output to suit the standard 250 mA of the original octal and B7G types *should not be used with 25 mA filament valves*.

A third alternative which may well become popular in the future is a small 'inverter' capable of delivering both LT and HT from a miniature rechargeable battery. These have been developed for military applications, for valves still have their place in the armed forces since they can survive nuclear radiation that would destroy transistorised equipment. It is hoped that these devices will come on to the civilian market in the near future.

As mentioned earlier, the circuitry of nearly all makes of popular 'all-dry' sets was very similar. In fact, in a number of cases two or more different brand names would employ a common chassis, made by such firms as British Radiophone or Plessey. As examples, the post-war Every-Ready model 'C' and the Cossor model 469 shared a Plessey chassis in which only one resistor, that for the bias in the HT negative, differed. Nearly all the 'personal' portables of between about 1946 and 1950 were fitted with an identical Plessey chassis. This commonality is certainly helpful to the repairer as generally speaking the same faults tended to appear on most brands. The same techniques are used as for any battery superhet, starting with a careful check on valve electrode voltages. Experience shows that even at the low HT voltages to be found, small paper condensers were just as vulnerable. Sets using the B7G diode-pentodes employed very high values of screen-grid resistor, typically as much as 6.8 MΩ, with anode load resistors of up to 1.5 MΩ. Service data of the day, based on the use of the AVO 'Seven', was apt to refer to the anode and screen voltage simply as 'very low' or even 'not measurable'. An AVO 'Eight' will measure them reasonably successfully but even then only a few volts will be registered on each. The safest thing here is to measure the resistance of the two resistors in case either or both has gone very high or o/c. Don't forget either the screen-grid decoupling condenser or that coupling the anode to the grid of the output valve. With respect to the latter, it is surprising how much anode current one of those miniature valves will draw with only a tiny positive voltage on its grid.

Chapter 27
Oddities

One of the many strange facets of vintage radio collecting is that one of the most sought-after sets which commands prices of up to £500 should happen to be what was just about the cheapest set on the market when it appeared in 1936. This was the Philco Model 444 'People's Set' (Figure 27.1), and the resemblance of its Bakelite cabinet to the front of the German Volkswagen or 'People's Car' may or may not have been fortuitous. It cost 6 guineas, then about two weeks' wages for a manual worker and half the price of many contemporary four valve mains sets. Its performance was adequate on the long aerials used in those days and the fact that large numbers of them are still in working order more than sixty years later attests to the fact that they certainly were not shoddily made. In fact, the circuitry of the 444 serves to illustrate many of the conventional features described in earlier chapters, plus some unconventional ones including an odd AVC system that reverses the normal rules.

The frequency-changer stage uses a 6A7 heptode (pentagrid) to produce an IF of 451 kc/s. This is coupled to the IF amplifier (a 78E) by a normal type of transformer, but the second IFT was, for some unknown reason, untuned. This can scarcely have been on economy grounds since the cost of a pair of trimmers could not have been all that much. The lack of tuning reduces severely the gain of the stage which, as this is a 'short' superhet appears to have been a perverse piece of design.

A double-diode-pentode acts as detector and output valve, and since no suitable 6.3 V heater type was then available Philco had Mazda make a special one just for the 444. Known as a PenDD61,

it was effectively an AC2PenDD with a modified heater which Mazda pretended never happened, for nowhere can it be found in that firm's technical or sales data.

The power supply stage uses a standard type of mains transformer with a directly heated full-wave rectifier (an 80) and HT smoothing by the field winding of an energised loudspeaker. Note that a 0.015 μfd condenser is wired from the live side of the mains input to the chassis of the set to suppress modulation hum; as the condenser is connected on the mains side of the on/off switch it is in circuit at all times. Even when new and in perfect condition this condenser would have passed sufficient current to give a nasty tingle to anyone touching the chassis, and after the passage of time it must be regarded as a definite hazard. Incidentally, many of the condensers in the 444 (as in other Philco models) were moulded into what appear to be four large terminal blocks with three soldering tags each. Three of these blocks contain pairs of condensers (C2 and C6, the AVC decouplers, C3 and C7, the cathode decouplers for the 6A7 and 78E, and C9 and C10, the IF by-pass condensers on the detector load) and the other holds just C16, the mod. hum condenser. They are notorious for developing leaks but only C2, C6 and C16 are likely to cause trouble.

In earlier discussion of 'short' superhets it was mentioned that the AVC was usually delayed heavily to permit maximum sensitivity on weak signals. Following American practice, the AVC in the 444 was tapped off from the detector load without benefit of delay, which must again count as being deliberately perverse. We now need to

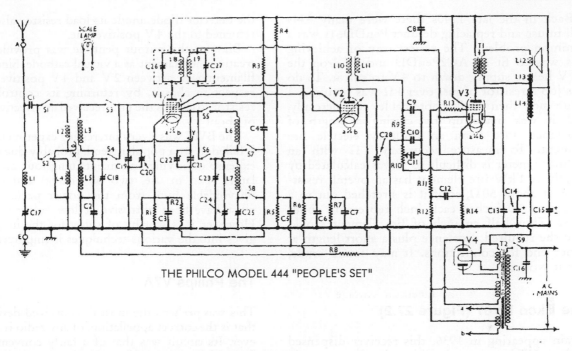

THE PHILCO MODEL 444 "PEOPLE'S SET"

Figure 27.1 The circuit for the Philco Model 444 'People's Set'

look at the AVC in detail as its curious circuitry can easily baffle newcomers when things go wrong.

It will be seen from the circuit that the standing bias on the 6A7 is provided by a 700 Ω cathode resistor. This provides about 7 V bias which is over twice the usual amount. Likewise, the 800 Ω cathode resistor of the 78E provides about 6 V bias which is again far more than is usual. The reason for the high cathode bias figures is that the detector load resistor, from which the AVC is derived, is returned to the cathode of the PenDD61, which itself is 5 V positive with respect to chassis. Thus the AVC line, and consequently the grids of the 6A7 and 78E, is itself 5 V positive in the absence of a signal. This offsets the cathode bias on the two valves so that the actual grid bias they receive is, respectively, 2 V and 1 V. Just why Philco went to all this trouble when it would have been far easier and better to have used a proper delayed AVC system with a separate diode is something we probably shall never know, but what is certain is that the strange design introduces another perversity almost guaranteed to fool the unwary.

Many times in this book have we spoken of the undesirability of permitting positive voltages to get onto the control grids of valves. In the 444 we have

to turn this upside down because the set simply will not work well unless there *is* positive voltage on the grids of the 6A7 and 78E. Without that, the standing bias on them reduces their gain to the extent where reception is restricted to only very powerful stations. This condition arises when the two AVC decoupling condensers C3 and C6 develop leaks; as the AVC line is fed via a 2 MΩ resistor only a small amount of leakage will drop the positive bias sufficiently to spoil the performance of the set. After the two condensers have been replaced it is as well to check the value of the 2 MΩ resistor in case it has gone high.

Replacing the output valve

The present writer had been over forty years in the radio trade before he saw a brand new PenDD61 in a box, because Mazda discontinued making them as soon as the production run of the 444 and its wooden-cased derivative, the Model 269, ceased. To digress for a moment, in view of the current worshipping of Bakelite artefacts, it is interesting to note that in 1936 it was very properly regarded as a cheap substitute for wood, the price of the 269 being half as much again as that of the 444.

Back in the late 1940s there were many 444s still in use and replacing defunct PenDD61s was a common problem. The easiest way of achieving this was to fit an AC2PenDD and to drop the 6.3 V heater supply down to 4 V to suit it. To do this job a resistor of just over 1 Ω was required, a value not then easily obtainable commercially, and it was made up by cutting a length of resistance wire from an ordinary electric fire element. To measure accurately 1 Ω with an ordinary meter is difficult so it was calculated by length. A 1 kW fire element has an overall resistance of about 50 Ω, so if it is stretched out to a length of 20 inches each inch represents about 2.5 Ω. Thus half an inch of fire element would give the required resistance plus a short length at either end for connections. It may sound crude, but it worked.

The Ekco BV67 (Figure 27.2)

Again appearing in 1936, this receiver dispensed with an HT battery and employed a vibrator pack, thus providing an excellent example of how a synchronous vibrator operates. It also illustrates the wiring of filaments in series–parallel, the use of a separate local oscillator in the frequency-changer stage and how an indirectly heated double-diode is used to provide delayed AVC.

The power for the BV67 was provided by a pair of heavy-duty 2 V accumulators connected in series to give 4 V. This was used directly to supply the vibrator pack, which provided a smoothed HT output of 135 V. Five valves were employed, two VP2 RF pentodes, a PM1HL triode, a 2D2 double-diode and a PM22D output pentode. The first three mentioned formed one group and the last two another group, each with the valve filaments wired in parallel. The two groups were then wired in series across the 4 V LT supply. The combined filament currents of the first group amounted to 460 mA and those of the second group to 390 mA, so a shunt resistor was wired across the latter to make up the difference.

The frequency-changer stage used one of the VP2 pentodes as a mixer with the PM1HL operating as local oscillator, its output being coupled into the suppressor grid of the pentode. The IF amplifier was conventional. The cathode of the 2D2 double-diode was returned to the positive side of the 4 V LT supply and the anode of the AVC diode was returned to chassis, thus providing a 4 V delay bias. To prevent this bias from reaching

the detector diode anode its load resistor also was returned to the 4 V positive line.

Bias for the output pentode was provided by treating the filament as a virtual cathode. Since the filament was between 2 V and 4 V positive with respect to chassis, by returning its control grid resistor to chassis the grid received an effective bias of about 3 V.

The BV67 was a very interesting experiment, but it failed to set a popular trend, probably due to the expense of buying and maintaining four accumulators (two in use, two on charge) and to the unavoidable hum from the power pack, which must have been obtrusive at low volume levels. However, it is worthy of study for the insight it gives into the various techniques it employed.

The Philips V7A

This was perhaps the most ill-conceived design, if that is the correct appellation, of any radio receiver ever. Its circuit was that of a fairly conventional four-valve plus rectifier superhet but it had no chassis, all the various components being fitted around the inside of a Bakelite cabinet. Some were bolted into position but others were held in by no more than pitch. The V7A must have been extraordinarily difficult and expensive in labour to produce and it presented service engineers with great problems when repairs were required. Fortunately the idea was very soon scrapped but even so, fair numbers of V7As remain in circulation to plague set repairers to this day.

The 'monoknob' receivers

Again a brainchild of Philips, a series of sets under that name and of its subsidiary Mullard appeared in the late 1930s. The outstanding feature of what again were reasonably conventional superhets was the use of a single control knob to change its wave bands, to tune it and to adjust its volume and tone. In fact it was a bit of a trick, because what appeared to be one large knob in the bottom centre of the cabinet was actually two, working concentrically. The front section worked the tuning and the rear section the wavechange, whilst the complete assembly could be moved up and down and side to side to control, respectively, the volume and tone. It all depended on a complicated system of Bowden control cables which again must have been costly to produce and install, and which are

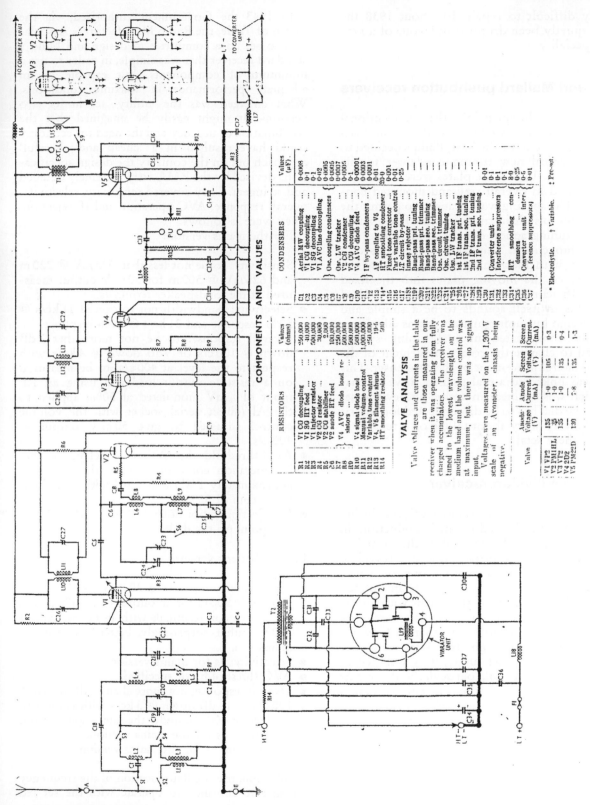

COMPONENTS AND VALUES

RESISTORS

		Values (ohms)
R1	V1 CG decoupling	250,000
R2	V1 SG HT feed	40,000
R3	V1 injector resistor	500,000
R4	V2 CG resistor	30,000
R5	V2 CG resistor	2,000
R6	V2 CG stabiliser	100,000
R7	V2 anode HT feed	250,000
R8		500,000
R9	V4 AVC diode load re-	500,000
R10	sistors	1,000,000
R11	V4 signal diode load	500,000
R12	Manual volume control	1,000,000
R13	Variable tone control	250,000
R14	V4, V5 filament shunt	19.5
	HT smoothing resistor	500

CONDENSERS

		Values (μF)
C1	Aerial MW coupling	0.0008
C2	V1 CG decoupling	0.1
C3	V1 SG decoupling	0.1
C4	V1 AVC line decoupling	0.02
C5	Osc. coupling condensers	0.0005
C6		0.0005
C7		0.0007
C8	Osc. LW tracker	0.0005
C9	V2 CG condenser	0.1
C10	V3 CG decoupling	0.1
C11	V4 AVC diode feed	0.00001
C12	IF by-pass condensers	0.0003
C13		0.0001
C14*	AF coupling to V5	0.01
C15	HT smoothing condenser	20.0
C16	Fixed tone corrector	0.001
C17†	Part variable tone control	0.01
C18	LT circuit by-pass	0.25
C19‡	Image rejector	—
C20†	Band-pass pri. tuning	—
C21†	Band-pass pri. trimmer	—
C22	Band-pass sec. tuning	—
C23	Band-pass sec. trimmer	—
C24	Osc. circuit trimmer	—
C25	Osc. circuit tuning	—
C26	Osc. LW tracker	—
C27	1st IF trans. pri. tuning	—
C28	1st IF trans. sec. tuning	—
C29	2nd IF trans. pri. tuning	—
C30	2nd IF trans. sec. tuning	—
	Converter unit	
C31	Interference suppressors	0.01
C32		0.1
C33		0.1
C34*	HT smoothing con-densers	8.0
C35*		0.25
C36	Converter unit inter-	0.5
C37	ference suppressors	0.01

* Electrolytic.   † Variable.   ‡ Pre-set.

VALVE ANALYSIS

Valve voltages and currents in the table are those measured in our receiver when it was operating from fully charged accumulators. The receiver was tuned to the lowest wavelength on the medium band and the volume control was at maximum, but there was no signal input.

Voltages were measured on the 1,200 V scale of an Avometer, chassis being negative.

Valve	Anode Voltage (V)	Anode Current (mA)	Screen Voltage (V)	Screen Current (mA)
V1 VP2	135	1.0	105	0.3
V2 PM1HL	135	1.0		
V3 VP2	135	1.0	135	0.4
V4 2D2				
V5 PM22D	130	7.8	135	1.3

Figure 27.2 The circuit of the Ekco BV67 receiver.
Note the use of an indirectly heated double-diode for detection and AVC.
The synchronous vibrator unit is shown on the far left. As it is very similar to that used in automobile receivers it is worthy of being studied in conjunction with the section dealing with these sets

extremely difficult to repair. By about 1938 the idea had quietly been dropped in favour of a new Philips speciality.

Philips and Mullard pushbutton receivers

Again ignoring the principle that conventional designs had become so because they had been proved in practice to be the best, Philips decided to produce a radical new tuning condenser for its pushbutton sets. In this the plates were circular in section, graded in overall size and mounted concentrically. The spacing was arranged to permit the moving plates to slide in and out of the fixed plates, so that the tuning action depended on lateral and not rotary movement. This condenser was used in conjunction with mechanical pushbuttons which simply pressed the moving plates into the fixed plates a sufficient distance to tune in the desired station. This in itself sounds to be a simple and sensible design but unfortunately it was allied to a desperately complicated manual tuning system which had to change the rotary movement of the knob into lateral movement of the condenser. This in turn was linked to a dial pointer drive mechanism which was unlinear in action and evil to set up. Since the dial has to be disconnected for chassis removal be very wary about taking on a repair job on one of these sets!

The wartime civilian receiver

The shortage of domestic radio receivers during the Second World War led to the introduction, in 1944, of what officially was called the 'Wartime Civilian Receiver', the official title not revealing that there were to be two versions, one for mains, the other for batteries. The public soon dubbed these as 'utility sets', bracketing them with all the other various household articles that were produced under the term 'utility' to spartan Government standards. Although the sets were not successful from the sales angle, it is well worth examining them in detail, particularly the mains version. The latter has some ingenious design points and illustrates the use of a Westector as detector and provider of AVC. Both mains and battery versions are totally unique in being the only domestic sets ever to have had their full technical specification published, enabling its performance to be checked against known criteria rather than against advertisement hyperbole.

In 1943 the Government authorised, via the Board of Trade, the Radio Manufacturers Association to set up a committee to design and build a standard receiver that was simple, in order to use a minimum of components, but capable of an adequate performance in wartime conditions. What emerged was the 'utility' set in its two versions. It might easily be imagined that the combination of urgency and the need for economy could have resulted in a 'cheap and cheerful' approach but on the contrary the design specifications devised by the RMA were extremely detailed and stringent. They covered sensitivity, selectivity, overall response, AVC operation and IF rejection. They were as follows.

For the mains version:

- Sensitivity: to be not less than 325 mV @ 220 m and 625 mV @ 500 m for 50 mW output measured at the loudspeaker terminals.
- Selectivity: bandwidth not to exceed 11 kc/s @ 50% response and 21 kc/s @ 10% response.
- Overall response: to be not more that 7 dB down @ 100 c/s or 9 dB down @ 4 kc/s with respect to the level at 400 c/s, to be measured on a resistive output load and using an RF input of $\frac{1}{2}$ mV modulated at 30%, applied to the A1 (direct aerial socket with the volume control adjusted to give 50 mW output @ 400 c/s.
- AVC threshold: the AVC to be delayed so that its operation commences when the output is approximately 1 W on a signal with a modulation depth of 50%.
- IF rejection ratio: not to be worse than 5:1 at any point on the dial.

For the battery version:

- Sensitivity: not less than 300 mV @ 200 m and 600 mV @ 500 m for an output of 50 mW across the speech coil terminals with an HT voltage of 120 V (rather surprisingly, better than for the mains set).
- Selectivity: as for mains version.
- AVC threshold: not specified.
- Overall response: not more than 10 dB down @ 100 c/s or 14 dB down @ 3 kc/s with respect to the level at 400 c/s, under the same input/output conditions as for the mains version.
- IF rejection: as for the mains version.

In addition, the oscillator section of the frequency changer to continue to operate with the set fed from a 60 V HT battery (i.e. 50% down on normal) via a 2.2 kΩ series resistor.

The cabinets used by all manufacturers to be as nearly as possible of the same appearance. Standard drawings to be prepared by the British Radio Cabinet Makers' Association and kept at the RMA offices. The tuning scales to be finished in the standard manner and to have the same appearance as the prototype.

General quality: 'In view of the difficulty of producing a sufficiently detailed specification to cover such points as loudspeaker performance or the durability of workmanship of the receiver, the RMA wishes to draw the attention of manufacturers to the fact that the sets will be so coded that defective apparatus can be traced to its source, and it is therefore in the interests of each manufacturer to adhere to the spirit of the specifications.

In other words, you have been warned!

Meeting the specifications

The technical design committee appointed by the RMA came up with what was a near conventional four valve superhet for the battery set and a highly individual three-valve plus rectifier superhet for mains. It is doubtful if much argument was required over the battery set, which in fact bore more that a passing resemblance to the Murphy B89 of 1940, shorn of its long waveband and variable tone control. On the other hand, the mains set must have provoked a great deal of discussion, much of which must have been due to the restrictions imposed on what valve types might or might not be used. The committee had concluded that the required performance could be obtained with a 'short' superhet, which in peacetime would have meant the use of a frequency changer, an IF amplifier and a double-diode-output pentode. Unfortunately, whilst the first two types of valve were in quantity production, the last, having no military application, had been abandoned 'for the duration' and the Board of Trade would not permit its manufacture solely for the Civilian Receiver. To get around this the committee opted for the 'Westector' miniature metal rectifier as detector, driving an ordinary high slope output pentode, which type was also in plentiful supply. It would be the first time that the Westector had been used on a large commercial scale since the early 1930s. Its adoption here, whilst solving one problem, would cause others to arise, as we shall see.

Valves and secrecy

The valves used in both battery and mains versions were to be as anonymous as the sets themselves and carried arbitrary type numbers devised by the British Valve Makers Association to cover groups of equivalents made by individual members, with the final figure of each number indicating the actual manufacturer, as follows: 1, Cossor; 2, Mazda; 3 Ferranti; 4, Osram; 5, Marconi; 6, Mullard; 7, Brimar. It seems odd that even at that critical time the illusion that Marconi and Osram valves were unrelated, instead of coming from the same factory, had still to be fostered. The types employed in the battery version were known as the BVA172, BVA142, BVA132 and BVA262; due to their unique bases they were patently the Mazda series TP25, VP23, HL23DD and Pen25 and as the numbers reveal were made only by that firm. The prototype mains set used the BVA273, BVA243, BVA264 and BVA211, or in other words the Ferranti 6K8G and 6K7G, the Osram KT61 and the Cossor 431U. Oddly enough, the BVA273 was not included in the list of valves used in production models, which were (including equivalents): BVA274/5/6 (BVA277 was listed but not produced); BVA243/6/7; BVA254/5/6/7; BVA211/4/5/6/6. A list of commercial equivalents is reproduced in Figures 27.4 and 27.5.

An opportunity missed

Other design aspects intended to minimise the use of components in short supply included the single medium wave band and the use of resistance capacity smoothing for the HT, the latter 'to avoid the use of large quantities of copper wire in . . . a smoothing choke'. It seems rather curious why this aim was not pursued further and more profitably by making the set AC/DC and eliminating the much more copper-hungry mains transformer. The cabinet, like that of the battery version, was to be of plain wood with no identification of any particular manufacturer.

The design of the prototype in detail

The aerial input was to the primary of a simple iron-dust cored RF transformer via a 500 pfd condenser, with another of that value in series with the bottom of the winding and chassis. The bottom of the winding was also strapped to that of the secondary, which provided a mixture of inductive and capacitive coupling, the first at the high end of the band and the second at the low end. The top of the secondary was connected directly to the control grid of the frequency changer (V1, BVA273) and the bottom taken via a surprisingly low resistance (680 Ω) to the AVC line. The local oscillator also used an iron-dust cored RF transformer with the top of the tuned winding taken directly to the grid of the triode section of V1, with another 500 pfd from the bottom end to chassis (shunted by a 47 kΩ resistor) acting as both grid condenser and padder. The anode winding was supplied from the HT via the same 6.8 kΩ resistor which supplied the screen grids of the frequency changer and the IF amplifier (V2, BVA 283). The grid of the latter was coupled to the anode of the triode-hexode by an iron-dust cored transformer operating at 460 kc/s.

A second iron-dust cored transformer coupled the anode of V2 to the detector (MR1, WX6). The Westector has different operating conditions to that of a thermionic diode, requiring a lower impedance input and a small standing current flowing through it to ensure conduction at low signal strengths. The first was achieved by making the IF transformer have a step-down ratio of 1.5:1, which had the unfortunate side-effect of lowering the 'Q' and thus the gain of the IF amplifier; to counteract this some close-wound coupling turns were added to the primary winding. The way in which the standing current through MR1 was achieved will be discussed later when the AVC system is described.

The AF from the detector was filtered by a 47 kΩ resistor and 100 pfd condenser before being coupled via a blocking condenser to the top of a 1 MΩ volume control. The tap was taken directly to the grid of V3 whilst the bottom end was returned to the junction of the two cathode resistors in order to establish a standing grid bias of 4.5 V. The output valve was conventionally coupled to the loudspeaker with a fixed tone control condenser across the primary of the output transformer.

The power supply stage employed a mains transformer with only two tappings to cover 195/250 V inputs, and a full-wave rectifier (BVA211). As mentioned earlier, resistance-capacity smoothing was used with a resistor between the HT– and chassis providing a bias voltage of approximately 10 V. The mains on/off switch was not conventionally ganged to the volume control but took the form of a single-pole toggle switch mounted at the rear of the chassis.

The AVC system

We have already seen that in any short superhet where the output valve grid is driven directly from the detector it is essential to delay the AVC to prevent its coming into operation until sufficient voltage is developed across the detector load to give full volume. In normal practice, when a double-diode-output pentode is employed one of the diodes is used as AVC rectifier. It would have been possible technically in the utility set to employ another Westector for the job but this was ruled out by the Board of Trade, possibly because of supply difficulties. This left as the only source of AVC bias the DC existing at the top of the detector load resistor, which, had it been used direct, would not have been delayed and would have restricted severely the performance of the set. This problem was solved by some rather complicated circuitry which included pressing the suppressor grid of V2 into service as an effective diode. The circuit needs to be studied carefully in conjunction with the following description (incidentally of three contemporary descriptions studied during research for this topic none appears entirely to be correct).

It will be seen that the negative end of the bias resistor R13 is taken to R8 (4.7 MΩ) and via this resistor to R7 (1.5 MΩ), the other end of which is connected to the suppressor grid of V2. The latter is also taken via R6 (1 MΩ) to the top of the detector load resistor R9 (330 kΩ) which, as mentioned earlier, returns to the cathode of V3. Thus the four resistors effectively form a potential divider between a point that is at –10 V and one that is at +12.5 V – but with the important addition of the path to chassis provided when the AVC 'diode' is conducting, as occurs under no signal conditions. It may then be regarded as a fairly low resistance, so that the centre of the potential divider is 'clamped' to chassis and effectively split into two sections, the upper at between +12.5 V and chassis, the lower at between –10 V and chassis. The voltage at the junction of R6 and R9 will be approximately +9 V, which is applied to the anode of MR1 via R5 (47 kΩ). Meanwhile, MR1 cathode returns to the junction of R11 and R15 where a voltage of 4 V with respect to chassis exists. Since the anode is more positive than the chassis the required steady current will flow through MR1.

The lower half of the potential divider provides approximately –2.5 V at the junction of R7 and R8 and this is applied to V2 and V1 as standing bias. When a signal arrives at MR1 a negative voltage will be developed at its anode that is in opposition to the positive voltage applied from the junction of R6 and R9, but until it exceeds –9 V it will not have any effect on the AVC system. However, as soon as this critical delay voltage is exceeded the 'diode' anode will become negative and it will cease to conduct, removing the 'clamp' to chassis. The negative voltage is then passed on as AVC bias to V1 and V2, but it is important to bear in mind that it will not reach their grids at the same potential, due to the dividing effect of R7 and R8. In fact, the voltage on the actual AVC line will be approximately 0.4 of that at the 'diode' when the latter is at –9.5 V and approximately 0.5 when it is at 15 V, with the proportion rising as the 'diode' becomes more and more negative.

The battery version

As mentioned earlier, this bore a strong affinity to the Murphy B89. It is a fairly straightforward four-valve superhet using a triode-pentode frequency changer, pentode IF amplifier, double-diode-triode detector and AF amplifier, and pentode output. One feature worth examining in detail, however, is the way the AVC bias is obtained. For some reason instead of using the second diode of the DDT as an AVC rectifier, bias was drawn via R7 (2.2 MΩ) from the top of the volume control (R9, 1 MΩ), which also acts as the detector diode load. The frequency changer and IF amplifier also derive their standing grid bias from the AVC line, which is returned via R8 (3.3 MΩ) to the HT– line, at a point approximately –3 V with respect to chassis. To prevent this voltage reaching the detector diode, the bottom of the volume control is returned to LT+, thus making R7, R8 and R9 into a virtual potential divider. Note that since the LT– is taken to chassis, the total voltage across the divider will be 5 V and not 3 V. The valves thus receive approximately –2.7 V standing bias from the junction of R7 and R8, whilst the diode anode receives about +0.7 V, so no AVC bias will be developed until the signal strength exceeds this.

Breaking the mould

The conception of every manufacturer (140 in all) involved in the production of the Utility Receivers building absolutely identical sets was hardly likely to have been achieved in practice and in fact many of them modified the design to a certain extent.

Whilst some changes are unlikely to puzzle anyone working on the sets, others could well cause problems, especially to those with lesser experience of them. Figure 27.3 to 27.6 show a complete list of all modifications for both mains and battery versions.

Likely problems in servicing

Mains receivers

Many years of experience in servicing these sets suggest that generally speaking trouble is confined to the detector and power supply stages. It cannot be emphasised too much that the voltages around the unusual combination of metal rectifier and suppressor grid for detection and AVC are critical and that discrepancies can ruin the sensitivity of the receiver. Particular attention needs to be paid to the 4.7 MΩ resistor which feeds negative bias to the suppressor of V2 and which is susceptible to going very high. The steady negative voltage developed across the Westector appears on C24, the coupling condenser to the volume control, and if this should leak there is a danger of the output valve receiving unwanted negative bias, in spite of the bottom of the control being returned to the junction of the bias resistors R11 and R12. One consolation is that although this may cause V3 to become less sensitive, at least it will not make it draw excess anode current as in most sets. To offset this it has been found that certain output valves have had a tendency to leak internally between cathode and grid and thus bring about overloading. Note that the voltage reading at the cathode (nominally 13 V) does not represent the actual grid bias; this should be measured between the junction of R11 and R12 and the cathode. It should be a shade over 4 V and any discrepancy should be investigated, commencing with trying a replacement valve. Another item worth checking is the bias condenser C13, by means of shunting another across it.

Apart from the normal chance of one or more of the smoothing condensers losing its capacity there is always a possibility that someone in the past may have replaced them incorrectly, for it is all too easy for the uninitiated to ignore negative bias and return the reservoir to chassis instead of to the centre tap of the HT winding on the mains transformer.

The mains transformers themselves, of whatever make, have a good record for reliability. This is probably due to the fact that standard types were employed of the same ratings as those used for

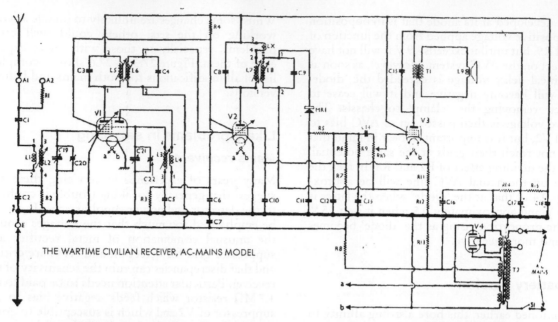

THE WARTIME CIVILIAN RECEIVER, AC-MAINS MODEL

COMPONENTS AND VALUES

CAPACITORS

		Values (μF)
C1	Aerial series capacitor ...	0·0005
C2	V1 hex. CG decoupling...	0·005
C3	1st IF transformer tuning capacitors	0·0001
C4		0·0001
C5	V1 osc. CG capacitor ...	0·0002
C6	Oscillator circuit tracker ...	0·0005
C7	V1 osc. anode decoupling	0·1
C8	V2 CG decoupling ...	0·1
C9	V1, V2 SG's decoupling ...	0·1
C10	2nd IF transformer tuning capacitors	0·0001
C11		0·0001
C12	IF by-pass capacitors ...	0·0001
C13		0·0001
C14	HT circuit by-pass ...	0·5
C15	AF coupling to V3 triode	0·005
C16	V3 triode to V4 AF coupling ...	0·005
C17	Fixed tone corrector ...	0·005
C18†	Aerial circuit tuning ...	0·000532
C19‡	Aerial circuit trimmer ...	0·00005
C20†	Oscillator circuit tuning	0·000532
C21‡	Oscillator circuit trimmer	0·00005

† Variable. ‡ Pre-set.

RESISTORS

		Values (ohms)
R1	A2 series resistor ...	47,000
R2	V1 pent. CG decoupling	680
R3	V1 osc. CG resistor ...	22,000
R4	V1 osc. anode HT feed	39,000
R5	V1, V2 SG's HT feed ...	47,000
R6	IF stopper ...	100,000
R7	V1, V2 fixed GB and AVC feed resistors ...	2,200,000
R8		3,900,000
R9	Manual volume control: V3 signal diode load	1,000,000
R10	V3 triode CG resistor...	3,300,000
R11	V1, V2, V3 HT feed	10,000
R12	V3 triode anode load ...	68,000
R13	V4 CG resistor ...	330,000
R14	V4 grid stopper ...	100,000
R15	V1, V2, V4 fixed GB resistor ...	390

OTHER COMPONENTS

		Approx. values (ohms)
L1	Aerial coupling coil ...	0·8
L2	Aerial tuning coil ...	3·0
L3	Oscillator reaction coil ...	0·8
L4	Oscillator tuning coil ...	1·7
L5	1st IF trans. { Pri.	7·0
L6	{ Sec.	7·0
L7	2nd IF trans. { Pri.	7·0
L8	{ Sec.	7·0
L9	Speaker speech coil ...	3·0
T1	Speaker input { Pri.	600·0
	trans. { Sec.	0·2
S1	HT circuit switch	—
S2	LT circuit switch	—

VALVE ANALYSIS

Valve voltages and currents given in the table below are those quoted as average values by the designers. They are based on readings taken on a receiver operating from an HT battery measuring 115 V on load, with the gang turned to maximum, but with no signal input.

Voltages were measured on the 400 V scale of a model 7 Avometer, chassis being negative.

In addition to the table, the following information is given: GB voltage (across R15) is −3 V; total HT current is 8·9 mA; total LT current is 0·45 A; V1 oscillator grid current (with gang at maximum) is 220 μA.

The valves are fitted with Mazda octal bases, whose pin connections are given in the diagrams beside the circuit diagram on this page.

Valve	Anode voltage (V)	Anode Current (mA)	Screen Voltage (V)	Screen Current (mA)
V1 172	85 / 58	Oscillator 0·36 / 1·3	46	0·71
V2 142	85	0·61	46	0·17
V3 132	52	0·5	—	—
V4 162	109	4·2	112	0·98

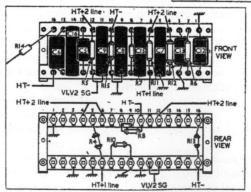

Figure 27.3 The circuit, component values (with tag strip details) and valve voltages for the battery version of the utility set

MANUFACTURERS' CODE NUMBERS and RECEIVER MODIFICATIONS

The following is a list of manufacturers concerned with the production of the Wartime Civilian Receivers, together with their code numbers, which precede the serial number. From the code number, dealers can ascertain to whom they should apply for spares. This information must be regarded as confidential to the Trade.

Below the list are details of modifications to be found in some manufacturers' versions grouped under their code numbers.

U1	...	Bush Radio Ltd.	U12	...	Burndept Ltd.	*U23	...	Plessey Co., Ltd.	U33 ... Roberts Radio Co., Ltd.
U2	...	E. K. Cole Ltd.	U12A	...	Vidor Ltd.	U24	...	Regentone Products Ltd.	U34 ... Radio Gramophone Dev. Co., Ltd.
U3	...	A. C. Cossor Ltd.	U13	...	Central Equipment Ltd.	U25	...	R.M. Electric Ltd.	U35 ... R.S.G. Radio Ltd.
U4	...	Gramophone Co., Ltd.	U14	...	Ferranti Ltd.	U26	...	Decca Record Co., Ltd.	U36 ... Beethoven Electric Equip. Co., Ltd.
U4A	...	Marconiphone Co., Ltd.	U15	...	Feigate Radio Ltd.	U27	...	Dulci Company.	U37 ... J. G. Graves Ltd.
U5	...	Ferguson Radio Corp. Ltd.	U16	...	Hale Electrical Co., Ltd.	U28	...	R. N. Fitton Ltd.	U38 ... Aron Radio & Television Ltd.
U6	...	General Electric Co., Ltd.	U17	...	Halcyon Radio Ltd.	U29	...	Portadyne Radio Ltd.	U39 ... N.H. Radio Products Ltd.
U7	...	Murphy Radio Ltd.	U18	...	Invicta Radio Ltd.	U30	...	Pamphonic Radio Ltd.	U40 ... Ace Radio Ltd.
U8	...	Philips Lamps Ltd.	U19	...	Lissen Ltd. (Ever Ready).	U31	...	Mains Radio Gramophones Ltd.	U41 ... Solectric Ltd.
U9	...	Pye Ltd.	U20	...	McMichael Radio Ltd.	U32	...	Kolster-Brandes Ltd.	U42 ... Whiteley Electrical Co., Ltd.
U10	...	Ultra Electric Ltd.	U21	...	Philco Radio & Tel. Corp. Ltd.				U43 ... Aerodyne, Ltd.
U11	...	A. J. Balcombe Ltd.	U22	...	Pilot Radio Ltd.				U44 ... R.A.P. Manufacturing Co., Ltd.

U1

An additional resistor of 100,000 Ω is connected between the A1 and E socket.

On component assembly there are several divergencies as compared with our sketches overleaf. Using same tag numbering, with "T" representing tag remote from, and "B" tag adjacent to, chassis, the arrangement is:

UPPER SKETCH		LOWER SKETCH	
Component	Tags	Component	Tags
R14	T16/B16	Westector	T3/B3
R4	T14/B14	R12	T4/B4
C6	T13/B13	R11	T4/B5
C7	T11/B11	R9	T5/B5
R8	T12/T10	R6	T7/B7
C10	T8/B8	R13	T10/B10
C14	T6/B6	R2	T12/B12
R5	T5/T3		
C12	T5/B4		
C11	T2/B2		
R15	T1/B1		

U2

An additional resistor of 47,000 Ω is connected between the A1 and E sockets.

The coupling coil LX shown on the second IF transformer is omitted.

The mains transformer does not conform to the type shown in our sketches, the connections being taken to two terminal strips.

U3

Aerial and oscillator coils L1, L2, L3, L4 are not fitted with dust-iron cores. Should C5 be renewed, tracking of the Oscillator section may sometimes be improved by softening wax securing Oscillator coil L3 and altering its position.

The connecting tags of coil units L1, L2 and L3, L4 are not numbered, but in the diagrams below they are given numbers to agree with those shown in our circuit diagram overleaf, as seen viewing the free ends of the units.

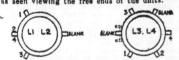

IF transformer coupling coil Lx is omitted. The IF transformers are adjusted to a frequency of 462.5 kc/s. Their core adjustments are all accessible without removing chassis.

BVA VALVE CODE

Claims for free replacement under guarantee of any valve in the Wartime Civilian Receiver must be made on the valve manufacturer whose name can be identified by reference to the final figure of code marking on the valve, as follows :—

1 .... Cossor
2 .... Ediswan (Mazda)
3 .... Ferranti
4 .... GEC
5 .... Marconiphone
6 .... Mullard
7 .... Standard Telephones

All applications for such replacements must be made in conjunction with a perperly completed BVA replacement form. This information is confidential to the Trade.

Tag numbers are not applicable, but the connections are as follows :

1st IF transformer (centre can, No. MC11572):
Blue lead to junction of C7, R2, R3, R7.
Black lead to grid cap of V2.
Brown lead to V1 anode.
Orange lead to HT+ line.

2nd IF transformer (end can, No. MC11573).
Orange lead to HT+ line.
Brown lead to V2 anode.
Black lead to MR1+.
Blue lead to junction of R10, R11, R12.

A 47,000 ohms anti-modulation hum resistor is fitted across A1 and E terminals.

Resistor R14 is composed of two 2,200 Ω resistors in parallel.

Capacitor C1 is mounted on the component assembly. R1, C2, R3, C5 are in slightly different positions from those shown by us.

Mains transformer unit differs from our sketch.

U4, U4A

The second IF transformer coupling coil Lx is omitted.

The low potential ends of capacitors C11 and C12 will in most cases be returned not to chassis as shown in our circuit diagram, but to the earthy side (tag 2 in our diagram) of L8.

U7

R8 may be mounted on the back of the component assembly, and R4 on the front, but they will be connected to the same pairs of tags as shown in our sketches.

An additional resistor of 27,000Ω is connected directly between the A1 and E sockets.

U8, U31, U33, U34, U37, U39

Adjustment of the aerial and oscillator inductances at the low frequency end of the band is not necessary. The coils are closely adjusted in the factory, then sealed.

The IF coils are of the usual Philips type. The secondary adjusting core is the upper one in each case. The extra coupling coil Lx is not used, but the detector MR1 is connected to a tap on L8 to reduce damping.

Deviation in coil resistances should be noted as follows: L1, 3Ω; L3, 6.5Ω; L4, 2Ω; T1 primary, 470Ω; T2 primary, 55Ω (total).

The speaker is mounted with the input transformer T1 on the right, viewed from rear.

The mains transformer is of a different construction from that in our chassis, with vertical connecting strips either side of the core

at one end. Viewed from this end, and reading from top to bottom, the connections are as follow: Left-hand strip: 1, HT secondary (end); 2, HT secondary (CT); 3, HT secondary (end); 4 and 5, HT heater secondary (4 V). Right-hand strip: 1, primary (250 V end); 2, primary (220 V tap); 3, primary (common end); 4 and 5, V1-V3 heater secondary (6.3 V).

Capacitors C16, C17, C18 may be three wet electrolytics, three dry or a combination of the two types. When dry tubular capacitors are used they are strapped beneath the chassis as in the under-chassis view. When wet types are used, C17 and C18 may be any value between 8 mfd and 32 mfd 350 V. C18 is mounted on the chassis between V3 and the gang. C16 and C17 are carried on a metal bracket on T2.

C3, C8 and C9 are each 103 pf (0.000103 μF); C4 is 97 pf (0.000097 μF); C15 is mounted under the chassis below V3.

R8 is a longer resistor than that shown in our sketch, and is therefore connected between tags 10 and 13, with tag 13 joined to tag 11. R14 is made up of two 2,200Ω 1 watt resistors in parallel; R15 is made up of two 4,700Ω 1 watt resistors in parallel.

The revised coil connections are shown below.

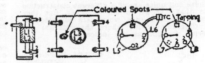

U14

The aerial coupling coil L1 shown in the circuit diagram is omitted, but a spare tag connected to the earthy end of the tuning coil L2 is provided for the aerial connection.

The IF transformers are variable-capacitance tuned, instead of variable-inductance as shown. In ganging, there is no necessity to adjust the core of L2.

U21

An additional resistor of 100,000Ω is connected between the A1 and E sockets.

The connections to L3 may be transposed. tag 4 becoming tag 3 and tag 3 becoming tag 4. L4 will also be reversed.

In the same way, the connections to Lx, tags 3 and 4, may be reversed.

REPLACEMENT VALVES FOR THE BVA NUMBERED TYPES

BVA VALVE NUMBER		COSSOR	MAZDA	EVER READY	FERRANTI	MARCONI OSRAM	MULLARD	PHILIPS	BRIMAR
V1*	274 275 276	OM10	—	ECH35	6K8G	X61M	ECH35	—	6K8G
V2	243 246 147	6K7G OM6	—	EF39	6K7G VPT62	—	EF39	—	6K7G
V3	264 265 266 267	—	—	EL33	—	—	EL33	—	6A96G
V4	211 214 215 216	43IU	UU6	A11D 511D	R4	U14 MU14	DW4/350 IW4/350	1561 1867	R2 R3

* Although two additional BVA numbers for V1—273 and 277—may appear in the instructions issued with the receiver, no valves of these types have been produced.

Figure 27.4 Receiver manufacturers' code numbers, modifications, valve manufacturers' code numbers and valve equivalents for the mains version of the utility set

MANUFACTURERS' CODE NUMBERS and RECEIVER MODIFICATIONS

The following is a list of manufacturers concerned with the production of the Wartime Civilian Receivers, together with their code numbers, which precede the serial number. From the code number, dealers can ascertain to whom they should apply for spares. This information must be regarded as confidential to the Trade.

Below the list are details of modifications to be found in some manufacturers' versions grouped under the code numbers to which they apply. Replacement valve types suitable for the Wartime Receiver are given in the table below against the BVA numbers which will be found on the original valves.

U1 ... ... Bush Radio, Ltd.	U14 ... ... Ferranti, Ltd.	U29 ... ... Portadyne Radio, Ltd.
U2 ... ... E. K. Cole, Ltd.	U15 ... ... Folgate Radio, Ltd.	U30 ... ... Pamphonic Radio, Ltd.
U3 ... ... A. C. Cossor, Ltd.	U16 ... ... Hale Electrical Co., Ltd.	U31 ... ... Mains Radio Gramophones, Ltd.
U4 ... ... Gramophone Co., Ltd.	U17 ... ... Halcyon Radio, Ltd.	U32 ... ... Keister-Brandes, Ltd.
U4A ... ... Marconiphone Co., Ltd.	U18 ... ... Invicta Radio, Ltd.	U33 ... ... Roberts Radio Co., Ltd.
U5 ... ... Ferguson Radio Corporation, Ltd.	U19 ... ... Lissen, Ltd. (Ever Ready).	U34 ... ... Radio Gramophone Dev. Co., Ltd.
U6 ... ... General Electric Co., Ltd.	U20 ... ... McMichael Radio, Ltd.	U35 ... ... R.S.C. Radio, Ltd.
U7 ... ... Murphy Radio, Ltd.	U21 ... ... Philco Radio & Tel. Corp. Ltd.	U36 ... ... Beethoven Electric Equip. Co., Ltd.
U8 ... ... Philips Lamps, Ltd.	U22 ... ... Pilot Radio, Ltd.	U37 ... ... J. G. Graves, Ltd.
U9 ... ... Pye, Ltd.	U23 ... ... Plessey Co., Ltd.	U38 ... ... Aren Radio & Television, Ltd.
U10 ... ... Ultra Electric, Ltd.	U24 ... ... Regentone Products, Ltd.	U39 ... ... N.H. Radio Products, Ltd.
U11 ... ... A. J. Balcombe, Ltd.	U25 ... ... R.M. Electric, Ltd.	U40 ... ... Ace Radio, Ltd.
U12 ... ... Burndept, Ltd.	U26 ... ... Decca Record Co., Ltd.	U41 ... ... Selectric, Ltd.
U12A ... ... Vidor, Ltd.	U27 ... ... Dulci Company.	U42 ... ... Whiteley Electrical Co., Ltd.
U13 ... ... Central Equipment, Ltd.	U28 ... ... R. N. Fitten, Ltd.	

U2

Capacitor C1 is mounted on the component assembly, while C6 is suspended in the wiring.

U3

The aerial and oscillator coils L1, L2 and L3, L4, are not fitted with dust-iron cores. Should C6 be renewed, tracking of the oscillator circuit may sometimes be improved by softening the wax securing L2 and L4 and altering their positions.

These coil units are of a different construction from that shown in our sketches, but they have connecting tags which are identified in the diagrams below, where the internal and external connections are indicated.

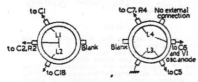

The IF transformers are also of a different construction from that shown in our sketches, and are so arranged that trimmers are accessible without removing the chassis from the cabinet. Their flexible connecting leads are colour coded, and they are connected to the following points in the circuit :—

1st IF transformer (centre can, No. MC11572/2):
Orange lead to HT+2 line.
Brown lead to V1 pentode anode.
Blue lead to AVC line.
Black lead to V2 top cap lead.

2nd IF transformer (end can, No. MC11574):

Orange lead to HT+2 line.
Brown lead to V2 anode.
Blue lead to R6, C12.
Black lead to V3 signal diode anode.
The IF transformers are adjusted to a frequency of 462.5 kc/s.

U4, U4A

The coil units are of a different construction from those shown in our sketches, and the IF transformer core adjustments are vertical screws reached from beneath the chassis in the case of the primaries and the tops of the cans in the case of the secondaries. No tags are fitted on the aerial and oscillator units, lead-out wires being continuations of the windings.

Some of the components beneath the chassis occupy positions different from those shown in our under-chassis view, and roughly an approximately equal number of components are distributed on either side of the component assembly. S1 and S2 are transposed as compared with those in our illustration.

U5, U13, U32, U40

The aerial coupling coil L1 is of the high-impedance type as against the low impedance one in our basic chassis, and it is returned directly to chassis instead of to the bottom of L2. Its DC resistance is 23.5Ω.

The IF transformer coils are air cored, and have mica pre-set trimmers whose adjustments are reached through holes in the tops of the cans. The DC resistance of all the IF coils is 7.5Ω each.

U7

The shorting link shown connected between tags 1 and 3 in our sketch of the L1, L2 coil unit is now connected instead between tags 2 and 4.

U8, U14, U31, U33, U34, U37, U39

Adjustment of the aerial and oscillator inductances at the low frequency end of the band is not necessary. The coils are closely adjusted in the factory, then sealed, and they should not be altered.

The IF coils are of the usual Philips type. The secondary adjusting core is the upper one in each case. The signal diode of V3 is connected to a tapping on L5 to reduce damping.

The following deviations occur in the DC resistance of the coils, as compared with the values in our table: L1, 3Ω; L3, 2Ω; L4, 6.5Ω.

Capacitors C3, C10 and C11 are 108 pf (0.000108μF) each, and C4 is 97 pf (0.000097μF). C17 is mounted beneath the chassis, underneath V4 holder.

Resistor R8 is a longer type than that shown in our sketch, and is therefore connected between tags 8 and 10 on the component assembly, and tag 8 is joined to tag 9.

At present the HT negative lead is yellow instead of black, and the speaker leads are green, but the maintenance of consistent lead colours is in any case subject to supply limitations. The connecting tags of the coil units are indicated in the diagrams below.

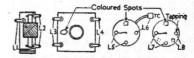

U10, U17, U35, U41

The principal difference in these chassis lies in the DC resistance values of coil windings. Those that differ from our " Other Components " are as follows: L1, 0.6Ω; L2, 2.6Ω; L3, 0.5Ω; L4, 2.7Ω; L5, L6, L7 and L8, 5.8Ω each; L5, 2.0Ω; L9, 2Ω; T1 pri., 550Ω; sec., 0.3Ω.

U21

The coil units in these chassis have numbered tags. Each coil has two tags, and in the following list the numbers of the two tags concerned follow the number of the coil, the first tag quoted being the one at the upper end of the coil as drawn in our circuit diagram, and the second the one at the lower end, as in the following example: L1, tag 1 (to C1), tag 2 (to C2, R2); L2, tag 3 (to V1 top cap lead), tag 4 (to C2, R2 and to tag 2).

Quoting in the same order, the rest are as follows: L3, tags 3 and 4; L4, tags 1 and 2; L5, tags 2 (to HT+2 line) and 1; L6, tags 1 and 2; L7, tags 2 (to HT+2 line) and 1; L8, tags 1 and 2.

Also in these chassis, the HT negative lead, which is given as black under " Battery Leads and Voltages," is slate or grey to distinguish it from the LT negative lead, which is black.

U28

The aerial and oscillator coils are air cored.

BVA VALVE CODE

Claims for free replacement under guarantee of any valve in the Wartime Civilian Receiver must be made on the valve manufacturer whose name can be identified by reference to the final figure of code marking on the valve, as follows :—

1 .... Cossor
2 .... Ediswan (Mazda)
3 .... Ferranti
4 .... GEC
5 .... Marconiphone
6 .... Mullard
7 .... Standard Telephones

All applications for such replacements must be made in conjunction with a properly completed BVA replacement form. This information is confidential to the Trade.

VALVE REPLACEMENT TABLE

VALVE	BVA NUMBER	REPLACEMENT (MAZDA)
V1	172	TP25
V2	142	VP23
V3	132	HL23DD
V4	162	Pen25

Figure 27.5 Receiver manufacturers' code numbers, modifications, valve manufacturers' code numbers and valve equivalents for the battery version of the utility set

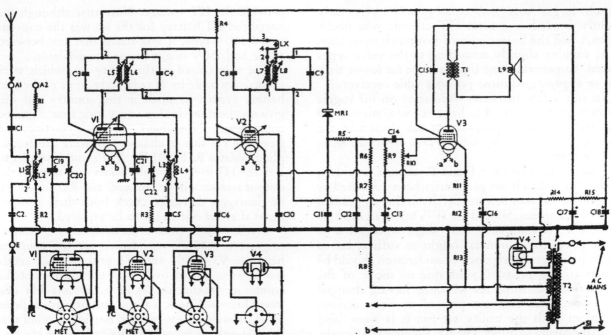

Circuit diagram of the Wartime Civilian AC Receiver. **MR1** is a Westector. The numbers at the coil ends indicate connecting tags shown in the sketches overleaf. In most chassis an anti-modulation hum resistor is connected between the A1 and E sockets.

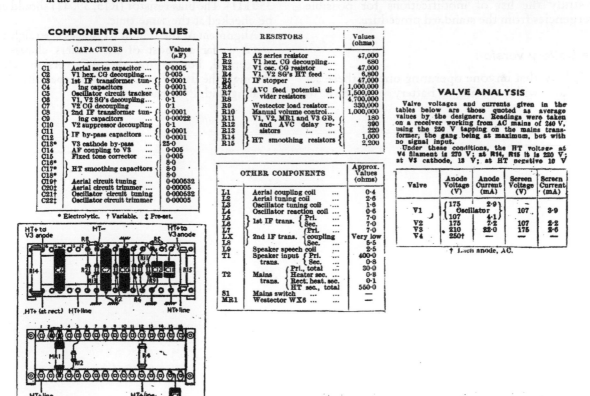

COMPONENTS AND VALUES

	CAPACITORS		Values (μF)
C1	Aerial series capacitor	...	0·0005
C2	V1 hex. CG decoupling	...	0·005
C3	1st IF transformer tun-		0·0001
C4	ing capacitors		0·0001
C5	Oscillator circuit tracker		0·0005
C6	V1, V2 SG's decoupling		0·1
C7	V2 CG decoupling		0·1
C8	2nd IF transformer tun-		0·0001
C9	ing capacitors		0·00022
C10	V2 suppressor decoupling		0·1
C11	IF by-pass capacitors		0·0001
C12			0·0001
C13*	V3 cathode by-pass	...	25·0
C14	AF coupling to V3	...	0·005
C15	Fixed tone corrector	...	0·005
C16*			8·0
C17*	HT smoothing capacitors		8·0
C18*			8·0
C19†	Aerial circuit tuning	...	0·000532
C20‡	Aerial circuit trimmer	...	0·00005
C21†	Oscillator circuit tuning		0·000532
C22‡	Oscillator circuit trimmer		0·00005

* Electrolytic. † Variable. ‡ Pre-set.

	RESISTORS		Values (ohms)
R1	A2 series resistor	...	47,000
R2	V1 hex. CG decoupling	...	680
R3	V1 osc. CG resistor	...	47,000
R4	V1, V2 SG's HT feed	...	6,800
R5	IF stopper	...	47,000
R6	AVC feed potential di-		1,000,000
R7	vider resistors		1,500,000
R8			4,700,000
R9	Westector load resistor	...	330,000
R10	Manual volume control	...	1,000,000
R11	V1, V2, MR1 and V3 GB,		180
R12	and AVC delay re-		390
R13	sistors		220
R14	HT smoothing resistors		1,000
R15			2,200

	OTHER COMPONENTS		Approx. Values (ohms)
L1	Aerial coupling coil	...	0·4
L2	Aerial tuning coil	...	2·6
L3	Oscillator tuning coil	...	1·6
L4	Oscillator reaction coil	...	0·6
L5	1st IF trans. { Pri.		7·0
L6	{ Sec.		7·0
L7	{ Pri.		7·0
LX	2nd IF trans. { coupling		Very low
L8	{ Sec.		5·5
L9	Speaker speech coil	...	2·5
T1	Speaker input { Pri.		400·0
	trans. { Sec.		0·8
	{ Pri., total		30·0
T2	Mains { Heater sec.	...	0·8
	trans. { Rect. heat. sec.	...	0·1
	{ HT sec., total		550·0
S1	Mains switch	...	—
MR1	Westector WX6	...	—

VALVE ANALYSIS

Valve voltages and currents given in the tables below are those quoted as average values by the designers. Readings were taken on a receiver working from AC mains of 240 V, using the 250 V tapping on the mains transformer, the gang being at maximum, but with no signal input.

Under these conditions, the HT voltage at V4 filament is 270 V; at R14, R15 it is 220 V; at V3 cathode, 13 V; at HT negative 10 V

Valve	Anode Voltage (V)	Anode Current (mA)	Screen Voltage (V)	Screen Current (mA)
V1	{ 175	2·9	107	3·9
	Oscillator			
	107	4·1		
V2	175	7·2	107	2·2
V3	210	22·0	175	2·6
V4	250†	—	—	—

† Each anode, AC.

Figure 27.6 The circuit, component values (with tag strip details) and valve voltages for the mains version of the utility set

much larger sets and consequently were only lightly loaded: the total HT current was under 45 mA and the heater current not much more than 1 A, varying slightly according to the valve types fitted. In general these figures were far lower than those applying to most pre-war table receivers.

In not a few instances the mains on/off toggle switch has been found to have become 'sticky' and not throwing over properly. If this should be the case it is worth trying some WD40 down the sides of the toggle.

Realignment of the RF and IF circuits is seldom called for unless it has previously been disturbed or if tests reveal that low sensitivity is due to gradual loss of alignment. Should the IFTs be only a little off their correct frequency only slight adjustment of their cores (or trimmers) ought to suffice, but if there is a large discrepancy consideration should be given to the possibility of one or more of the condensers shunting the windings having changed in value. This has not proved to be a serious problem with the utility set but it is something worth bearing in mind. The RF alignment is simple, with only the single wave band to worry about, but do study the list of modifications for possible divergencies from the standard procedure.

The battery version

It is likely that anyone operating one of these set nowadays will be using a battery eliminator to provide the HT voltage. Note that although the suggested HT battery for the set was the conventional 120 V type, it was stated that any between 108 V and 150 V might be used as available.

The valves used in the utility set, which were exclusively made by Mazda, were some of the best battery types ever made in this country and will give excellent service for a very long time if treated considerately. Note that C14 (0.5 μfd) is connected from the HT line to chassis, following the decoupling resistor R11 (10 kΩ). If it should happen to leak the HT to all valve electrodes save those of the output tetrode will be reduced, and R11 may itself be damaged; therefore check both these components if either one needs to be replaced.

Credit must be paid to the designers for using as original a 1000 VW condenser for the coupling between V3 anode and V4 grid. As V3 anode works at only just over 50 V, the chances of the condenser leaking ought to be remote but one never knows! Check it if the output valve appears to be drawing excess current, bearing in mind that there should an actual negative voltage on the grid via the stopper (R14, 100 kΩ) and the leak (R13, 330 kΩ). The bias resistor (R15, 390 Ω) should also be checked at the same time.

Alignment of the IF and RF stages is straightforward as far as most of the makers chassis are concerned, but as for the AC version, check through the list of modifications especially those dealing with Pye-made sets.

Chapter 28

Repairing FM and AM/FM receivers

To a large extent the same techniques as used for AM-only receivers will apply on both AM or FM to the main chassis of the receiver, bearing in mind that on FM the IF is increased to 10.7 Mc/s and the detector works in a different way. The AF amplifier and output stages are similar to those of AM sets only in respect of extending the treble response. More expensive sets were fitted with special high frequency 'tweeter' loudspeakers alongside the ordinary unit. Experience has shown that the chief causes of trouble other than those to which any receiver is heir have been breakdowns on the AM/FM switching due to the presence of HT on the wafers and a greater susceptibility to self-oscillation in the IF amplifier stage.

For the first problem you may be lucky enough to find some unused switch contacts that can be pressed into service, otherwise some kind of replacement will have to be made. Rather than change the entire switch unit, which may be difficult or impossible with press button types, you may well be able to fit a separate small switch on the rear of the chassis.

IF instability is usually due to a faulty decoupling condenser on the HT feed to the anode or screen grid of the IF amplifier – or to someone having twiddled the cores in the transformer. We shall look at realignment in a moment.

By far the greatest problem with VHF tuner units has proved to be deterioration of an ECC85 or UCC85. The best way of proving this is by substitution, but if you haven't a spare to hand you could try the old service engineers' trick of gently tapping the valve with the handle of a small screwdriver. It is surprising how often this treatment will restore the performance, if only temporarily. Don't overdo the tapping!

If there appears to be trouble within the tuner unit you may or may not be able to check voltages, etc., because the usual small size and near totally enclosed construction makes it extremely difficult to get at the valve bases and components. Each tuner will have to be considered on its particular merits.

Be warned that the sensitivity on FM of many sets was not much better than poor, and a good outside aerial is called for unless you live in a very good reception area.

Realigning FM and AM/FM receivers

Contrary to what might be expected, an FM signal generator is *not* necessary. Although most manufacturers issued their own particular instructions for alignment the following procedure usually will serve well enough. With combined AM/FM sets carry out the AM alignment first, if necessary, in exactly the same way as for ordinary AM-only sets.

On FM the 10.7 Mc/s IF AM signal needs to be injected into the grid of the VHF frequency changer, which more than likely will be inaccessible. In this case, try partially removing the almost invariable screening can from the valve so that it no longer is in contact with the chassis, then clip the live generator output lead to it. You may have to use a fairly high output from the generator to make this work but it usually will do the trick.

De-tune the secondary of the last IFT (i.e. the winding that feeds the ratio detector) by unscrewing its core outwards until it is level with the outside of the can. This will render it sensitive to AM signals. Connect a multimeter on its 10 V AC range across the loudspeaker and tune all the IF cores save the last for maximum output. Reduce the output from the signal generator as necessary to maintain about 1.25 V. Now tune the last core for *minimum* output, which should coincide with the point of maximum sensitivity to FM signals.

Remove the signal generator input, replace the can on the ECC85/UCC85 and try the set on an aerial to see if stations can be received satisfactorily. Sometimes, against the rules for AM alignment, it may be necessary to fine tune the last IF core by ear to achieve completely undistorted sound.

On the RF side you really need an FM signal generator to carry out detailed alignment. Fortunately in most cases this will not be necessary and a little careful adjustment on actual stations should enable you to position them correctly on the dial and to realise maximum sensitivity. When a VHF tuner has been mistuned badly in the past it will be necessary to obtain the manufacturer's servicing data and to follow it faithfully.

Bear in mind that UK valve FM receivers covered only from about 87 Mc/s to 100 Mc/s, although it may be possible by a little adjustment of the local oscillator trimmer or core to raise the upper limit sufficiently to receive Classic FM, if so desired.

In all cases, seal cores and trimmers with wax or paint after alignment is completed Note that this was done in the radio trade not only to prevent accidental movement but also to reveal if the customer or a 'knowledgeable' friend interfered with them.

Special note: be warned that a number of cases have been recorded of Murphy A252 AM/FM receivers having left the factory with a 1000 pfd decoupling condenser on the HT feed to the VHF tuner omitted, suggesting that there may have been a bad batch which escaped the factory quality control. Check for the absence of this condenser in the event of oscillation being experienced on the tuner unit or the AM frequency changer when acting as a VHF IF amplifier.

Chapter 29

Public address and high fidelity amplifiers

The use of microphones and loudspeakers to address public gatherings was pioneered in the early 1920s and rapidly became widespread, especially as more powerful output valves and loudspeakers became available. When nowadays young people talk casually of 160 W amplifiers in motor cars, it is interesting to look at the output powers considered suitable in the valve era, when watts really meant watts and was not a meaningless tag. Around 1930 a book published on the work of wiring cinemas for talking pictures stated that between 3 W and 5 W was perfectly sufficient for medium to large halls, in conjunction with very efficient giant horn loudspeakers. Even then some patrons complained that the sound was too loud! Twenty years later 25 W was considered adequate for most general purpose PAs and in the early 1960s the present writer installed a complete 25-loudspeaker music system in a noisy factory using an amplifier with this output rating.

The term 'high fidelity' (usually abbreviated to 'hi-fi') came into use in the 1930s, when attention was being turned to amplifiers and loudspeakers that should reproduce speech and music, especially the latter, with as much faithfulness as possible to the original sounds. The amplifier is, of course, only a link in the chain between the sound, a microphone and a broadcast or gramophone record. As we have seen earlier, the frequency range of AM broadcasting was necessarily restricted in accordance with the channel spacing of transmitters, and well-recorded 78 r.p.m. records, with a range of between 50 c/s and 8000 c/s could comfortably provide better sound quality when reproduced properly. This superiority was extended further when the Decca Record Co. introduced its 'FFRR' (Full Frequency

Range Recording) system in 1947, making it possible for 78s to reproduce from 20 c/s to 15 000 c/s. To complement the records Decca introduced new magnetic pick-ups with similarly extended frequency response and some excellent radio-gramophones and record players. When long playing (LP) records arrived a few years later the upper frequency response was enhanced still further, calling for even better pick-ups and amplifiers. Throughout the late 1940s and 1950s many well-known designers and manufactures produced a succession of continually improved amplifiers, although it must be said once the question of providing adequate frequency response had been settled much of the claimed advances lay in the reduction of distortion. This was carried to the point where improvements were being claimed in fractional percentage figures which makes it doubtful if even a highly trained musician could have told the difference. In fact, it used jokingly to be suggested that many hi-fi buffs spent more time listening for faults that in enjoying the music or song provided by records!

As regards output powers, G.A. Briggs, a leading authority on sound reproduction, observed that in most domestic listening conditions no more than 0.25 W was required, but amplifiers needed to have a much greater reserve power to preclude distortion on very loud passages of music. Ten watts maximum would have been considered more than adequate.

Design features

The design of PA and hi-fi amplifiers follows similar lines, bearing in mind that the former do not need to have as wide a frequency response as the latter, the circuitry of which is likely to be somewhat more complex. Designing amplifiers is an exceedingly involved subject, involving as it does considerations of such things as harmonic and intermodulation distortion and of different classes of operation such as Classes A1 and A2, and AB1 and AB2. In quoting examples we shall confine ourselves to designs using what are almost certainly the best known and most popular output valves, the PX4 and PX25 directly heated power triodes and the KT66 beam-tetrode. The first two date from the late 1930s and the third from 1938, and all remain extremely desirable today, it being not unusual to see good second-hand pairs of these valves changing hands at between £50 and £150.

Output powers

The PX4 is capable of giving up to 13.5 W in Class A push-pull with an anode voltage of 300 V, this being more than adequate for domestic purposes. The distortion at maximum output is 2.5%, and is proportionally lower at less output. If the anode voltage is reduced to 250 V the maximum output is 9 W with a distortion of 2%. Both these outputs are achieved with virtual cathode biasing via separate centre-tapped heater windings for each PX4.

A pair of PX25s Class A will give 15.5 W with an anode voltage of 400 V or 20 W with 500 V applied. The respective distortion figures are 2.5% and 2%; note that the figure is smaller for the larger output. This is achieved with virtual cathode biasing. If fixed bias is employed and the valves operated in Class AB1 with fixed grid bias the output with 525 V nominal on the anodes is raised to 26 W, at the cost of slightly more distortion – 14%.

The KT66 is capable of giving much greater power outputs than the PX25 but with somewhat higher distortion figures. For instance, in Class A push-pull with cathode bias and 250 V on the anodes, a pair will give 17 W, with a distortion of 4%. At 390 V on the anodes the power is raised to no less than 30 W, with 6% distortion. The use of Class AB1 with fixed grid bias and a nominal HT of 510 V the power rises still further to 50 W, with a better distortion figure of 5%.

The high power applications of the KT66 are largely confined to PA amplifiers. For hi-fi work the valves may be connected as triodes (anode strapped to screen grid) to give less distortion although at much reduced power. In Class A with

cathode bias and 250 V on the anodes a pair will give 4.5 W with a distortion of 2%. Raising the anode voltage to 400 V results in an output of 14.5 W with 3.5% distortion.

The Osram PX4 push-pull amplifier (Figure 29.1)

The first valve (L63) operates as a 'concertina' phase splitter. As both the 10 000 Ω anode feed resistor and 1000 Ω cathode bias resistor are by-passed the respective load resistors, both 20 000 Ω, are balanced. The anti-phase signals are fed to the grids of two more triodes operating as conventional amplifiers and used to drive the output valves by choke-capacity coupling; note the use of a centre-tapped choke. There are 10 000 Ω stoppers in each PX4 grid and 100 Ω stoppers in each anode. Each PX4 has its own heater winding on the mains transformer, centre tapped to return to chassis via a 1000 Ω bias resistor by-passed by a 50 μfd condenser.

The Osram PX25 (Class A) amplifier (Figure 29.2)

Only the output stage is shown, with the valves driven by the centre-tapped secondary of a transformer the primary of which is in the anode circuit of a preceding driver valve, typically another L63. Ten thousand ohms grid stoppers are again fitted but the anode stoppers are omitted. Virtual cathode bias is again employed with centre-tapped heater windings for each PX25.

The Osram PX25 (Class AB1) amplifier

The driver valve (L63) again works into a transformer, this time with two separate secondary windings for the PX25 grids. This enables the negative grid bias to be applied individually and to be balanced by the two 50 000 Ω potentiometers shown below the lower PX25. In the initial setting-up procedure each PX25 would be tried on its own and the bias altered until each drew the same anode current. Note the D41 double-diode shunted across the grid circuits of the two PX25s. This unusual feature is intended to provide a low impedance path to earth should potentially harmful positive voltages appear on the grids, and it is not likely to be encountered in commercial amplifiers. The negative bias voltage is obtained from a tapping on one side

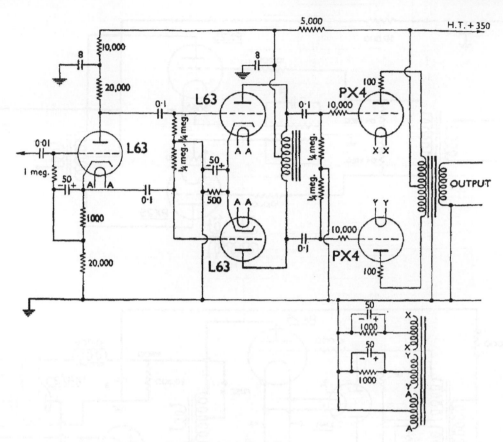

Figure 29.1 The circuit of an amplifier using PX4s in push-pull

of the main HT winding on the mains transformer, taken to a half-wave rectifier working 'backwards' with the output taken from its anode to provide –70 V after smoothing by a 50 000 Ω resistor and two 0.1 μfd condensers.

Note that the HT smoothing uses a 'swinging' choke, which helps to stabilise the output voltage. The 20 000 Ω bleed resistor connected from HT+ to chassis maintains a necessary minimum current through the choke.

The Williamson KT66 amplifier (Figure 29.3)

The Williamson amplifier was conceived especially to take advantage of the new ffrr records, and its performance shows how that aim was fully realised. Negative feedback from the secondary of the output transformer to the cathode of V1 helps the achievement of a very nearly constant frequency response from 10 c/s to 20 000 c/s.

The first AF amplifier valve (V1, L63) is directly coupled to the concertina phase splitter (V2, L63), the anode of the first being connected to the grid of the second. In normal circumstances the presence of the 100 V anode voltage of V1 on the grid of V2 would destroy it, but due to its operation with a large value cathode load resistor (R5, 22 000 Ω) the cathode voltage is 105 V, so the grid is actually at –5 V with respect to it. Anti-phase outputs from V2 are coupled via 0.05 μfd condensers to the grids of the two driver triodes (V3, V4, L63). These in turn are resistance-capacity coupled to the grids of the two output valves (V5, V6, KT66), and the inclusion of the potentiometer (R12, 25 000 Ω) between the two anode load resistors for V3 and V4 enables very accurate balancing of the drive voltages. Used in conjunction with the potentiometer (R17, 100 Ω) and variable resistor (R21, 100 Ω) in the common cathode circuit of V5 and V6 extremely good overall balance of the output stage may be achieved.

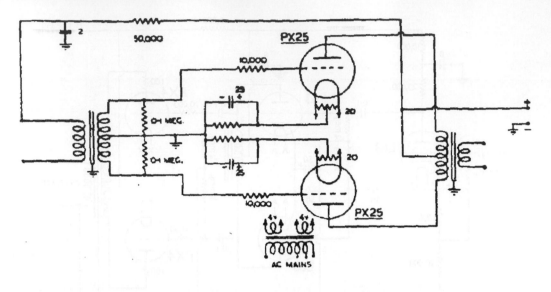

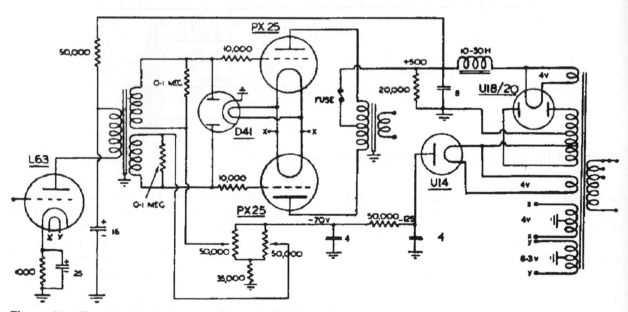

Figure 29.2 Top: the circuit of an amplifier using PX25s in Class A push-pull. Bottom: the circuit of an amplifier using PX25s in Class AB1 push-pull. Note the separate rectifier used to provide negative grid bias

It will be seen that V5 and V6 are operated as triodes by having their screen grids and anodes strapped via 100 Ω resistors. Used in this manner they produce some 15 W of output power at very low distortion.

Had the KT66s been operated as beam tetrodes the output power of this amplifier easily could have been doubled at the price of somewhat reduced frequency response and a higher per-

centage of distortion, neither of which would have been of great significance for PA work.

Preamplifiers (Figure 29.4)

It was usual for a hi-fi amplifier to be fed from a discrete pre amplifier incorporating facilities for tone control and for switching in different signal

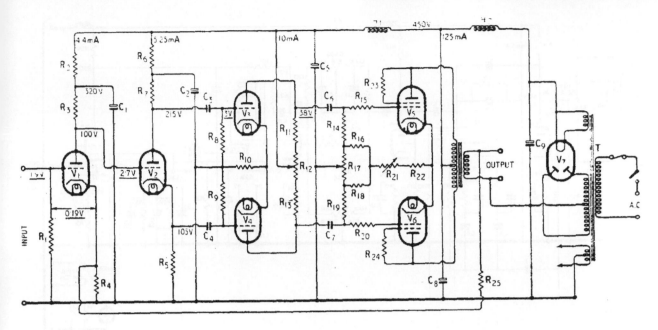

R_1	1 MΩ ¼ watt ± 20 per cent	R_{15}, R_{20}	1,000 Ω ¼ watt ± 20 per cent	C_9	8 μF 550 V. Wkg.
R_2	33,000Ω 1 watt ± 20 „	R_{16}, R_{18}	100Ω 1 watt ± 20 „	C_9	8 μF 600 V. Wkg.
R_3	47,000 Ω 1 watt ± 20 „	R_{17}, R_{21}	100Ω 2 watt wire-	CH$_1$	30 H at 20 mA (Min.)
R_4	470 Ω ¼ watt ± 10 „		wound variable.	CH$_2$	10 H at 150 mA (Min.)
R_5, R_6, R_7	22,000 Ω 1 watt ± 10 „	R_{22}	150Ω 3 watt ± 20 „	T	Power transformer.
R_8, R_9	0.47 MΩ ¼ watt ± 20 „	R_{23}, R_{24}	100 Ω ½ watt ± 20 „		Secondary 425-0-425 V.
R_{10}	390 Ω ¼ watt ± 10 „	R_{25}	1,200 √speech coil impedance,		150 mA (Min.) 5 V. 3A, 6.3
R_{11}, R_{13}	39,000 Ω 2 watt ± 10 „		¼ watt.		V. 4A. C.T.
R_{12}	25,000 Ω 1 watt wire-	C_1, C_2, C_5	8 μF 450 V. Wkg.	V_1 to V_4	L63
	wound variable.	C_3, C_4	0.05 μF 350 V, Wkg.	V_5, V_6	KT66.
R_{14}, R_{19}	0.1 MΩ ¼ watt ± 20 „	C_6, C_7	0.25 μF 350 V, Wkg.	V_7	U52.

Figure 29.3 The circuit and component values of the Williamson amplifier. (Reproduced by courtesy of *Wireless World*)

sources. There might, in addition be high or low note filters and the means of compensating for the different kinds of recording techniques used for 78 and LP records, British and American. General duty PA amplifiers usually had a built-in pre-amplifier which included simplified tone controls and 'mixing' facilities whereby inputs from say, a microphone and a gramophone pick-up might be faded in and out as required.

A very well-known manufacturer of hi-fi amplifiers once remarked that he built his equipment to have a flat frequency response over a very wide range and that technically speaking no tone controls were either necessary or desirable, but he had to include them because the public demanded them. This is a very valid point which must have struck other designers but nevertheless all hi-fi preamplifiers incorporate some kind of treble and

bass cut or boost controls, many of them being based on original designs by P. Baxandall.

A good example of a commercial preamplifier is provided by the Pye model PF91A (1955), which used two ECC40 double-triodes. Inputs were provided for a microphone, crystal or magnetic pick-up cartridges and inputs from a radio unit or a reel-to-reel tape recorder. The first valve (V1A) acted as a straight amplifier, but its output was coupled to the second valve (V1B) via an 'equalisation' network which modified the response to suit the different inputs.

The output from V1B was resistance-capacity coupled to a third triode (V2A) via treble and bass controls which could modify the response to each over wide limits. V2A's output was again RCC coupled into a switched filter unit which could limit the overall high note response to 4 kc/s, 7 kc/s

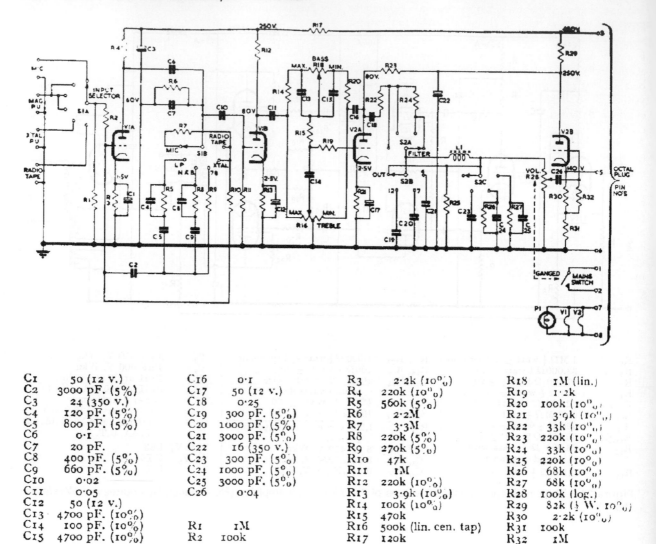

Figure 29.4 Top: the circuit of the Pye PF91A preamplifier. Bottom: the condenser and resistor values

C1	50 (12 v.)	C16	0·1	R3	2·2k (10%)	R18	1M (lin.)
C2	3000 pF. (5%)	C17	50 (12 v.)	R4	220k (10%)	R19	1·2k
C3	24 (350 v.)	C18	0·25	R5	560k (5%)	R20	100k (10%)
C4	120 pF. (5%)	C19	300 pF. (5%)	R6	2·2M	R21	3·9k (10%)
C5	800 pF. (5%)	C20	1000 pF. (5%)	R7	3·3M	R22	33k (10%)
C6	0·1	C21	3000 pF. (5%)	R8	220k (5%)	R23	220k (10%)
C7	20 pF.	C22	16 (350 v.)	R9	270k (5%)	R24	33k (10%)
C8	400 pF. (5%)	C23	300 pF. (5%)	R10	47k	R25	220k (10%)
C9	660 pF. (5%)	C24	1000 pF. (5%)	R11	1M	R26	68k (10%)
C10	0·02	C25	3000 pF. (5%)	R12	220k (10%)	R27	68k (10%)
C11	0·05	C26	0·04	R13	3·9k (10%)	R28	100k (log.)
C12	50 (12 v.)			R14	100k (10%)	R29	82k (½ W. 10%)
C13	4700 pF. (10%)			R15	470k	R30	2·2k (10%)
C14	100 pF. (10%)	R1	1M	R16	500k (lin. cen. tap)	R31	100k
C15	4700 pF. (10%)	R2	100k	R17	120k	R32	1M

or 12 k/cs (a fourth switch position cut out the filters completely).

The output from the filter unit was taken to the grid of the fourth valve (V2B) which was operated as what is called a cathode follower. In this the anode is taken directly to HT and the output obtained from a load resistor in the cathode circuit. This results in 100% negative feedback being applied to the stage, endowing it with excellent frequency response with very low distortion. In addition the input impedance is high and the output impedance low, making it useful for working into a main amplifier via long connecting cables which otherwise might be susceptible to hum pick-

up. The price for all these advantages is a stage gain of less than unit – typically 0.9 – but this is of no importance when the preceding stages have an abundance of gain.

Incidentally, Pye's advertised claim for the PF91A and its associated PF91 main amplifier (which used KT66s strapped as triodes) was an output of 12 W with less than 0.1% distortion at 1000 c/s, and with an overall frequency response of between 2 c/s and no less than 160 kc/s! Although, of course, far beyond the upper limit of audible sound, the very high treble response was considered to be valuable for the correct reproduction of transients in music.

Repairing amplifiers

Fortunately the technical niceties of amplifiers are academic as far as repair work is concerned because it only involves the original designer. All you need to remember is that amplifiers are only extended version of the output stages used in radio receivers, the servicing of which we have discussed. As with radio receivers the majority of faults will be found to be due to leaky condensers, abnormally high resistors and bad previous repair work. There is absolutely no need to fear that amplifiers must be difficult to repair, provided they are tackled logically, stage by stage in the manner already advocated in this book, with the knowledge of what the various valve electrode and other voltages ought to be. At this point it must be mentioned that we ought to differentiate between grid bias provided by resistors in the cathode circuit of the valve and that provided by a source of negative bias, so from now on we shall refer to the first as cathode bias and only to the second as grid bias.

It is usual for there to be at least two stages preceding the output stage to provide initial amplification of signal from gramophone pick-up or radio and then to act as a phase splitter to drive the output valves. There may, however, be several valves in each of those stages, especially in high power amplifiers where the output valves require a considerable amount of signal voltage to drive them. In many cases the main amplifier is itself preceded by a 'preamplifier' which as well as building up the signals also provides for the switching in of different inputs and for controlling the tone. We shall look at these units later, after we have examined main amplifiers.

Power supply sections

These follow the same lines as for radio receivers except that the HT voltages involved are likely to be considerably higher. Whilst some small power amplifiers may manage with an HT line of 300 V, larger units may well have lines of 400 V or even more. The amount of anode current drawn by the output valves in the bigger amplifier also is higher, with the HT winding on the mains transformer and the rectifier commonly being capable of supplying 250 mA. Choke-capacity smoothing is the norm although you may occasionally encounter the swinging choke method. More often than not there will be several stages of smoothing providing different levels of voltage for the various valve

stages. Older high power amplifiers often employed large paper condensers with working voltages of around 600 V in the reservoir and first smoothing positions; it is not unusual to find these still in perfect order after fifty and more years of use.

On the LT side there will usually be more windings than are found in ordinary radio mains transformers. When the PX4 (4 V @ 1 A) and PX25 (4 V @ 2 A) are used in the output stage each may have its own filament supply with one or more 6.3 V windings for the other valves. KT66s require 6.3 V @ 1.27 A each, usually supplied by a single winding but with one or more others to supply the other valves. A large rectifier such as the U52 draws 2.25 A @ 5 V, so the overall heater wattage of an amplifier may well amount to 40 W. With the HT supply wattage easily reaching 100 W, the mains transformer is likely to be vary large and very expensive to replace. It thus behoves you to take extra special care to ensure that such items as grid coupling condensers are changed at the least sign of leakage.

HT fuses

To their credit, many amplifier makers fitted a fuse in the return circuit of the HT winding centre tap to HT–. In the event of heavy HT current being drawn this offers far better protection to the HT winding than does a fuse in the mains transformer primary, which may not blow until the winding has been irreparably damaged. You are strongly recommended to install an HT fuse if none is already fitted. Use a 'quick-blow' fuse which will blow long before there is a chance of the winding being overheated.

HT– switches

Amplifiers used intermittently, as for occasional public address announcements, are usually silenced between periods of operation by opening the return circuit from the centre tap of the HT winding to HT– by means of a switch. This permits all the valve heaters to run at normal temperature and gives an 'instantly-on' facility. If HT should inexplicably be absent in such an amplifier, check that the switch is operating properly.

Negative bias supplies

Larger amplifiers may have a small separate winding to provide about 50–100 V for a negative bias line. Half-wave rectification is usually employed with resistance-capacity smoothing. The current drain is almost infinitesimally small so the rectifier, which may well be a small metal type, will give very long service. If it should have to be replaced bear in mind the advice on the subject given earlier in this book.

Output transformers

These again will be larger than in radio receivers due to the heavy anode currents they have to carry and the need for good low frequency handling. The secondary windings may be wound for various values of loudspeaker impedance. At one time the usual impedance for large hi-fi speakers in this country was 15 Ω, until the American 16 Ω became generally adopted. The two may be interchanged without more than academic problems. PA amplifiers often have a special '100 V' (500 Ω) output which can be carried over long lengths of connecting cables without incurring losses. Each loudspeaker connected into the 100 V line has its own step-down transformer, normally with a multi-tapped secondary enabling from about 0.5 W to 5 W to be fed to the speaker as required by its situation. Be warned that without modification ordinary 15 Ω or 16 Ω speakers cannot be used directly on 100 V lines, nor 100 V speakers used on 15 Ω or 16 Ω outputs.

Faults on and around output valves

Because the output valves in amplifiers are generally more powerful than in ordinary radio sets, and are capable of passing heavy anode currents, it is even more essential that they shall run with the correct grid bias. As always, check for leaky grid coupling condensers, and if offenders have to be changed, check any cathode bias resistors and by-pass condensers. In some amplifiers a single cathode resistor serves for both valves, in others there are individual resistors. In either case the values will have been chosen with care and must be adhered to if replacements have to be fitted. Some amplifiers used wire-wound resistors because it is easier to achieve close tolerance with this type. They have an added advantage that they will stand a considerable amount of abuse should they be forced to pass heavy current. It is possible to find examples that have run so hot that the ceramic covering has burned off and the resistance wire is blackened, yet they remain electrically sound. In fact, this ability to stand overloading does make it possible for the

cathode voltage to rise to such an extent that any associated by-pass condenser receives far too much voltage across it and fails explosively. The evidence of such an event is unmistakeable, with the underside of the amplifier chassis peppered with bits of tinfoil and electrolyte.

Whether bias resistors have to be replaced or not, always check the cathode voltages on the output valves. When separate bias resistors are used markedly different voltages on each cathode will point to one or other of the valves drawing either too much or too little anode current. To cross-check that this is in fact due to a valve fault and not to associated components, change the valves over in their sockets and check the voltages again. Obviously, when a common cathode resistor is employed the voltages must be the same, so remove one valve at a time and check the cathode voltage of that which remains. Then reverse them in their sockets and recheck the voltages. If these tests reveal that one of the valves must be faulty, replace it with a new one and check again. It is in fact desirable that both valves should be replaced at the same time, if this is practicable from either a supply or financial viewpoint.

Preceding stages

Amplifiers tend to have a good deal of HT decoupling which implies that there will be a number of electrolytic condensers that are potential sources of trouble, should they go either short circuit or open circuit. Short circuits will, of course, make themselves obvious by removing HT from the valves they help to feed, whilst open circuits are liable to cause instability and maybe even self-oscillation. If shorted condensers have to be replaced, don't forget to check the associated HT feed resistors.

It is essential to check that anode resistors, particularly in the case of phase-splitting stages, are of the correct value, as discrepancies will cause imbalances in the drive to the output valves. Use close tolerance replacements of ample wattage rating. Problems with valves are not so likely to be associated with low emission as with heater-cathode leakage causing hum and with microphony. Replacement is the only cure for both.

Preamplifiers

Again the usual advice about checking components applies, with the additional note that in some preamplifiers the required accuracy of anode feed resistors is such that close tolerance (1%) types are fitted. The same must be used if replacements have to be made.

Most preamplifiers will be connected to the main amplifiers via a multi-core cable and plugs and sockets at either end. When the two units are fitted permanently in, say, a cabinet there ought to be no danger of intermittent contacts developing but the fact remains that they do happen on occasion. Beware particularly of poor earthing of screened cables and of screen to inner shorts on the same.

Be very wary of using a mixture of two makes of preamplifier and main amplifier, even if they do appear superficially to be compatible. HT and heater supplies are usually drawn from the main amplifier and the writer once saw the mains transformer in an amplifier burn out due to the fact that the preamplifier pressed into service had one side of the valve heaters taken directly to chassis, whereas the amplifier itself had a centre-tapped heater winding. The result of this was that half of the heater winding was shorted out at the pre-amplifier end of the connecting cable. The resistance of the latter stopped this from being a dead short and the two units actually worked for several hours before the transformer erupted in an expensive pall of evil smelling smoke.

Common abbreviations

Notes: this list is not exhaustive but it does cover most of the terms used in the vintage radio era and which will be found in literature and receiver circuit diagrams of the period. In some cases alternatives have been given for abbreviations but others might well be found to have been used. In particular there seems to have been a fairly free interchange of upper and lower case, or of normal or italic letters in different publications (sometimes even in the same!) but as with certain other ambiguities the context will usually show which particular interpretation should be used. Valve makers' data books used a large number of abbreviations, of which are shown here only the few that are also used generally in vintage radio.

The writer makes no apology for omitting more modern abbreviations which are inappropriate to vintage radio.

A or AE   Aerial. Note: the old spelling used a diphthong, viz. Ærial.
AE   See above.
A or an   Anode. The context will usually prevent confusion with aerial.
'A' battery   The American term for an LT battery.
AC   Alternating current.
AC/DC   Receiver suitable for either AC or DC mains.
AF   Audio frequency
AFC   Automatic frequency control.
A or Amp   Ampere.
Ah   Ampere hour. In vintage radio this is used exclusively to indicate the capacity of an accumulator.
AM   Amplitude modulation.
Amp   Amplifier.

Amps   Amperes.
ANT   Antenna, the American term for aerial.
AVC   Automatic volume control.
AVO   The trade name of the best-known test meter, supposedly derived from Amps, Volts, Ohms and used by radio engineers since the 1920s. Now virtually a generic term for testmeters.

'B' battery   The American term for an HT battery.
BC   Broadcast (band), the American term for medium waves.

C   Condenser or capacity.
'C' battery   An American term for a grid bias battery.
cps, c/s or cs   Cycles per second.
CRT   Cathode ray tube
CW   Continuous wave.

dB   Decibel.
DC   Direct current.
DDT   Double-diode-triode (valve).
Det   Detector.

E   Earth.
E Used to indicate voltage, e.g. $E_a = 90$ V.
EHT   Extra high tension.
EMF   Electromotive force.

f or f Frequency.
FC   Frequency changer (valve).
FM   Frequency modulation.

g   Grid (of valve). May be followed by a number.
gc   Conversion conductance (in frequency-changer valve).

GB   Grid bias.
gm   (Valve) mutual conductance.
GND   Ground, the American term for earth.

H   Henry.
HF   High frequency.
HT   High tension.

I or *I*   Used to indicate current (see below).
I_a Anode current (valve).
IF   Intermediate frequency.
IFT   Intermediate frequency transformer.
I_{g2} Screen grid current (valve).
I_k Cathode current (valve).

K   Cathode (of valve).
kc, kcs or kc/s   Kilocycles per second.
$k\Omega$ kilohm (= $1000\,\Omega$). See also $M\Omega$ for American usage.

γ Wavelength, e.g. γ = 300 metres.
L   Inductance.
LF   Low frequency.
LO   Local oscillator (in superhet).
LS   Loudspeaker.
LT   Low tension.
LW   Long wave.

μ (Valve) amplification factor. Also used generally to indicate a subdivision by 1 000 000.
M   metres.
mA, ma or M/a   Milliamp – one thousandth of an ampere.
Ma/V   Milliamps per volt (in valve).
μA   Microamp.
MCW   Modulated continuous wave.
mfd or μfd   Microfarad.
MIC or mic   Microphone.
mmfd or $\mu\mu$fd   Micro-microfarad (= picofarad).
MSW   Medium short waves (band: also known as trawler band).
mV   Millivolt – one thousandth of a volt.
μV   Microvolt – one millionth of a volt.
MW   Medium waves.
$M\Omega$ Megohms (= millions of ohms). Note: in American radio circuits M indicated not meg but mil and $1\,M\Omega$ = $1000\,\Omega$.

Neg   Negative, as in voltage.

O   Sometimes used to indicate ohms instead of Ω when the printer did not have a set of Greek letters!

OC or o/c   Open circuit – used as adjective, noun or verb in respect of a break in a circuit.
Osc   Oscillator.

P   Plate (in valve) or Power.
PA   Public address.
PB   Pushbutton.
PD   Potential difference. Essentially the same as voltage but is handy for describing the change of voltage from one point in a circuit to another.
Pen   Pentode.
pfd   Picofarad (= $\mu\mu$fd).
PM   Permanent magnet.
Pos   Positive, as in voltage.
Pot   Potentiometer.
PU   Pick-up (gramophone).

QPP   Quiescent push-pull.

r_a or r_A Anode resistance = impedance of valve.
R or *R*   Resistance or resistor.
RA   Recommended value of load resistance or impedance for valve.
RCC   Resistance-capacity coupled.
Rec   Receiver (sometimes Rectifier).
Rect   Rectifier.
RF   Radio frequency.
R_g Recommended or actual value of resistor used to return grid of valve to chassis or cathode.
R_{g2} Recommended or actual value for resistor used to supply voltage to screen grid of valve.
R_k Recommended or actual value of cathode bias resistor for valve.
RMS or r.m.s   Root mean square (value of AC voltage).
RT or R/T   Radio telephony.

S or Sw   Switch.
SC or s/c   Short circuit. May be used as adjective, noun or verb in respect of a breakdown in insulation in a circuit.
SG   Screen grid.
Sig. gen.   Signal generator.
Spkr.   (Loud) speaker.
SW   Short wave.

TI or t.i.   Tuning indicator.
Trans.   Transformer.
TRF   Tuned radio frequency.

UHF   Ultra high frequency.
USW   Ultra short waves.

V   Valve.

V or v (following numeral)   Volt(s).

V_a Recommended or measured anode voltage in valve.

VC or vc   Volume control.

V_f Filament voltage (valve).

V_{g2} Recommended or measured screen-grid voltage in valve.

Vh   Heater voltage (valve).

VHF   Very high frequency.

VT   Vacuum tube, the American term for valve.

W   Watt(s).

W/C or wc   Wave change (switch).

X Used to indicate reactance, e.g. $X = 400\,\Omega$.

Z Used to indicate impedance, e.g. $Z = 400\,\Omega$.

Ω Ohms. Note: prior to about 1930 the lower-case φ was used to indicate ohms and Ω was used as for millions of ohms or megohms. See also MΩ.

Greek and Latin prefixes used to indicate the number of electrodes in a particular valve.

Note: this does not indicate the number of base connections.

Diode   Two electrodes.
Triode   Three electrodes.
Tetrode   Four electrodes.
Pentode   Five electrodes.
Hexode   Six electrodes.
Heptode*   Seven electrodes.
Octode   Eight electrodes.

Two or more prefixes may be combined to indicate multiple valves, e.g. triode-hexode.

Ranges of frequencies commonly used in vintage radio

Audio frequency (AF) is normally taken as the range of frequencies audible to the human ear, i.e. from about 16 c/s to 16 000 c/s†. In vintage radio low frequency (LF) may be taken to mean the same.

High frequency (HF) may be taken as a general term for anything above about 16 000 c/s, i.e. above AF. The alternative radio frequency (RF) is more

*Also known in America as pentagrid.
† Some sources suggest that certain individuals may be able to hear frequencies as high as 30 kc/s.

or less synonymous in vintage radio, e.g. HF amplifier is the same as RF amplifier. In receivers designed specifically for commercial or military communications extra low frequency (ELF) may be used for frequencies below about 30 kc/s.

Intermediate frequency (IF) as used in super-heterodyne receivers is not in itself a term for a specific group of frequencies but for the particular frequency used in a particular. However, the choice is limited and 99.9% of the IFs used in vintage AM receivers lie within the ranges of 100 kc/s to 140 kc/s or 450 kc/s and 470 kc/s. The odd exceptions included 175 kc/s, 265 kc/s and 360 kc/s, but others may have been used. Vintage FM receivers nearly all used a 10.7 mc/s IF but again the occasional oddity may be found.

The definition of very high frequency (VHF) and of ultra high frequency (UHF) has changed considerably over the vintage years. Before about 1940 anything over 30 mc/s was considered to be VHF, whilst above 60 mc/s was UHF. By about 1944 VHF covered up to around 100 mc/s and UHF up to around 200 mc/s. Ultra short wave (USW) covered more or less anything above 30 mc/s; for instance it was usual to refer to the original BBC 405-line television service on 45 mc/s as being USW.

Some obsolete radio terms which may be encountered in old literature

Anode battery   An HT battery.

Gramophone attachment   A pick-up.

Grid leak   A high value resistance used to return the grid of a valve to filament, cathode or chassis.

Mansbridge condenser   A type of paper condenser noted for having a good capacity to size ratio.

Note magnifier   An AF amplifier.

Power valve   An output valve.

Telephone condenser   A tone control condenser shunted across a loudspeaker.

V   Indicates the detector (it may have stood for valve) and is a formula once used to describe receivers, e.g. a 1-V-2 set had four valves, one prior to the detector and two following it.

An 0-V-0 was a one-valve set.

Wet battery   An accumulator.

X   Used widely as an abbreviation for the first few letters of a radio term which could be guessed from what followed. For instance, Xtal meant Crystal, Xmitter meant Transmitter and Xformer meant Transformer.

Some colloquialisms used in vintage radio

Belt   Electric shock.

Bike   A cycle, e.g. the UK mains frequency of *fifty bikes*.

Bottle   Valve.

Burn up   The disastrous failure of a component resulting in its being reduced to ashes; often the outcome of a dead short (q.v.).

Cs and Rs   Condensers and resistors.

Dead short   A zero-resistance short circuit of a potentially calamitous nature (*cf.* Short).

Deck   Chassis.

Diddlyode   Diode.

Fils   Filament(s).

Freak   Frequency changer.

Gain pot   An RF gain control (*cf.* Pot).

Genny   Signal generator.

Iffy   IF transformer.

Leak, leaky   An unintentional resistance path of potentially harmful nature, e.g. *a leaky condenser* is one that permits the passage of DC through it. Not to be confused with the completely intentional grid leak (q.v.).

Lining up (job)   Realigning the tuned circuits of a receiver, esp. a superhet.

Mike   Microphone.

Mikey (component, esp. a valve)   Emitting a ringing sound when tapped or subjected to vibration.

Mod. hum   Modulation hum, a 50 c/s or 100 c/s hum that appears to be tuned in with a radio signal.

Neon   A screwdriver incorporating a small neon lamp to give indications of voltage, esp. for checking the chassis of AC/DC receivers.

Open   Open circuit.

Pot   Potentiometer. Not to be confused with tea pot, one of the most essential items in any workshop.

Puffer   Small condenser of picofarad value.

Reccy   Rectifier.

Short   Short circuit, e.g. *a short on the HT line*, but not necessarily of a calamitous nature, such as in *high resistance short* (*cf.* Dead short).

Soak (test)   Leaving a receiver running in the workshop for a very long time to see if anything fails.

Stopper   A low value resistor inserted into the grid, screen grid or anode circuit of a valve to prevent self-oscillation.

Terminal (screwdriver)   Used in certain areas only to indicate a narrow bladed screwdriver for use with electrical terminals.

Tranny   Transformer, e.g. *mains tranny, auto-tranny*.

Trolly   Electrolytic condenser.

Wob, Wobbulator (frequency modulated oscillator).

Part 3

Appendix 1

Intermediate frequencies

This list covers models going back to the first commercial superhets and has been submitted, where possible, to the makers for checking. Frequencies thus: 465, 473, are alternatives, but 123–127 indicates the circuits should be staggered over the band indicated. Sometimes the frequencies for each circuit in a 'staggered' set are shown thus: 127–123–123–127.

ACE	kHz
RG3	470
RG5	427
RG6	427
S6	125
SH6	125
RG7	427
RG8	470
RG9	427
AW35	470
AW53	427
AW53B	427
AW73	427
AC85	470
AW94	427
AW115	470
AW563	427
AC939	450
A50	465

AERODYNE	
Aerogram	125
Aeromagic	125
Cardinal	125
Falcon	125
Silver Wing	125
Swallow	125
42	125
47	125
50	125
53	125
54	465
56	125
58	125
63	465
73	125
100	117.5
105	465
110	117.5
115	465
135	465
290	465
291	465
295	465
300	117.5
301 AC	460
302 AC/DC	460
305	117.5

ALBA	
AC superhet	473
5V Bat. SH, 1934	473
Clipper	470
35	470
40 Universal	370
57 AC and Universal	117.5
67	473
68 AC and Universal	117.5
78	473
79 AC and Universal	117.6
90	465
98	365
230	117.5
315	117.5
320	460
330	117.5
335	460
340	470
450	117.5
455	460
461 AC	460
462 AC/DC	460
510	470
540	117.5
550 AC and Universal	117.5
605	465
610	470
615	117.5
620	117.5
625	117.5
635	460
640	117.5
650	117.5
660 AC and Universal	117.5
670	117.5
698	365
710	470
725	117.5
730	470
740	117.5
745	470
755	470
770 AC and Universal	117.5
790	465
798	365
805	465
810	470

815	117.5
820	117.5
825	117.5
830	470
835	460
845	470
850	117.5
855	470
870 AC and Universal	117.5
880 AC and Universal	117.5
890 AC and Universal	117.5
905	465
910	117.5
920	117.5
930	460
970	117.5
990 and Universal	117.5

ALLWAVE	
Standard Superhet, 1935	465
Standard Superhet RG	465
Tallboy RG, 1935	465
Ambassador 6778	465

AMPLION	
Radiolux Superhet	110
Radiolux Superhet RG	110

ARMSTRONG	
5V 7 stage	110
5V 8 stage	110
RF/PP	465
RF/PR	465
2B/PR	118
2B/T	118
3NBP/8	457
3NBP/8 Late model	470
3NBP/10	427
3WT/PB	427
AW3/PB	427
AW/38	465
4B/PR	118
4B/T	118
AW/36	465
AW/59	470
AW93PP	427
RF94PP	427

AW125PP	470
SS10	465
3NWT	450
3NBP/T	427
U3NBP/T	427

ATLAS	
758	117.5
A13	126
A17	126
A24	126

BEETHOVEN	
Baby Grand	450.5
Little Prodigy AC	450.5
Little Prodigy, Bat.	450.5
Twin Speaker, All-electric Superhet	118
AC40	450.5
AC42	450.5
B43	450.5
AC77	118
B88	118
PBA201	450.5
AD303	450.5
AD404	450.5
RG717	118
AC720	450.5
B730	450.5
AC740	450.5
PBB750	450.5
AD770	450.5
PBA780	450.5
RG827	118
PBA820	450.5
B848	450.5
AC852	450.5
909	450.5
909AC	450.5
RG938	450.5

BELMONT	
520	465
525	465
530 AC-DC Midget	456
541	456
544	465
545	465

BELMONT – *cont.*

555	465
556	465
570P	465
600	465
625	465
650	465
700	465
720	465
721	465
746	465
755	465
760P	465
770	465
780	465
781	465
800	465
820	465
821	465
845	465
856	465
860P	465
900	465
1100	465
1150	465

BENSON

AWP Midget Portable	470

BLUE SPOT

Aristocrat	465
A67	465
A68	465
A69	465
AC5	110

BRUNSWICK

BCA/01	456
BCA/1	456
BCW/01	456
BGA/01	456
BGA/1	456
BGA/1E	456
BGA/2	456
BGA/3	456
BGCA/01	456
BGCA/	456
BGCA/3	465
BGU/01	456
BPU/1	465
BTA/01	456
BTA/1	456
BTA/1E	456
BTA/2	456
BTA/3	465
BTB/1	456
BTU/01	456
39CGM	465
39EH	120
39TGM	465
40	465
40U	465
42	380
42D	380
43D	380
45	380

47	380
47U	380
50	465
51	465
54	465
56	465

BTS

Trophy 5	465

BURGOYNE

AW47	473
AWS	473
AWS/G	473
BSH	117.5
DTG	473
Dragon	437
Dragon AC Recorda-graph	473
Dragonette	473
Superhet 5, B5	117.5

BURNDEPT

Ethodyne 209	473
Universal Trans.	130
Universal Superhet	473
201	130
203	473
209	473
210	473
211	473
218	130
225	130
226	130
229	130
231	130
233	450
257	130
259	473
266	473
267	473
276	473
281	473
285	473
290	473
292	450
298	473
299	473
303	473
309	450
312	473
313	473
314	473
315	473
316	473
317	473
318	473
319	473
323	465

BURRELL

4Y Superhet	110

BUSH

DAC1	123
SAC1	123
SUG1	123

SB1	123
TG1	123
SB3	123
SAC4	123
SB4	123
BP5	123
SAC5	123
SAC6	123
SAC7	123
DAC21	123
DUG21	123
SAC21	123
SB21	123
SAC25	123
SAC31	123
SAC35	123
SUG31	123
RG33	465
SSW33	465
SUG33	465
RG37	465
SSW37	465
SUG37	465
RG41	465
SW41	465
BA43	465
DAC43	465
DUG43	465
RG43	465
SUG43	465
SUG43G	465
SW43	465
SB44	123
SW45	465
PB50	465
BA51	465
DAC51	465
DUG51	465
PB51	465
RG52	465
RG52G	465
SUG52	465
BA53	465
DAC53	465
PB53	465
RG53	465
PB55	465
SUG55	465
PB60	465
BA61	465
PB61	465
SUG61	465
DUG62	465
BA63	465
DAC63	465
PB63	465
RG63	465
RG63 Auto	465
RG64	465
RG64	465
SUG64	465
PB65	465
SUG65	465
BP70	465
BA71	465
DAC71	465

DAC73	465
DUG73	465
PB73	465
SUG73	465
RG64 Auto	465
AC81	465
PB83	465
DAC81	465
BA81	465

CAC

Austin Superhet AC	110
Austin Bat. 5	110

CAMEO

AC Cameo	430
All Wave	430
Atom	430
Bookcase RG	430
Cameo	430
Cameogram	430
Emergency	430
Super Midget 4	430
ABX	430
ARP	430
AWP	430
P	430
RP	430
RP9	430
TW	430

CIVILIAN WAR-TIME RECEIVERS

Battery model	460
AC model	460

CLIMAX

AC5	115
AC-DC5	115
S4AC	121
S5	115
534	111

COLUMBIA

356	128–125–125–125
357	125
358	125
380	125
621	125
631	128–125–125–125
640 and 640A	125.2
1006	125

COSSOR

31	465
32	465
33	465
34	465
35	465
37	465
AD41	465
46	465
47	465
53	465
55	465
56	465

COSSOR – *cont.*			CROSLEY			110	465	AW88	126.5
57		465	Roamio		455	120	465	C88	126.5
61	(SW 1363)	465	A358		455	180	130	UAW88	126.5
62, 62B		465	5C2		181.5	190	130	ADT95	110
63		465	538BT		450	350	130	BT95	110
64, 64B		465	638T		450	400	130	ACT96	130
66, 66A		465	848C		450	405	130	AC97	126.5
67		465	848CU		450	500	130	RG97	126.5
67A		465	848R		450	510	130	AW98	126.5
70		465	848RU		450	520	456	BAW98	126.5
71, 71B		465	848T		450	530	456	ARG107	126.5
72		465	848TU		450	540	456	AW108	460
73		465	1058AR		450	550	130	RG109	126.5
74		465	1058T		450	919	130	AW119	126.5
77, 77B		465				1010	130	UAW119	126.5
81		465	**DECCA**			1111	130	P150	465
82		465	Twin S/het R/GAC6		183	1616	130	PB179	465, 480
85		465	AW3		380	4040	130	PBU179	465, 480
338 and 348 SW only		1563	AW3P		380	4141	130	PB189	126.5
364		128	AW4341		465	4242	130	PBU189	126.5
365		128	AW6		465	4343	130	PB199	480
366		128	AW6V		465			PB279	465, 480
366A		128	Decca-Brunswick 6V			**DRUMMER**		PB289	126.5
374		128	RG (Med W only)		183	M45	117.5	C389	126.5
375		465	AW7		465			ARG399	480
375U		465	AW8		465			RG489	126.5
376B		128	AW9		465	**EKCO**		C501	126.5
385		465	AW10		465	C25	110	TRG502	126.5
394		465	AWD47		380	RG25	110	PB505	477
395		465	AWG16		465	SH25	110	PBU505	477
396		465	ML		465	AC64	110	PBU506	477
397		465	MLB		465	DC64	110	PBU506	477
398		465	ML4		380	AD65	110	PB507, 508	477
438	(SW 1363)	465	ML5 and 42		380	B67	126.5	(If red serial No, 465)	
438U	(SW 1363)	465	ML6		465	BV67	126.5	C509	465
439		465	ML6U		465	AW69	126.5	CU509	465
456AC		465	MLD/3		380	BAW69	126.5	PB510	126.5
456B		465	MLD/5		380	C69	126.5	C511	126.5
464AC		465	PC/AW		465	CU69	126.5	PB515	126.5
483	(SW 1363)	465	PC/ML		465	UAW69	126.5	RG516	126.5
484	(SW 1363)	465	PG/AC		465	AW70	126.5	EX401	126.5
484U	(SW 1363)	465	PG/AW		465	UAW70	126.5	EXU401	126.5
485		465	PG/ML		465	BAW71	126.5	EX402	480
535		128	PG/U		450, 465	AC74	110	A21	477
538		465	PT/AC5		456	B74	110	B25	477
583		465	PT/AW		465	DC74	110		
584		465	PT/BS		465	AD75	480	**EVER READY**	
584U		465	PT/M		125, 465	AC76	130	5001	127
598		465	PT/ML		465	AD76	130	5002	127
634		128 or 134	PT/ML/B		465	AC77	126.5	5003	127
635		128 or 134	PT/ML/U		450, 465	AD77	126.5	5004	127
736		128	PT/U		465	CT77	126.5	5005	127
737		128	PAW5		465	CTU77	126.5	5006	127
836		128	UAW78		465	BAW78	460	5007	127
837		465	Double Decca MB5		380	BV78	460	5008	127
3733 SW only		1563	Portrola		130	C78	460	5011	465
3764		465	Portrola AC/DC 1939		465	UAW78	460	5014	465
3774		465	44		465	RG84	110	5019	127
3783 SW only		1563	55		465	AC85	110	5025	465
3863		465	56		465	B86	130	5029	455
3884		465	Prestomatic		465	AC86	130	5030	455
3952		465	66		465	AD86	130	5031	455
3974		465	77		465	B86	110	5032	455
3974A		465	88		465	RG86	130	5033	455
6864		465	88U		465	AW87	460	5034	455
6874		465	99		465	CTA87	460	5036	455

EVER READY – *cont.*

Model	Page
5038	455
5040	455
5101	452
5103	452
5104	452
5105	473
5117	452
5118	452
5122	473
5132	473
5203	452
5214	452
5215	452
5216	452
5218	452
5219	452
5221	455
5247	452
5263	452
5347	452
5380	452
5381	452

EMERSON

Model	Page
301	455
330	455
331	455
332	455
336	455
351	455
453	455
376	455
400	455
414	455
415	455
419	455
421	455
422	455
425	455
439	455
441	455
461	455
463	455

FERGUSON

Model	Page
Fergusonic	470
Fergusonic AC-DC Batt.	470
101, 101U, 101UX	470
Fergusonic Mains Minor	470
103B	470
104	470
104AC6	470
106	470
350	465
365	465
366	465
378	465
378U	465
545	470
501	465
502	465
502C	465
502RG	465
503	465
503C	465
503CT	465
503RG	465
503RGT	465
503T	465
601	465
602	465
602C	465
602RG	465
603	465
603C	465
603CT	465
603RG	465
603RGT	465
603T	465
701	465
702	465
704	465
705	465
715	465
771	465
772	465
773	465
774	465
775	465
777	465
801	465
802	465
804	465
805	465
815	465
881	465
882	465
884	465
885	465
901	470
901U	470
902	470
902U	470
903B	470
904	470
904U	470
905	470
906	470
907 Fergusonic AC-DC	470
908	470
909	470

FERRANTI

Model	Page
1933/4 Gloria	125
1933/4 Lancastria	125
1933/4 Arcadia	125
1933/4 Battery S/het	125
1934/5 Arcadia AC S/het	125
1934/5 Gloria	125
1934/5 Gloria Arcadia	125
1934/5 Lancastria	125
1934/5 Lancastria Portable	125
1934/5 Lancastria Reflex	125
1934/5 Universal	125
1936/7 Magna AC	125
1935/6 Gloria Auto-gram	125
1935/6 Nova AC	125
1935/6 Nova AC-DC	125
1935/6 Lancastria AC	125
1935/6 Lancastria AC-DC	125
1935/6 Arcadia Cons and R/G	125
1935/6 Arcadia AC	125
1935/6 Nova Batt.	125
1936 All Wave S/het	125
1936/7 Nova 2-wave AC	125
1936/7 Nova All-wave	125
1936/7 Nova All-wave Batt.	125
1936/7 Nova All-wave AC-DC	125
1936/7 Magna All-wave AC	125
1936/7 Magna AC-DC	125
1936/7 Arcadia All-wave AC	125
1936/7 Arcadia Console	125
1936/7 Arcadia R/G	125
A1	135
48B	125
49B	125
139	450
141	450
239	450
241 and 341	450
512AM	450
513	450
513AM	450
514	450
514PB	450
515PB	450
539	450
617PB	450
837	450
1037	450
1037U	125
1137	450
1137B	125
1137U	125
1237B	125
1737	125
1737 Gram	125
2037	125
2337	450
145	450

GAROD

Model	Page
930	456
931 Consoles	456
930D	456
931D	456
930KC	456
931KC Gram	456
1240	456
1240E	456
1240LC	456
1650	456
1650 Console	456
1650LC Gram	456

GEC

Model	Page
3358	110
3440	107
344	107
3442	107
3443	107
3444	107
3445	107
3446	125
3448	107
3449	107
3460	125
3466	125
3480	125
3484	125
3488	125
3540	125
3541	125
3542	125
3544	125
3545	107
3548	125
3550	125
3551	125
3558	125
3558R	125
3566	125
3640	125
3645	125
3646	125
3650	125
3651	125
3658	125
3659	125
3740	125
3745	125
3746	125
3748	125
3750	445
3754	125
3758	445
3760	445
3762	445
3766	445
3780	445
3781	445
3782	445
3788	445
3789	445
3846	456
3850	456
3855	456
3856	456
3857	456
3860	456
3862	456
3865	456
3866	445
3867	456

GEC – *cont.*

Model	IF
3868	456
3880	445
3882	445
3888	445
3889	445
3890	456
3892	456
3910	456
3918	456
3940	456
3942	445
3946	456
3950	456
3955	456
3956	456
3957	445
3960	456
3964	456
3965	456
3966	445
3967	456
3968	456
3969	456
3970	456
3972	445
3978	445
3977	445
3979	445
4010	445
4018	445
4040	456
4045	456
4046	456
4050	456
4051	456
4054	456
4055	456
4056	456
4058	456
4059	456
4060	456
4065	456
4066	456
4070	456
4135	456
4141	456
4157	445
4166	445
4172	445
4173	445
4177	445
4177U	445
4178	445
4178R	445
4179R	445
4227	430
4237	445
4247	445
4242	445
4262	445
4267	445
4337	445
4342	445
4347	445
4362	445
4367	445
4641	456
4650	456
4652	456
4655	456

GILBERT

Model	IF
Kumfe 2	122.5
A50	110
A63	122.5
C64	122.5
Gordon Elf	430
Cameo AC, DC5	430

HMV

Model	IF
146	456
147	456
166	465
340	456
341	456
381	123–127–125–125
404	128–123–128–125.5
425	123–127–125–125
438	128–123–128–125.5
440	128–123–128–125.5
441	125
442	123–127–123–127
443	123–127–123–127
444	125
445	123–127–125–125
446	123–127–125–125
456	465
457	465
458	465
459	125
462	125
463	127–123–123–127
464	126.5–121.5–121.5–126.5
467	120–114–117–117
469	465
470	128–125–125–125
479	465
480	460
481	460
482	465
485	460
485A	460
486	465
487	465
488RG	460
489	465
490	465
491	465
492	465
493	465
494	465
495	465
496	460
497	460
498	460
499	465
505	128–123–128–125.5
512	128–123–128–125.5
523	128–125–125–125
524 and 524A	120–114–117–117
531	125.2
532	125.2
532C	125.2
540	128–123–128–125.5
540A	125
541	125
542	128–123–128–125.5
545RG	123–127–125–125
546	123–127–125–125
570	123–127–123–127
580	122–112–117–117–117–117
581RG	460
582	460
622RG	123–127–125–125
632	123–127–125–125
642	465
645	123–127–125–125
650	465
651	465
653	465
654	465
655	465
656	465
657	465
658	465
659	465
660	465
661	465
663	465
664	465
665	465
666	465
668	465
670	465
671	465
682	465
800 Auto RG	125
801 Auto RG	460
1100	465
1101	465
1102	465
1103	465
1104	465
1105	465
1106	465
1107	465
1111	485
1112	485
1113	465
1200	465
1300	465
1301	465
1350	465
1351	465
1351A	465
1354	485
1355	485
1400	465
1403	465
1404	465
1406	465
1500	465
1501	465
1504	465
1600	465
1601	465
1750	465
5211	485
5212	485
5311	485
5312	485

HALCYON

Model	IF
Briton	130.5
Briton Radiogram	130.5
Nine-Stage AC	110
Royal County	130.5
Seven-Stage Universal	110
Royal Gram and Auto	130.5
A57	130.5
A581	130.5
A5820	130.5
AC5	110
AC6	110
AC7	110
AC7G	110
B691	130.5
GA33	130.5
GR37	130.5
MS6	465
RGA581	130.5
RGCA581	130.5
RGCU6801	130.5
RGU6801	130.5
U57	130.5
U537	130.5
U571	130.5
U573	130.5
U5820	130.5
U6801	130.5
701	110
4501	110
4501G	110
4501GA	110
4701	110
4701G	110
4701GA	110
6701	110

HARTLEY-TURNER

Model	IF
MA	130
MA7	130
MA12	130
MA25	130
RF41	465
RGMA	130
RGNA	130

HAYNES

Model	IF
All Superhet Models	110

HIGGS

Model	IF
A48R (and G)	465
A49R	450
AW49C	450
AW49R	450
AW57R (and G)	465
AW58R (and G)	465

HIGGS – *cont.*

Model	Value
AW59C	450
AW59R	450
AW69B	450
AW69G	450
AW69PB	450
AW69PG	450
AW99C	450
AW99G	450
BW49PR	450
BW49R	450
UW57R (and G)	465
UW58R (and G)	465
UW59C	450
UW59R	450
UW69B	450
UW69G	450
UW69PB	450
UW69PG	450

INVICTA

Model	Value
Multi Mains	110
SF	119
SHB	127
UP2	467
A29	465
B29P	465
A40RG	465
A40	465
A40C	465
B39	465
B40	465
U40	465
AC/45	127
AC/45D	127
AC/45RG	127
45U	127
45U/RG	127
A46PB	465
AC47	127
47B	127
47U	127
47U/RG	127
AC48/RG	127
A49	465
A49C	465
A49PB	465
A49RG	465
AR49	465
B29P	465
B39	465
B49	465
U49	465
AW56	465
AW57	465
AW57/RG	465
300	465
310	465
310/RG	465
330	465
360	465
390	465
400	465
430	465
431	465
440	465
450	465
460	465
461	465
500	465
502	465
503	465
510	465
520	465
570	465
580	465
635	110
635 (D)	110
635/RG	110
650/NJ	465
800	469
820	465

KOLSTER-BRANDES

Model	Value
285	114
286	114
339	130
345	130
346	130
365	130
365A	130
378KB	130
381	130
383	130
398	130
405	130
405A	130
422	130
425	130
426	130
427	130
427A	130
428	130
428A	130
444	130
444A	130
540	130
550	130
560	464
580	130
590	464
592	464
610	130
630	464
632	464
640	464
642	464
650	464
652	464
660	464
666	130
666A	130
666B	130
666C	130
670	464
720	130
730	464
735	464
740	464
740P	464
750	464
760	464
770	464
800	464
808	464
817	464
820	464
830 and OA1	464
831	464
835 and OA1	464
840	464
850	472
860	464
865	464
870	464
875	464
880	464
885 and OA1	464
888 A, B, C	130
890	472
935	130

LISSEN

Model	Value
All Purpose	452
Arundel	452
Balmoral	452
Caernarvon	473
Carisbrooke	452
Conway	452
Dover	455
Edinburgh	473
Glamis	452
Kenilworth	452
Lancaster	452
Marquis	455
Richmond	452
Salisbury	452
Skyscraper 7 (Fixed)	126
Winchester	452
York	452
8060	126
8108	127
8109	127
8110	127
8111	127
8114	465
8116	127
8117	127
8121	127
8125	127
8128	127
8129	127
8169	127
8214	465
8301	465
8302	465
8304	455
8317	455
8318	455
8319	455
8321	455
8322	455
8329	455
8401	452
8402	473
8414, 8415	452
8417	452
8418	452
8421	473
8422	473
8424	455
8432	473
8458	452
8468	452
8480	452
8481	452
8482	452
8485	452
8503	452
8514	452
8515	452
8518	452
8521	455
8529	452
8533	452
8539	452
8547	452
8563	452
8467	452

MAJESTIC

Model	Value
4V Midger Superhet	465
15/15B	175
20	175
21	175
22	175
23	175
25	175
44	465
49	465
52	175
60	175
61A	175
120	175
150	175
151	175
153	175
154	175
155	175
156	175
163	175
194	465
261A	125
261B	125
264	125
265	125

MARCONIPHONE

Model	Value
209	123–127–125–125
219	123–127–125–125
222	465
223	456
224	456
234	456
236	456
239	123–127–125–125
249	123–127–125–125
255	125
256	128–125–125–125
257	456
258	128–125–125–125
262	128–123–128–125.5
264	125
269	125

MARCONIPHONE – *cont.*	
272	128–123–128–125.5
273	125
274	128–123–128–125.5
276	120–114–117–117
278DC	128–123–128–125.5
279	127–123–123–127
280DC	128–123–128–125.5
286	128–123–128–125.5
Q286	125
287	125
288	128–123–128–125.5
289	123–127
290	120–114–117–117
291	120–114–117–117
292	122–112–117–117–117–117
296	123–127
297	125
298	123–127
336	125–125–122–128
345	460
346	460
347	460
363	460
365	460
366	460
367	460
382	465
392	465
395	465
534	465
535	128.8–122.8–126.5–122.5
536	128.8–122.8–126.5–122.5
537	465
538	465
539	465
549	465
556	465
557	465
559	465
561	465
562	126.5–121.5–121.5–126.5
563	465
564	465
566	465
567	465
571	465
572	465
573	465
575	465
576	465
851	465
852	465
853	465
855	465
857	465
857T	465
858	465
859	465
861	465
864	465
865	465
866	465
868	465
869	465
870	465
871	465
872	465
873	465
874	465
878	465
879	465
880	465
881	465
882	465
883	465
884	465
885	465
888	465
890	465
891	465
892	465
893, 893A	465
895	465
91	465
918	485
919	485
920	485
921	485
922	465
950	465

McCARTHY	
BPS6	127
BS4	127
BS4/AW	465
BS6/AW	465
PP6/AW	125
PP8/AW	465
PP9/AW	473
PP9/AW De Luxe	473
RF6/AW	465
RF6/AW/AC-DC	465
RF7/AW	465
RF7/AW/AC-DC	465
RGS6	465
S5/AW/AC-DC	465
S5/AW/AC	465
S6/AW	465
S6/AW/AC-DC	465
S8/AW	465
S6/AW/AC-DC De Luxe	110

McMICHAEL	
AC Mains Superhet	428.5
Superhet Trans	110
Twin Spker S/H (Early Models 406/410)	428
135	128.5
135U	128.5
137	465
137U	465
138	465
235	128.5
235	128.5
361	128.5
362	465
363	128.5
364	128.5
365	128.5
366	128.5
368	128.5
369	128.5
371	128.5
371U	128.5
372	128.5
373	128.5
374	128.5
375	465
378	128.5
380	460
380U	460
381	465
382	465
385	460–465
386	460
386RC	460
386RG	460
386U	460
388	465
389	465
390	465
391	465
393	465
394	465
396	465
396RC	465
396RG	465
398	465
399	465
401	128.5
401U	128.5
435	128.5
535	128.5
802	465
803	465
808	465
903	465

McMURDO-SILVER	
Masterpiece VI	465
Masterpiece Chippendale	465
Masterpiece Olympic	465
Masterpiece Sheraton	465
15–17 Chassis	465
15–17 Georgic	465
15–17 Homeric	465
15–17 Olympic	465

MILNES	
Diamond (SW 3.1 MHz)	465
Emerald (SW 3.1 MHz)	465
Mercury	450
Onyx	450
Pearl (SW 3.1 MHz)	465
Ruby (SW 3.1 MHz)	465
Saturn	450
Venus	450
466 (SW 3.1 MHz)	465
466 B (SW 3.1 MHz)	465
589 (SW 3.1 MHz)	465

MULLARD	
MAS2	128
MAS3	128
MBS3	128
MUS3	128
MAS4	128
MBS4	128
MUS4	128
MAS5	128
MUS5	128
MAS5	128
MBS6	128
MUS6	128
MAS7	128
MBS7	470
MAS8	128
MAS12	128
MUS12	128
MAS15	470
MUS15	470
MAS17	470
MUS17	470
MAS18	128
MUS18	128
MAS20	128
MUS20	128
MBS23	128
MAS24	470
MUS24	470
MU35	115
MAS82	470
MUS82	470
MAS90	128
MUS90	128
MAS94	128
MAS97	128
MUS97	128
MAS103	470
MAS104	128
MAS137	128
MUS137	128
MAS139	470
MXS139	470

MURPHY	
A4	117–120
D4	117–120
B5	117
A8	120
A24	117
A24C	117
A24RG	117
B24	117
D24	117
D24C	117
D24RG	117
B25	117–119
A26	117–119
A26C	117–119
A26RG	117–119
D26	117–119
D26C	117–119

MURPHY – cont.							
D26RG	117–119	B81	465	34B	460	D531RG	475
B27	119	B81	465	A52BG	451	D531W	475
A28	119	A90	465	A53BG	451	A537BG	451
A28C	119	A90RG	465	U53BG	451	A537CG	451
A28RG	119	D90	465	54	460	A537RG	451
D28	119	D90RG	465	56	125	B537	451
D28C	119	B91	465	70	260	C537BG	451
D28RG	119	A92	465	70A	260	U537BG	451
A30	119	B89	465	71	125	V537	451
A30C	119	B93	465	98/1	460	V537BG	451
A30RG	119	B95	465	98/2	451	W537	451
D30	119	B97	465	99	451	P538BG	470
D30C	119	AD94	465	116N	460	A539BG	475
D30RG	119	SAD94S	465	116Q	460	A539PB5	475
AD32	119	SAD94L	465	116RX	470	A539RG	475
B33	119	A96	465	116S	460	580	451
A34	119	A98	465	116X	460	581	451
A34RG	119			200X	175	582BG	451
D34	119	**ORR**		237	125	582CG	451
D34RG	119	Multimains	110	238	125	582RG	451
A36	119	SF	119	247E	125	583BG	451
D36	119	SHB	127	248	125	583CG	451
A38C	119	AC/45	127	248E	125	583RG	451
D38C	119	45/U	127	255	451	584BG	451
A40	119	AC/47	127	256	125	584CG	451
A40C	119	47/U	127	260	125	584RG	451
A40RG	119	AW/56	465	261	125	620	125
A46	465	AW/57	465	263	125	A637BG	451
A46C	465	635	110	264	125	A637CG	451
A46RG	465	635D	110	265	125	A637RG	451
D46	465			269BG	451	A638ARG	470
D46C	465	**PETO SCOTT**		269CG	451	A638BG	451
D46RG	465	Trophy 5	465	269RG	451	C638	470
B47	465	Trophy 6	465	271	125	C638BG	470
B47A	465	Trophy 8	465	280	460–451	CA638	470
B47C	465	S51	465	282	451	U638BG	470
A48	465	H51	465	281A	125	U647BG	451
A48RG	465			281F	125	U647VG	451
D48	465	**PHILCO**		281G	125	680	460
D48A	465	Empire 5	451	282	451	D732	475
D48RG	465	Empire 7	451	290	451	D732BG	475
A50	465	Empire 20	470	295	451	D732CG	475
A50C	465	4 Star BG	125	322	470	A847BG	470
D50	465	5 Star BG	125	A421	451	A938CG	470
D50C	465	5 Star CG	125	U427	451	1237	125
A52(SW 1st 3.1 MHz,		A1	470	P429	470	1260	125
2nd 465 kHz)	465	A1RG	470	U429BG	475	1263	125
B69	465	B1	465–468	444	451	1280	460–451
A70	465	BEF1 and BEF2	465	450	451	1280X	460–451
A70C	465	U1	465	471CG	451	1281	125
AtoRG	465	A2	465	471RG	451	1281A	125
D70	465	B2, B29	451	D521	475	1281F	125
D70C	465	U2	465	D521B	475	1281G	125
D70RG	465	A3	451	D521W	475	1281Q	125
B71	465	A4	451	S521	475	1282X	451
B71A	465	BP4	465	S521B	475	1582ARG	451
A72	465	C4	125	S521W	475	1583ARG	451
A72RG	465	C4T, C4S	125	X521	475	1584ARG	451
D72	465	A5	451	X521B	475	U1647	451
D72A	465	C5	470	X521W	475	U1647RG	451
A74	465	A7	451	A527	451	A1847RG	470
A74C	465	A9	451	C527	451	A2258	470
A76	465	A9RG	451	P527	451	A2258ARG	470
A78C	465	16	460	U527	451	2620	460
A78RG	465	16B	460	D531	475	2620A	460
				D531B	475	2620E	460

PHILCO CAR RADIO		555U	128	B34	451	RGU650	456
806 + 806T	260	575A	115	C35	451	U650	466
801T	260	580A	115	RG35	451	CU690	456
803T	125	584A	115	T35	451	RGAU690	456
M522T	475	585HU	115	PT36	451	RGU690	456
M522S	475	587HU	115	PTC36	451	U690	456
Transitone 10 + 10T	260	588A	115	B43	451		
Transitone 5	460	588U	115	53	451	**PORTADYNE**	
Transitone 11 + 11T	260	597A	128	C53	451	Jubilee Superhet	112
C4T	125	597U	128	RG53	451	Jubilee 5V Battery	
C4S	125	599A	128	C63	451	Superhet	112
K728T	125	617L, 617A	128	T63	451	A36	112
K728S	125	650A	470	U106	456	A37	112
L728T	125	650U	470	CU225	456	A38	456
L728S	125	660A	470	RGAU225	456	A39	450
K628T	125	660U	470	RGU225	456	A52	112
K628S	125	680A	128	RU225	456	A53	456
L628T	125	691A	128	U225	456	A58	450
L628S	125	691U	128	C335	451	A59	112
K628TC	125	698A	128	RG335	451	A64	456
K627TS	125	698U	128	T335	451	A72	112
6	260	699A	128	B344	456	AC55	112
9	260	699U	128	CB344	456	B36	112
12	260	701AX	128	C350	451	B37	112
		702A	128	T350	451	B42	112
PHILIPS		702U	128	Y353	456	B48	112
V5A	128	711A	128	CU355	456	B49	450
V7A	128	714B	128	RU355	456	B72	112
V7U	128	716B	128	U355	456	J/AC	112
206H, 206A	470	727A	128	CU357	456	J/AC-DC	112
212B	128	727U	128	LM357	456	J/RG	112
218B	128	735A	128	RGAU357	456	MS5	450
219B	128	735L	128	RGU357	456	PA6	112
225B	470	745A	128	U357	456	PB6	112
228B	470	745U	128	CU385	456	PB/AC	450
229B	470	747A	128	LM385	456	PB39AC	450
241	115	747AX	128	RGAU385	456	PB39U	450
243	115	748A	128	RGU385	456	PBC/AC	450
245	115	748U	128	U385	456	PBC/U	450
246	115	753A	470	T401	451	PBS/AC	450
247	128	753U	470	T405	451	PBS/U	450
248	128	771A	475	EX405	451	PB/U	450
249	128	771Y	475	T455	451	RG2/AC	450
250	128	785AX	128	RGAU475	456	RG2/U	450
258	470	787AX	128	RGU475	456	RG3/AC	450
259	470	790A	128	U475	456	RG3/U	450
260	470	790U	128	530	451	RG6/AC	450
261	470	791A	128	BT530	451	RG6/U	450
262	470	791U	128	BTC30	451	RG7/AC	450
263	470	792A	128	BT532	451	RG7/U	450
264	470	792U	128	BTC32	451	RG/AC	112
265, 265B	470	794A	128	C533	451	RG/PB/AC	450
268	470	794U	128	RG530	451	RG/PB/U	450
269	470	795A	128	T533	451	RG/S	450
361A	475	795U	128	CU535	456	RGS/U	450
361U	475	797A	128	LM535	456	S/AC	112
362A	475	797U	128	RGAU535	456	S/B	112
362U	475	805A	470	RGU535	456	SB4	450
470A	128	805X	470	U535	456	SP/U5	450
470U	128	855X	470	BL550	456	SW5	450
480A	128			RG583	456	SU/5	456
480L	128	**PILOT**		RGA583	456	TA38	450
520A	115	Armchair Console	456	C633	456	TU38	450
522A	115	Little Maestro	451	T633	456	U38	456
539A	115	Major	451	CU650	456	U39	450
555A	128	Twin Miracle	451	RGAU650	456	U53	456
						U58	450

PYE

Model	Value	Model	Value
Empire AW (1936)	465	T6	127
International 5V	462	T7	127
International Console	462	T9	127
New Baby Q	469	T9C	127
New Baby (all dry)	469	T9RG	127
Nipper	467	T10	465
BS6	462	T10A	465
CAW	465	T12	127
CR/AC	114	T17	127
CR/DC	114	T17RG	127
CR/RG/AG	114	T18	127
E/AC	114	T18C	127
E/B	114	T18RG	127
E/DC	114	T20	127
MP	462	T21	127
MP/B462		T37	127
MP/C	462	T37RG	127
MP/RG	462	T60	127
MP/U	465	T61	127
MP/UC	465	26E	465
P	462	26ERG	465
P/AC	114	26ERG Auto	465
P/B	114	62E	465
PP/AC	465	62EV	465
PP/B	467	802	465
PP/U	465	803	465
PS	465	805	465
PS/B	465	805RG	465
PS/C	465	805RG/Auto	465
PS/RG	465	806	465
PU	465	806RG	465
QAC2	127	809	465
QAC3	465	810	465
QAC5	465	811	465
QAC38	465	812	465
QB	467	812RG	465
QB3	465	823	127
QPAC	465	824	127–131–123
QPB	465	826	465
QPU	465	830	465
QU	467	834	465
QU3	465	834C	465
Q43C (Model with AFC)	465	834RG	465
Q43RG (Model with AFC)	465	835	465
Q49C	465	841RG	465
Q49RG	465	842	465
Q49FC	465	901	462
RS4	462	906 'International'	462
S	114	930E	462
SE/AC127		931U	462
SE/AC/RG	127	946	462
SE/B	127	951	462
SE/DC	127	952	462
SE/U	127	Baby Q Senior (all dry)	469
SE/U/RG	127	Baby Q Senior (acc. model)	467
SP/AC	127	BS6B	462
SP/B	127	BS6V	462
S/RG	114	RS4 First Model (Serial Nos, MBH, MCH, MDH, MEH)	465
S/RG/Auto	114	E40RF	465
TC/RG	114	E42	165
TP/AC	127	E140RF	465
TP/B	127		
T4	127		

Model	Value
V41RF	465
G10	463
G10C463	
G10RG	463
15A	465

REGENTONE

Model	Value
AC-DC Transportable	465
Bennet Bantam	456
Bennet Transportable	456
Dickson All Dry	450
Permeability	456
Transportable 5	465
Transportable 6	465
World Wide 5	456
5V S/het with round cans	110
– Otherwise	123
AC/47	110–123
AC/56	110–123
AC/56U	110–123
AC/57	110–123
AC/57U	110–123
AW/S	465
RG66	110–123
RG66U	110–123
USP59	465
R55	465
R55A	465
AW44	465
AW66	465
U33	465
U33X	465

REMCO

Model	Value
All Models	465

RGD

Model	Value	Model	Value
166	465	702AC	110
166U	465	702DC	110
196	465	703AC	110
196U	465	704AC	110
296	465	704C	110
296U	465	704DC	110
356	465	704RG	110
356U	465	705	460
516AC	460	718AC	460
522	460	718AC-DC	465
535	460	722	465
623AC	460	723AC	465
623DC	460	723AC-DC	465
625AC	460	727	465
625DC	460	739AC	465
628AC	460	739AC-DC	465
628DC	460	A739AC	465
630AC	460	A739AC-DC	465
630DC	460	743	465
643DC	460	748	465
645AC	460	878	460
645DC	460	880	460
658	460	901AC	110
660	460	901DC	110
700AC	110	925	465
700DC	110	930	465
701AC	110	948	465
701DC	110	955	465
		1015	465
		1129	465
		1135	465
		1153	465
		1155	465
		1175	465
		1201	110
		1202	110
		1203	460
		1204	110
		1220	460
		1221	465
		1295	465
		3611	465
		5311	465
		5511	465
		7511	465
		1046G	465
		1046	465

RI

Model	Value
4V Batt. Superhet	118
Superhet RG	118
Airflo	118
Duotone	118
Moderne	118
Ritz	118
Ritz AC S/het	118
Ritz Micrion Batt 5	118
Ritz Twin Speaker	118

ROBERTS

Model	Value
Up to 1939:	
M5A	430
M4D	430
From 1939 onwards:	
M5A	465
M4D	465

ROBERTS – *cont.*
1946

P5A	465
P4D	465

ROGERS MAJESTIC

11/6	456
11/8	456
11/8X	456
11/9	456
11/9DX	456
11/11	456
11/11X	456
12/6	456
12/7	456
12/9	456
13/8	456
13/8C	456
13/10	456
13/15	456
14/8C	456
14/8R	456
14/8T	456

SELMER

Truvoice 5	450
139	450
140	450
1239	450

SPARTAN

401	465
501	465
510	465
511	465
519	465
520	465
521	465
530	465
531	465
540	465
541	465
548RG	345
548T	345
559	465
610	465
61	465
619	465
620	465
62	465
629	465
630	465
631	465
639	465
640	465
641	465
648AG	345
648C	345
648RG	345
649	465
650	465
651	465
719	345
748AG	345
748C	345

748T	345
1268AG	456

SPENCER

All Models	430

STANDARD

S40	130
S60	130

SUNBEAM

22	456

SUNRAY

55	110
99	465

TELSEN

6V Superhet	117.5
3435/BH	117.5
3435/BV	117.5
3435MH	117.5
3435/MV	117.5

TEMPOREX

R3	465
R3G	465
R3U	465
R3UG	465

TRUPHONIC

AT5	127
AW5	127
AW5A	127
AW5B	127
AW5C	127
AW5T	127
AW6	456
B4	127
BB4	127
BW5	127
BW5B	127
CA6	127
CU6	127
MA5	465
MA5RG	465
MA5T	465
MU5	465
MU5RG	465
MU5T	465
MA6	465
MA7	465
MA8	465
MA8RG	465
MU5	465
MU5RG	465
MU5T	465
NA5	127
NAC5	127
NAW5	127
NU5	127
NUW5	127
PA5	465

PAT5	127
PU5	465
PUT5	127
RGAW5C	127
RGA6	127
RGUW5	127
RGU5C	127
RGU6	127
UT5	127
UW5	127
UW5B	127
UW5C	127
NW5T	127
U6	127
UW6	127

ULTRA

1934 Panther AC	456
Tiger M AC-DC	456
M22	470
22AC	456
22 Batt	456
22 DC	456
M23	470
25AC	456
25DC	456
26AC and AC-DC	456
44	456
47	456
48	456
49	456
50	456
P60	510
P61	510
P62	510
P63	460
P70	510
88	456
95	456
96	456
97	456
99	456
101	456
102	456
103	456
105	470
106	470
115	456
116	456
121	456
122	456
123	456
125	456
133	456
134	456
140	456
150	456
201	470
202	470
203	470
204	470
205	470
206	470
207	470
208	470
209	470

210	470
301	470
302	470
303	470
304	470
305	470
306	470
307	470
308	470
309	470
310	470
315	470
316	470
320	470
330	470
400	456
500	470
401	470
402	470
405	470

VARLEY

AC Superhet 4	110
AP46	110
AP48	110
Square Peak Mains Superhet	110

VIDOR

220	473
221	473
227	130
237	130
258	130
275	473
277	473
280	473
284	473
288	450
291	473
300	473
301	473
302	473
308	450
322	465
323	465
351	456

WAR-TIME CIVILIAN RECEIVER
See *Civilian War-time Receivers*

WB

4VA/wave Superhet	128
5VA/wave Superhet	128
394B	128
395	128

ZENITH

5S29	252.5

ZETAVOX

ST/AC	125

USA Receivers Imported by Board of Trade

ADMIRAL
- 67M5 — 455
- 76P5 — 455
- 77P5 — 455
- 78P6 — 455
- 79P6 — 455
- P6XP6 — 455
- 4202B6 — 455
- 4203B6 — 455
- 4204B6 — 455
- 4220D5 — 455

ANDREA
- 35H5 — 455

EMERSON
- 301 — 455
- 310 — 455
- 311 — 455
- 318 — 455
- 320 — 455
- 330 — 455
- 331 — 455
- 332 — 455
- 336 — 455
- 343 — 455
- 349 — 455
- 350 — 455
- 351 — 455
- 353 — 455
- 363 — 455
- 376 — 455
- 389 — 455

- 400 — 455
- 402 — 455
- 413 — 455
- 414 — 455
- 415 — 455
- 418 — 455
- 419 — 455
- 421 — 455
- 422 — 455
- 424 — 262
- 425 — 455
- 426 — 455
- 427 — 262
- 428 — 262
- 433 — 455
- 439 — 455
- 440 — 455
- 441 — 455
- 461 — 455
- 463 — 455
- 465 — 455
- 465A — 455
- 467 — 455

FADA
- 115 — 455
- 148 — 455
- 200 — 455
- 203 — 455
- 205 — 455
- 209 — 455
- 215 — 456
- 220 — 455
- 252 — 455
- PD41 — 456
- PL23 — 456

- PL41 — 456
- 169W — 456
- 215T — 456

GE
- HJ612 — 455
- J54W — 455
- L513 — 455
- L541 — 455
- L543 — 455
- L570 — 455
- L571 — 455
- L572 — 455
- L574 — 455
- L600 — 455
- L604 — 455
- L613 — 455
- L621 — 455
- L643 — 455
- L651 — 455
- LB673 — 455
- LB700 — 455
- LB702 — 455

MOTOROLA
- 51X16 — 455
- 51X19 — 455
- 61X17 — 455
- 61L11 — 455

PHILCO
- PT3 — 455
- PY87 — 455
- PT88 — 455
- PT95 — 455
- 321T — 455

- 42–327T — 455
- 42–842T — 455

RCA
- 1X — 455
- 6X2 — 455
- 14X — 455
- 34X — 455
- 35X — 455
- 36X — 455
- 45X12 — 455
- 15X — 455
- 55X — 455
- 16X2 — 455
- 16X3 — 455
- 16X11 — 455
- 16X13 — 455
- 26X1 — 455
- 26X3 — 455
- 26X4 — 455
- 26BP — 455
- 26X21 — 455

STROMBERG-CARLSON
- 500H — 455

WESTINGHOUSE
- 12X4 — 455
- 13X8 — 455
- WR13X8 — 455
- WR62K1 — 455
- WR62K2 — 455

ZENITH
- 5G603M — 455
- 6G601 — 455

When the Copenhagen wavelength plan was implemented in March, 1950 a recommended IF of 470 kc/s was adopted by some manufacturers almost immediately, and by others in the next few years. Some clung to their old favourites for a long time, in some cases even changing from 470 kc/s to non-standard IFs. Where 'post CP' is quoted, confirmation of model dates may be obtained by examination of the tuning dials in conjunction with the information given on page 128.

	kc/s
Ace:	
C53, 701	472
Alba:	
C112	455
Post CP	470
Allander	
A410	465
Ambassador:	
650H	420
849, 4756	452
MU540, 3540, 3541, 5243, U3645, 545H, PA145, PA146	465
Coronet	470
AM/FM chassis	470 & 10.7 Mc/s

Amplion:	
HU160	465
Argosy:	
See *Regentone*	
Armstrong:	
FM56	10.7 Mc/s
Banner:	
See *Sobell*	
Beethoven:	
A415	450.5
U3048, A3348, A3050PRG, BP2040, A1188, A2030ARG, U2038, U2030C, U3030	465
B84 and CP	470

Berec:
See *Ever-Ready*

Bush:

BP10, RG4, PB83, RG3, AC1, DAC1, AC2, DAC2, SUG3, DUG3, DAC90, AC91, DAC91, SUG91, AC81, DAC81, AC11, DAC11, RG11A, DRG11A, DAC90A, PB22, DAC22, SUG26, DAC10	465
DAC10 above S/no. 39501	470
DAC90 above S/no. 69591	470
AC11 S/nos. 60501–61500 inclusive	470
Post CP	470
VHF41, RG46, VHF54, VHF55 (FM)	19.5 Mc/s
Subsequent FM models	10.7 Mc/s

Champion:

Planet, 822, 834, 838, 781A, 781B, 844, 862	465
820, 825, 830, 851, 887, 951/3	470
856, 880, 840, 841, 859	470 & 10.7 Mc/s
836	10.7 Mc/s

Cossor:

469	452
480KU, 487U, 487AC, 480K, 490K	465, 450
474AC, 463AC, 464AC	465
All subsequent models (AM)	470
All subsequent models (AM/FM)	470 & 10.7 Mc/s

Decca:

Double Decca ML	380
Doucette, Beau Decca, RG98, RG102	472
RG100, 66, RG103	460 & 10.7 Mc/s
All subsequent models (AM)	472
All subsequent models (AM/FM)	472 & 10.7 Mc/s

Defiant:

Post CP (AM)	470
Post CP (AM/FM)	470 & 10.7 Mc/s

Denco:

DR23, DR18, DR21B, DR22, DR21, DR21U	465
DR17	1600

Eddystone:

504, 556, 556B	450 & 2.5 Mc/s
670, 659, 740, 710B, 840	450
640	1600
750, 88B (Double Superhet)	1620 & 85
680, 680X	450 & 1 Mc/s
820 (AM/FM)	462 & 10.7 Mc/s

Ekco:

A104, A144, U109, U143, B53, A52, U199, A160	460
MBP99	455
ARG168, A33, P63, CR117, ARG233, CR152	465
A147, MBP149, U122, U159, U195	470
A239	460 & 10.7 Mc/s
All subsequent models (AM)	470
All subsequent models (AM/FM)	470 & 10.7 Mc/s

Emerson:

E601, E600	470

Etronic:

All pre-CP models	465
All post-CP models	470

Ever-Ready:

A, C/E, C/A, K, T	452
B2	455
B	465
Later K models and all subsequent	470
AM/FM models	470 & 10.7 Mc/s

Ferguson:

289A, 299RG, 299RG(S)	475
All other AM models	470
All FM or AM/FM models	470 & 10.7 Mc/s

Ferranti:

145, 146, 147, 148, 149, 194, 147S, 347, 447, 248, 547, 546, 746, AR15	465
All subsequent models (AM)	470
All subsequent models (AM/FM)	470 & 10.7 Mc/s

Ford:

(Car radios) Manual models	470
Push-button models	480

GEC:

All pre-CP models	456
All post-CP models	470

Grundig:

AM models	468
AM/FM models	468 & 10.7 Mc/s

HMV:

1604, 1119	465
Post-CP	470

Invicta:

15, Mk. 1 & 11, 32, 33, 43, 52, 58, 75	420
51, 42, 12, 20, 200, 200W, 200C, 91, 92, 93, 10, 11, 40, 60, 65, 73, 31, 25, 30	465
42 (S/nos. 657001–657100 inclusive)	420
14, 26, 36, 55, 57 and all subsequent AM/FM and FM only models	470 & 10.7 Mc/s

KB:

AR30	464
DR10, CG20, CR20, FR15, DR15, ER15, FR10, ER10	465
GR40, FP151, GR10/1, FB10, HG35, HG30, HR10, FP11, KR20, LR30, PR10, LR10, LG30, LG35, PGT10	422
AR50	472
KR20FM, LR10FM, PG20, PG30, MR10, MG30, QB20	422 & 10.7 Mc/s
BR20, MP151, PP3/1, OP21/1, RB10, PP11, PP21, PP251, EG50M, EG50TM, EG35, BM20, FG25, FG20, FG50, FG50TL	470
PB10FM (and all models incorporating FM)	10.7 Mc/s
All subsequent AM models	470

Masteradio:

All models until 1954	465
All models after 1954	470

McCarthy:

All AM only models	460
AM/FM chassis 855	460 & 10.7 Mc/s

McMichael:

All models prior to CP	465
All AM models after CP	470
AM/FM models	470 & 10.7 Mc/s

Marconiphone:

P20B, P20BX	365
T24DAB	365, 360
T21A, T18DA, T18DAX, T18DAXM, T14A, ARG14A, C10A, T10A, T10AW, T11DA, T11DA/B, T11DA/AG, RG11A, T19A, ARG 19A, ARG27A, T26A, RG22A, P17B, P17BX, T15DA, 23 series	465
T29A, ARG29A, ARG31A, T25DA	470, 465
ARG30AE and all subsequent models	470
FM IF in all models	10.7 Mc/s

Mullard:

All models	470

Murphy:

U144, A100, B141, A130, A168M, BUi83	465
A182, A182R, U182, A188C, A172R, A192, U198H, and all subsequent	470
FM IF in all models	10.7 Mc/s

Pam:

All models	470

Pennine Ranger:

E54/3	450

Peto-Scott:

S51, H51, H52, SU51, HU52	456
BP41, BP50, ARG64T, R54, ARG65 ARG63	465
ARG68 (AM/FM)	470 & 10.7 Mc/s

Philco:

All pre-CP models	465
All post-CP models	470
FM IF in all models	10.7 Mc/s

Philips:

All models except 681A	470
681A (medium/long waves)	452
(short waves)	3 Mc/s ± 250 kc/s & 452

Pilot:

All pre-CP models	451
All post-CP models	470
FM IF in all models	10.7 Mc/s

Portadyne:

AG33, AG44	465
AM/FM chassis	465 & 10.7 Mc/s

Portagram:

Unimidge	465

Pye:

P61	420
Other pre-CP models	465
Post-CP models	470
TCR1000, TCR2000	480

Radiomobile:

Pre-CP models	465
Post-CP models	470

Rainbow:

750, 739, 536	465

Raymond:

Pre-CP models	465
Post-CP models	470

R.M.:

Gnome	430
PW461	465

RI:

493	450

Roberts:

RP4, RMB	472
R55, CR, and subsequent models	470

Regentone:

All pre-CP models	465
All post-CP models	470
FM IF in all models	10.7 Mc/s

RGD:

All pre-CP models	465
All post-CP models	470
FM IF in all models	10.7 Mc/s

Sobell:

All pre-CP models	465
All post-CP models	470
FM IF in all models	10.7 Mc/s

Strad:

553, 20RG	465

Ultra:

Minstrel, Symphonic, Leader, Ultragram, Twin, U405, T402, T406, T401	470
U626	471, 465
All subsequent until Thorn chassis introduced	471
FM IF in all models	10.7 Mc/s

Vidor:

CN351, CN359, CN353, CN396, CN396A, CN385, CN379, CN578, CN361, CN360, CN360(a)	456
CN393, CN392, CN381A, CN381B	456
Later versions of last four models	475
CN409, CN411, CN414, CN417, CN420, CN420A, CN421	475
All subsequent models	470

Notes:

Cossor models 470AC, 487AC, 480K, 490K, 487U. Certain versions used 450 kc/s IFs to eliminate re-radiation problems in the Midlands and North of England. These were marked with a letter M on the cabinet back.

Ekco MBP99. Models for use in the south of Britain used the 455 kc/s IF, those for northern areas 460 (cf. Cossor models above). Appropriate code letter, S or N, printed on inside of receiver back.

Marconi models T29A, ARG29A, ARG31A. Early versions may have the 465 kc/s IF, but should be re-aligned to 470 kc/s when serviced.

Valve characteristics and base connections

The valve characteristic tables on the following pages are as complete as possible. Many types, although used in thousands of sets, are no longer made. Alternatives can be chosen from a study of the characteristics, and of the bases.

The tables are subdivided into frequency changers, triodes, etc., and these appear approximately in the order in which they are used in receivers.

Order of presentation

Within each subdivision valves are listed in the alphabetical order of the manufactures' names (Brimar, Cossor, Dario, etc.). Thereafter they are dealt with in ascending order of filament or heater voltage, which in general means that battery valves come first, followed by AC-only types, then those for AC/DC use, and finally any special types. Thus valves will not necessarily be found to be in alphabetical or numerical order of type numbers.

Abbreviations in descriptions where these are not self-evident are: D, diode; DD, double-diode; S, screen-grid; VS, variable-mu screen-grid; P, HF pentode; VP, variable-mu HF pentode; DDT, double-diode triode; DDP, double-diode LF pentode; Pen, LF pentode; DPen, diode pentode; DTetrode, diode LF tetrode; DT, diode triode.

Valve base diagrams are drawn to show the connections when looking at the base with the valve inverted.

In the tables the base used is indicated by a code reference contained in Base column. The first number and letter indicate the group of diagrams to which reference should be made, and the final number, the base diagram in that group. For example, 4B3 indicates that the valve has a 4-pin British base, and that the third diagram in the section shows the pin connections.

Basing diagram groups in addition to 4B are: 5B, meaning 5-pin British; 7B, 7-pin British; 9B, 9-pin British; 8S, British side contact; OM, Mazda Octal; O, International Octal; 7C, 7-pin Continental; OF, footless.

Interpreting receiving valve nomenclature

You may have watched an old hand at work on a vintage radio set and have wondered how it was he knew just which valve did what and with what other valve it might be replaced. You may assume that this knowledge will have been built up over so many years of experience that you will never be able to emulate it but this is only partly true. Although that old hand will have been learning all the time, for no one ever ceases to learn (or shouldn't, anyway), most of what he knows about valves will have been picked up very early on in his career, because no service engineer could function properly without it. If you apply yourself to the following basic information it will give you a jolly good grounding, and by subsequent regular browsing through valve data you too will be able to store up a wealth of valve knowledge.

Obviously, it would be impossible to pay attention here to anything like every valve type made by every valve manufacturer during the vintage years, so as far as British valves are concerned here we confine ourselves to the most

popular types made by the four largest manufacturers. Further study of valve data will make
you aware of coding used by smaller firms.

As regards European valves, from about 1934 all
manufacturers adopted the same type numbers,
which makes things much simpler. American
manufacturers always employed this unanimity.

British types

Fortunately most of the British valves produced
from around 1927 were coded in fairly simply
ways which are easily understood. We shall look at
the 'big four' manufacturers, A. C. Cossor Ltd.,
The Marconi-Osram Valve Company Ltd, The
British-Thomson-Houston Co. Ltd. (Mazda) and
Mullard Ltd, in alphabetical order.

Cossor used to prefix its 2 V battery valves with
two or three digit figures which indicated the
voltage and current ratings of the filament. These
figures were followed by one, two or three letters
which indicated the type of valve: D or DD stood
for a diode or double diode; L for a low-impedance
triode, H for a high-impedance triode and HL for
a medium impedance triode; P for an output triode
(sometimes with the prefix X for 'extra power'); B
for a double output triode (Class 'B'); SG or VSG
for, respectively a 'straight' or variable-mu screen
grid; OT for an output tetrode; PT for an output
pentode; SPT and VPT for a 'straight' or variable-
mu HF pentode; PG for a heptode ('PentaGrid');
and TH for a triode-hexode.

Some instances: A 210HL meant a medium
impedance triode with a 2 V, 0.1 A filament,
220DDT meant a double-diode-triode with a 2 V,
0.2 A filament, 230OT an output tetrode with a
2 V, 0.3 A filament and so on. Note: the 2P and
2XP were output triodes with 2 V, s A filaments for
mains, not battery operation.

Early mains valves initially followed a different
pattern, with simply the prefix letter M indicating
mains operation which meant in this case a 4 V
heater. In addition the abbreviation Pen was used
to indicate an output pentode and PEN an HF
pentode (clever studd this). Thus an MVS/PEN
was a variable-mu HF pentode an an MP/Pen an
output pentode. In the days when mains valves
were made for DC only operation they wre
indicated by the prefix D taking the place of the M.
Around 1933 a similar type of nomenclature as for
battery types was adopted but with the letter M for
mains inserted after two digits representing the
heater voltage and current. Thus a 41MPT was a
mains output pentode with a 4 V, 1 A heater. At the

same time valves for AC/DC mains operation
were introduced and three digits were employed to
indicate voltage and current, e.g. the 402Pen was an
output pentode with a 40 V, 0.2 A heater.

Rectifiers had three digits followed by SU for a
half-wave type and BU for a full-wave (sometimes
the B was omitted).

Marconi-Osram initially used a similar range of
letters as Cossor to represent valve types but when
beam tetrodes (or 'kinkless' tetrodes as M-O called
them) were introduced for output stages the letters
KT were employed to indicate them and N was
substituted for PT for output pentodes. At the
same time W was introduced to indicate variable-
mu HF pentodes and KTW for variable-mu HF
beam tetrodes. Z was used in either case for
'straight' valves. A digit was placed after the letters
to indicate the filament or heater voltage in battery
types and mains types for AC operation, viz., 2 for
2 V, 4 for 4 V and 6 for 6 V. Another digit might be
added to represent a development on a basic types,
e.g. KT2, KT21, KT24. Valves suitable for AC/DC
operation always had the digit 3 after the type
letters to indicate the 0.3 A heater current, again
possibly followed by a development digit. Incidentally, most of the earlier types had 13 V heaters to
make them suitable also for car radio receivers in
vehicles with 12 V electrical systems.

All rectifiers, whether half- or full-wave had the
identifying letter U, very occasionally prefixed by
M for mains. Any digits after the letter(s) have no
significance as regards heater voltage except in the
case of 3, which again indicates a 0.3 A type. As
before, a development digit may follow.

Mazda also used much the same system of type
letters as Cossor and no difficulty should be
experienced in interpreting them. Battery types
had digit(s) after the letters to indicate the filament
voltage, viz. 2 for 2 V and 14 for 1.4 V, followed by
development digits. Mains valves with 4 V heaters
and British bases all had the prefix AC. When
Mazda introduced its own version of an octal base
the prefix was dropped an the digit 4, followed by
a development digit was placed after the letters, e.g.
VP41. AC/DC valves with British bases had digits
after the letters to indicate heater voltage and
current, e.g. VP1320 had a 13 V, 0.2 A rating.
VP1321 and VP1322 were developments of the
type. When only three digits followed, e.g. VP233
the first two indicated the heater voltage (i.e. 23 V)
but the third did not indicate the current; instead it
signified that the valve had a Mazda octal base.

Rectifiers used U for a half-wave type and UU
for full-wave. Following digits did not signify

heater voltage in AC-only types were were used for this purpose in AC/DC types, following the same pattern as explained immediately above. Note that the prefix M was used only for mercury-vapour rectifiers, not found in domestic equipment.

Mullard's early battery valves all had the prefix letters PM, to indicate the special patented type of filament used. These were followed by one or two digits which do *not* indicate the filament voltage, since anything between 1 and 22 might be used for 2 V types, and 24 for 4 V types intended for mains operation. Any letters after the digits sometimes indicated the purpose of the valve but cannot be relied upon to give useful information.

With respect to mains valves, the position improved in the early 1930s when, once again, a system of letters similar to that of Cossor's was introduced. These were always placed first and followed by a digit or digits to indicate the heater voltage, exact in the case of AC-only types, approximate in the case of AC/DC types. Thus, a valve such as the VP4 could be relied upon to have a 4 V filament but the PEN26 was rated at 24 V and the PEN36C at 33 V. The final letters A or B, when used, indicated a development on an earlier type, whilst C indicated that a British base was fitted in lieu of a side contact type (e.g. a VP13C versus a VP13). Mullard made a series of directly heated high-power output triodes with 2 V, 4 V or 6 V filaments. The code letters were simple, being ACO (presumably for AC Output) in all cases but the filament voltage coding was either ingenious or perverse, being 42 for 2 V, 04 or 44 for 4 V and 64 for 6 V.

Rectifiers for AC-only operation carried the letter W to indicate full-wave, preceded by the letter D for directly heated or I for indirectly heated. Just to be awkward, one supposes, just one particular rectifier had the letters FW – for full wave? A digit following the letters indicated only a development type and not the filament/heater voltage. Sometimes a final number was used to indicate the maximum anode voltage, e.g. DW4/350 or FW4/500. With the fiendish cunning one begins to associate with valve manufacturers, Mullard rectifiers for AC/DC or 'universal' operation bore the code letters UR, followed by a development digit and, where appropriate, the suffix C to indicate a British base.

Brimar British-based types

Standard Telephones and Cables Ltd., introduced its Brimar range of valves in the mid-1930s, the name being an acronym for BRItish Made American Range. Thus most of the firm's valves were exactly the same in type numbers and characteristics as the indigenous American types to be discussed below. However, a range of British-based types also was produced for AC-only and AC/DC operation for which an arcane coding system was devised and which may at first be thought to be meaningless. It consists of one or two digits followed by the letter A or the letter D, followed again by a single digit. One might wonder, for instance, why 7D3 should be allocated to an output pentode, 9D2 to variable-mu HF pentode and 20D2 to a triode-hexode frequency-changer, especially when the respective heater voltages are 40 V, 13 V and 20 V. Likewise, why 11A2 for a double-diode-triode and 15A2 for a heptode, especially when both have 4 V heaters?

The first clue is provided by the letters A and D. It will be seen that all the 4 V heater types incorporated the former and the high voltage heater types the latter. From this we may infer that A means AC and D means AC/DC. as regards the first two digits, the coding can be broken down to:

1 Half-wave rectifier
4 Triode
7 Output pentode
8 Straight HF pentode
9 Variable-mu HF pentode
10 Double diode
11 Double-diode-triode
15 Heptode
20 Triode-hexode

The final digit is a development number. For some reason no type number was allocated to full-wave rectifiers for AC-only operation and they were known simply at R1, R2, etc.

Continental valves

Prior to about 1934 there were so many different methods in use for coding valves that it would be impossible even to begin to innumerate them. Fortunately, when side contact types were introduced in that year a new system of identification was adopted which continued, with only minor additions and alterations until the end of the valve era. It consisted of up to four letters, the first one indicating the type of filament or heater voltage and the rest indicating the actual job of the valve. Then came one or two digits indicating the type of base used, with every occasionally a final letter to indicate a development type. The system represents the apotheosis of valve nomenclature since any valve may be identified at a glance for its function, heater voltage or current and type of base.

Filament/heater code letters:

A   4 V (AC-only)
C   0.2 A (AC/DC)
D   1.4 V (for dry batteries)
E   6.3 V (AC-only and AC/DC)
G   5 V (AC-only rectifiers)
H   0.15 A (AC/DC)
K   2 V (for accumulators)
P   0.3 A (AC/DC)
U   0.1 A (AC/DC)
V   0.05 A (AC/DC)

Type letters:

A   Single diode
B   Double diode
C   Triode
E   Secondary-emission valve (very rarely used)
F   HF pentode
H   Hexode or Heptode
K   Octode
L   Output pentode
M   Tuning indicator
N   Gas-filled triode ('thyratron')
Y   Half-wave rectifier
Z   Full-wave rectifier

Combinations of letters were used to indicate multiple valves, e.g. EBC33 for a double-diode-triode, ECH35 for a triode-hexode.

Base numbers:

1 to 9   5-pin side contact ('V' base) or 8-pin side contact ('P' base). On rare occasions, American octal.
10 to 19 Continental octal or 'footless' base (these are diminutive metal-envelope valves)
20 to 29 'Loctal' (all-glass, 8-pin lock-in base)
30 to 39 American type octal
40 to 49 Small 8-pin all-glass lock-in base (B8A)
50 to 59 9-pin all-glass lock in base (B9A) plus unusual types such as miniature diodes and triodes without conventional bases
60 to 79 Not used for domestic types
80 to 89 Small all-glass 9-pin (B9A)
90 to 99 Small all-glass 7-pin ('button-base' or B7A)

American valves

Prior to about 1934 the type numbers of American valves were just numbers which have no special significance and appear to have been 'picked out of the air'. For instance, a 43 is an output pentode with a 25 V heater whilst a 44 is a variable-mu RF pentode with a 6.3 V heater. 807 is a high-power output tetrode with a 6.3 V heater and 1625 is a 'magic eye' also with a 6.3 V heater. Something, at least, is to be inferred from type numbers which incorporate one or more digits followed by one or two letters, a digit and then (sometimes) a final letter because normally the first figure indicates – more or less accurately – the voltage rating of the filament or heater. For instance, 1 indicates a voltage of anything between 1 V and 2 V, whilst 2 indicates 2.5 V. This sort of inconsistency is endemic and many others will become apparent to anyone studying US valve data. 6 indicates 6.3 V and so, perhaps surprisingly, does 7, as applied to 'Loktal' or 'Octalox' valves. This is because originally the latter types had with 7 V heaters to suit a nominally 6 V car battery on charge, but they later were changed to 6.3 V. In the same way, both 12 and 14 indicate 12.6 V, the Loctals originally having been rated for a 12 V car battery on charge. Where heater voltages of over 12.6 V are concerned the type numbers mostly indicate the ratings accurately.

In many cases a valve with the prefix 12 for a 12.6 V heater will otherwise be electrically identical to one with the prefix 6 for 6.3 V. Thus a 12K8 triode-hexode is the same otherwise as a 6K8 and 12Q7 double-diode-triode as a 6Q7. There are scores of other examples. However, a 12A7 is a combined rectifier and output pentode whilst a 6A7 is a heptode. Again, a 12B8 is a combined triode and HF pentode whilst a 6B8 is a double-diode-HF pentode, its 12.6 V equivalent being a 12C8. Thus, unless you know your valves thoroughly it pays not to take chances but to look them up in the data before plugging them in.

A single letter following the first figure appears to have no particular significance and cannot be relied upon to indicate the function of a valve. For instance, 'A' may be found in such diverse types as the 1A3 diode, 1A5 output pentode and 1A7 heptode. Likewise, 'K' is found in the 6K6 output pentode, 6K7 variable-mu HF pentode and 6K8 triode-hexode.

One superficially reliable indication is that when the first figure is followed by two letters and the first of these in turn is 'S', it indicates that the valve is single-ended, that is, it does *not* have one of its electrodes brought out to a top-cap. In a lot of cases an 'S' valve will be electrically identical to one without this letter, but once again you cannot rely upon its being so. Whilst the 6SK7 is undoubtedly the single-ended version of the 6K7 with top-cap grid (they are both HF pentodes), a 6SN7 is a voltage-amplifying dual triode whilst a 6N7 is a high-power Class 'B' output valve, itself lacking a top cap. Again, if you are not sure, check first!

It is sometimes suggested that the digit following the letter indicates the number of base pins in use and that this may be used to deduce the purpose of a valve. In some cases this appears to work, but seems to be due to happy coincidence rather than a hard-and-fast principle. Certainly there are far too many discrepancies for it to be relied upon. One has only to look at the 6F8 (seven pins in use) and the 6SL7 (eight pins in use) to see that it can sometimes be the reverse of true. Incidentally, the 6SL7 provides another example of what we were saying about 'S' types above, for it certainly is *not* a single-ended version of the 6L7 heptode. For a comprehensive exposition of all the possible discrepancies mentioned above, let's take the case of the 12SA7, which is the 12.6 V equivalent of the 6SA7, but the latter is *not* the single-ended equivalent of the 6A7 mentioned above (and both the 12SA7 and the 6SA7 use all eight base pins!).

We are on somewhat safer (but not entirely firm) ground when we come to the final letter (if present) because with octal valves it indicates the type of envelope. The absence of a letter means a metal envelope, G a large glass envelope and GT a small ('glass tubular') type. Thus you may have a 6V6, a 6V6G and a 6V6GT, all electrically identical. The snag is that the system does not apply to non-octal valves such as our old friend

the 6A7, which has a small 7-pin base (sometimes incorrectly referred to as a 7-pin UX base – all true UX bases have only 4 pins). Here's a strange thing, though: there used to be a 6A7S, which had a permanent external shield connected to the cathode. Its characteristics were similar to that of the 6A7 but oddly enough the two were not normally interchangeable.

If all this has left you with the feeling that you may never fully understand American valve nomenclature, don't worry, you don't need to. As long as you can memorise something abou the fifty or so types which you are likely to come across in domestic radio repairs, that will be sufficient. If this piece can help you in this aim, and has persuaded you not to make guesses about unfamiliar valves, it will have served its purpose.

Valve bases are indicated in the table by a code in which O means Octal, OL, Loctal, and UX, the earlier American pin-type connections; figure in front of letters UX indicates number of pins in the base.

N.B. Where bias voltages are given as, for example −1.5 −30 V, it indicates that the valve in question has variable-mu characteristics and that its gain is fully controlled by applying a bias voltage in the range −1.5 V to −30 V, cut-off occurring at the latter figure. When the value of a bias resistor has to be calculated, it should provide the minimum voltage shown. This should be borne in mind if and when substitutions have to be made involving valves of similar but not identical characteristics, since it may not be possible to obtain full performance if the minimum bias is too high for the new type, whilst too low a value will cause it to be over-run. Little difficulty should be encountered where valve gain is further controlled by manual or automatic variation of the bias but the action may be sharper or less responsive in effect.

Table 1. Frequency changers

Make	Type	Description	Base	Fil. volts	Fil. amps	Anode volts	Screen volts	Oscil-lator volts	Conv. condct. Mbos	Bias volts
Brimar	20A1	Triode Hexode	7B38	4.0	1.2	250	80	150	650	−1.5−30
	15A2	Heptode	7B35	4.0	0.65	250	100	200	550	−3−40
	15D1	Heptode	7B35	13.0	0.2	250	100	200	550	−3−40
	15D2	Heptode	7B35	13.0	0.15	250	100	200	550	−3−40
	20D2	Triode Hexode	7B39	13.0	0.15	250	100	100	350	−3−30
	6A8G	Heptode	−	6.2	0.3	250	100	200	550	−3−40
	6K8G	Triode Hexode	−	6.3	0.3	250	100	100	350	−3−30
Cossor	210DG	Bigrid	5B1	2.0	0.1	150	−	−	190	0
	210PG	Pentagrid	7B7	2.0	0.1	150	80	150	450	0
	210SPG	Pentagrid	7B7	2.0	0.1	150	80	150	450	0
	210PGA	Pentagrid	7B7	2.0	0.1	150	80	150	450	0
	220TH	Triode Heptode	7B11	2.0	0.2	150	150	100	200	0
	41MDG	Bigrid	5B16	4.0	1.0	200	−	−	150	0
	41MPG	Pentagrid	7B35	4.0	1.0	250	100	100	1300	−1.5
	41STH	Triode Hexode	7B37	4.0	1.0	250	100	100	600	−1.5
	4THA	Triode Hexode	7B37	4.0	1.5	250	100	100	850	−2.0
	OM8	Octode	O54	6.3	0.2	250	50	200	550	−2.0
	OM10	Triode Hexode	O58	6.3	0.2	250	100	250	700	−2.0
	13PGA	Pentagrid	7B35	13.0	0.2	250	100	200	520	−3.0
	202MPG	Pentagrid	7B35	20.0	0.2	250	100	100	1300	−1.5
	202STH	Triode Hexode	7B37	20.0	0.2	250	100	100	600	−1.5
	302THA	Triode Hexode	7B37	30.0	0.2	250	100	100	850	−2.0
	4TP	Triode Pentode	7B45	4.0	1.4	200	200	−	4500	−5.0
Dario	BK22	Octode	7B7	2.0	0.14	135	45	−	250	0−12
	BH12	Hexode	7B5	2.0	0.135	135	60	−	1400	−1.5
	TK24	Octode	7B35	4.0	0.65	250	70	−	600	−1.5
	TCH24	Triode Hexode	7B37	4.0	1.45	250	100	−	−	−2.5−25
	TB5013	Octode	8S28	13.0	0.2	200	70	−	600	−1.5−25
	TCH229	Triode Hexode	7B37	21.0	0.2	200	70	−	1200	−1.5
Ever Ready	K80A	Octode	7B7	2.0	0.1	135	70	150	200	0
	K80B	Octode	7B8	2.0	0.13	135	45	135	270	−0.5
	A36A	Triode Hexode	7B37	4.0	1.0	250	70	−	1000	−1.5
	A36C	Triode Heptode	7B37	4.0	1.45	250	100	−	750	−2.5
	A80A	Octode	7B35	4.0	0.65	250	90	90	600	−1.5
	ECH3	Triode Hexode	8S29	6.3	0.2	250	100	−	650	−2.0
	ECH35	Triode Hexode	O58	6.3	0.3	250	100	−	650	−2.0
	CCH35	Triode Hexode	O58	7.0	0.2	250	100	−	650	−2.0
	C36A	Triode Hexode	7B37	21.0	0.2	250	70	−	1000	−1.5
	C36C	Triode Hexode	7B47	29.0	0.2	250	100	−	750	−2.5
	C80B	Octode	7B35	13.0	0.2	200	90	−	600	−1.5
	C36B	Triode Hexode	7B37	29.0	0.2	200	150	100	1000	−1.5
Ferranti	VHT2A	Heptode	7B7	2.0	0.1	150	70	70	−	−1.5
	VHT4	Heptode	7B35	4.0	1.0	250	100	100	650	−3.0
	VHTA	Heptode	7B35	13.0	0.2	250	100	100	−	−
	VHTS	Heptode	7B35	13.0	0.3	250	100	100	650	−3.0
Hivac	TP230	Triode Hexode	9B2	2.0	0.3	150	70	150	325	0−12
Marconi	X14	Heptode	O10	1.4	0.05	90	45	90	250	0
	X21	Heptode	7B9	2.0	0.1	150	70	70	240	−
	X22	Heptode	7B9	2.0	0.15	150	70	110	350	0
	X23	Triode Hexode	7B9	2.0	0.3	150	60	100	250	−1.5
	X24	Triode Hexode	7B9	2.0	0.2	150	60	100	250	−1.5
	MX40	Heptode	7B35	4.0	1.0	250	150	150	500	−3
	X42	Heptode	7B35	4.0	0.6	250	100	50	490	−3
	X41	Triode Hexode	7B35	4.0	1.2	250	80	120	640	−1.5
	X41C	Triode Hexode	7B35	4.0	1.2	250	80	120	640	−1.5
	X61M	Triode Hexode	O58	6.3	0.3	250	100	200	620	−3.0
	X63	Heptode	O58	6.3	0.3	250	100	200	490	−3.0
	X64	Hexode	O58	6.3	0.3	250	150	−	310	−6.0
	X65	Triode Hexode	O58	6.3	0.3	250	100	150	225	−3.0

Table 1. Frequency changers – *continued*

Make	Type	Description	Base	Fil. volts	Fil. amps	Anode volts	Screen volts	Oscil-lator volts	Conv. condct. Mbos	Bias volts
Marconi –	X30/32	Heptode	7B35	13.0	0.3	250	80	150	800	−3.0
continued	X31	Triode Hexode	7B35	13.0	0.3	250	80	150	640	−3.0
Mazda	FC141	Pentagrid	OM8	1.4	0.05	90	90	−	250	0
	TP22	Triode Pentode	9B2	2.0	0.25	150	150	150	500	−19.5
	TP23	Triode Pentode	7B10	2.0	0.25	150	150	150	400	−1.5
	TP25	Triode Pentode	OM5	2.0	0.2	150	150	150	225	−1.5
	TP26	Triode Pentode	OM5	2.0	0.2	150	150	150	550	−
	ACTP	Triode Pentode	9B5	4.0	1.25	250	250	200	700	−5
	ACTH1	Triode Hexode	7B38	4.0	1.3	250	250	250	750	−3
	ACTH1A	Triode Hexode	OM26	4.0	1.3	250	250	150	870	−3
	TH41	Triode Hexode	OM26	4.0	1.3	250	250	150	879	−3
	TP1340	Triode Pentode	9B5	13.0	0.4	250	250	200	700	−5
	TH2320	Triode Hexode	7B38	23.0	0.2	250	250	150	750	−3.0
	TH2321	Triode Hexode	7B38	23.0	0.2	250	250	150	640	−3
	TH233	Triode Hexode	OM26	23.0	0.2	250	250	150	640	−3
	TP2620	Triode Pentode	9B5	26.0	0.2	250	250	200	650	−5
Mullard	DK1	Heptode	8S10	1.4	0.05	90	90	45	250	0
	TH2	Triode Hexode	7B10	2.0	0.23	135	60	−	430	−5.0
	FC2	Octode	7B8	2.0	0.1	135	70	70	200	0
	FC2A	Octode	7B8	2.0	0.13	135	45	45	270	−0.5
	TH4	Triode Hexode	7B37	4.0	1.0	250	70	−	1000	−1.5
	TH4A	Triode Hexode	7B37	4.0	1.45	250	100	100	750	−2.0
	TH4B	Triode Heptode	7B37	4.0	1.45	250	100	−	750	−2.5
	FC4	Octode	7B36	4.0	0.65	250	90	70	600	−1.5
	ECH3/33	Triode Hexode	8S29/058	6.3	0.2	250	100	−	650	−2.0
	ECH35	Triode Hexode	O58	6.3	0.3	250	100	−	650	−2.0
	EK2	Octode	8S28	6.3	0.2	250	200	50	550	−2.0
	EK3	Octode	8S28	6.3	0.72	250	100	−	650	−2.5
	CCH35	Triode Hexode	O58	7.0	0.2	250	100	−	650	−2.0
	FC13	Octode	8S28	13.0	0.2	200	90	70	600	−1.5
	FC13C	Octode	7B36	13.0	0.2	200	90	70	600	−1.5
	TH13C	Triode Hexode	7B37	13.0	0.31	250	70	130	1000	−1.5
	TH21C	Triode Hexode	7B37	21.0	0.2	250	70	−	1000	−1.5
	TH22C	Triode Hexode	7B37	29.0	0.2	250	150	100	−	−
	TH30C	Triode Heptode	7B38	29.0	0.2	250	100	−	750	−2.5
Osram	X14	Heptode	O10	1.4	0.05	110	60	−	250	−
	X21	Heptode	7B9	2.0	0.1	150	70	−	240	−
	X22	Heptode	7B9	2.0	0.15	150	70	150	350	0
	X23	Triode Hexode	7B10	2.0	0.3	150	60	150	250	−1.5
	X24	Triode Hexode	7B10	2.0	0.2	150	60	150	350	−1.5
	MX40	Heptode	7B35	4.0	1.0	250	100	250	500	−3.0
	X41	Triode Hexode	7B37	4.0	1.2	250	70	250	640	−1.5
	X42	Heptode	7B35	4.0	0.6	250	100	−	490	−
	X73M	Heptode	O58	6.0	0.16	250	80	250	500	−3.0
	X61M	Triode Hexode	O58	6.3	0.3	100	−	−	620	−
	X62	Triode Hexode	O58	6.3	1.27	250	120	250	1750	−1.5
	X63	Heptode	O54	6.3	0.3	250	100	250	400	−3.0
	X64	Hexode	O35	6.3	0.3	250	150	−	310	−6.0
	X65	Triode Hexode	O58	6.3	0.3	250	100	250	225	−3.0
	X30/32	Heptode	7B35	13.0	0.3	250	100	−	800	−
	X31	Triode Hexode	7B37	13.0	0.3	250	80	150	640	−1.5
	X71M	Triode Hexode	O58	13.0	0.16	250	100	−	520	−
	X75	Triode Hexode	O58	15.0	0.16	250	100	250	225	−3.0
Record	OC2	Octode	7B8	2.0	0.13	135	45	135	270	−1−12
	AC/OC4	Octode	7B37	4.0	0.65	250	70	90	700	−1.5−25
	AC/TH4	Triode Hexode	7B37	4.0	1.0	300	80	150	1000	−1.5−25
	OC/13	Octode	7B36	13.0	0.2	200	70	90	600	−1.5−25
	OC/13L	Octode	8S28	13.0	0.2	200	70	90	600	−1.5−25
	TH/21DA	Triode Hexode	7B37	21.0	0.2	200	80	150	1000	−1.5−25

Table 1. Frequency changers – *continued*

Make	Type	Description	Base	Fil. volts	Fil. amps	Anode volts	Screen volts	Oscil-lator volts	Conv. condct. Mbos	Bias volts
Triotron	O202	Octode	7B8	2.0	0.13	135	45	–	250	0–12
	O406	Octode	7B36	4.0	0.63	250	70	–	600	–1.5
	TH401	Triode Hexode	7B37	4.0	1.0	300	150	–	750	–2.0
	O1307	Octode	7B36	13.0	0.2	200	70	–	600	–1.5–25
Tungsram	VX2	Hexode	7B5	2.0	0.135	135	60	–	300	–1
	V02/S	Octode	7B9/8S11	2.0	0.13	135	45	135	270	–
	TH4A/B	Triode Heptode	7B38	4.0	1.5	275	100	100	750	–2.5
	TX4	Triode Hexode	7B37	4.0	1.0	250	80	150	1000	–1.5
	V04/S	Octode	7B36/8S28	4.0	0.65	250	70	90	600	1.5–25
	V06S	Octode	8S28	6.3	0.2	250	50	200	450	–2–25
	VX6S	Hexode	8S29	6.3	0.2	250	150	–	350	–3–35
	6E89	Triode Hexode	O58	6.3	0.3	250	100	150	650	–2
	6TH8G	Triode Hexode	O58	6.3	0.6	250	100	150	1000	–1.5–25
	ECH11	Triode Hexode	OF5	6.3	0.2	230	100	150	650	–2
	ECH2	Triode Heptode	8S29	6.3	0.85	250	100	100	750	–2.5
	ECH3/33	Triode Hexode	8S29/058	6.3	0.3	250	100	150	650	–2
	ECH35	Triode Hexode	O58	6.3	0.3	250	100	150	650	–2
	EK2	Octode	8S28	6.3	0.2	250	50	200	550	–2
	EK3	Octode	8S28	6.3	0.65	250	100	100	650	–2.5
	V013/S	Octode	7B36/8S28	13.0	0.2	250	70	90	600	1.5–25
	TX21	Triode Hexode	7B37	21.0	0.2	250	80	150	1000	–1.5
	TH29/30	Triode Heptode	7B38	29.0	0.2	275	100	100	750	–2.5
	MH1118	Heptode	7C4	10.0	0.18	250	100	200	520	–2.5

Table 2. Screen-grids and HF pentodes

Make	Type	Description	Base	Fil. volts	Fil. amps	Anode volts	Screen volts	Bias volts	Anode current (mA)	Screen current (mA)	Bias res. ohms	Slope mA/V
Brimar	8A1	P	5B19/7B23	4.0	1.0	200	80	-1.5	3.5	0.7	200	4.0
	9A1	VP	5B19/7B23	4.0	1.0	200	80	-1.5–30	5.0	1.0	200	4.25
	8D2	P	7B30	13.0	0.2	250	100	-3	2.0	0.5	1000	1.25
	9D2	VP	7B30	13.0	0.2	250	125	-3–40	10.5	2.6	200	1.65
	6J7G	P	–	6.3	0.3	250	100	-3	2.0	0.5	1000	1.25
	6K7G	VP	–	6.3	0.3	250	125	-3–40	10.5	2.6	200	1.65
Cossor	215SG	S	4BS	2.0	0.15	150	80	-1.0	1.25	–	–	1.1
	220SG	S	4B5	2.0	0.2	150	80	-1.0	1.4	–	–	1.6
	220VSG	VS	4B5	2.0	0.2	150	80	-2.5	2.25	–	–	1.6
	220VS	VS	4B5	2.0	0.2	150	80	-2.5	1.0	–	–	1.6
	210VPT	VP	4B8/7B4	2.0	0.1	150	80	-1.5	2.9	7.5	–	1.1
	210VPA	VP	4B8/7B4	2.0	0.1	150	150	-3.0	2.2	–	–	1.1
	210SPT	P	4B8/7B4	2.0	0.1	150	80	-1.5	1.2	–	–	1.3
	220IPT	P	7B28	2.0	0.2	150	80	-1.5	2.5	–	–	1.0
	MSG/HA	S	5B17	4.0	1.0	200	100	-1.5	2.1	–	600	2.0
	41MSG	S	5B17	4.0	1.0	200	80	-1.5	0.8	–	1500	2.5
	MSG/LA	S	5B17	4.0	1.0	200	100	-1.5	5.2	–	250	3.75
	MVSG	VS	5B17	4.0	1.0	200	100	-1.5	7.8	7.5	V	2.5
	4TSP	P	7B23	4.0	1.0	250	250	-3.0	12.0	–	–	8.0
	4TSP	P	7B23	4.0	1.0	250	250	-3.0	12.0	–	–	8.0
	MS/PEN	P	5B19/7B23	4.0	1.0	200	100	-1.5	5.0	–	–	2.8
	MS/PEN A	P	7B23	4.0	1.0	200	150	–	9.0	5.0	200	4.0
	MVS/PEN	VP	5B19/7B23	4.0	1.0	200	100	-1.5	4.3	–	V	2.2
	MS/PEN B	P	7B26	4.0	1.0	200	100	-1.5	5.0	–	V	2.8
	MVS/PEN B	VP	7B26	4.0	1.0	200	100	-1.5	4.3	–	V	2.2
	OM5	P	047	6.3	0.2	250	100	-2.0	3.0	–	V	1.8
	OM6	VP	047	6.3	0.2	250	100	-2.5	6.0	–	V	2.2
	13VPA	VP	7B26	13.0	0.2	200	100	-3.0	7.0	–	V	1.8
	13SPA	P	7B26	13.0	0.2	200	100	-3.0	2.3	–	–	1.25
	DVSG	VS	5B17	16.0	0.25	200	100	-1.5	7.5	–	V	2.5
	DS/PEN	P	5B19	16.0	0.25	200	100	-1.5	4.7	–	–	2.3
	DVS/PEN	VP	5B19	16.0	0.25	200	100	-1.5	5.5	–	V	2.0
	202VP	VP	7B23	20.0	0.2	250	100	-1.5	4.3	–	V	2.2
	202VBP	VP	7B26	20.0	0.2	250	100	-1.5	4.3	–	V	2.2
	202SPB	P	7B26	20.0	0.2	250	100	-1.5	4.8	–	–	2.8
	4TPB	P	7B26	4.0	1.0	250	200	-3.0	12.0	–	–	8.0
	41MPT	VP	7B23	4.0	1.0	250	200	-1.5	12.0	–	–	4.8
	42MPT	P	7B23	4.0	2.0	250	250	-3.0	34.0	–	–	8.5
	42PTB	P	7B26	4.0	2.0	250	250	-3.0	34.0	–	–	8.5
	41MTS	Split anode P	7B43	4.0	1.0	250	100	–	–	–	–	–
	4TSA	"	7B44	4.0	1.0	250	100	–	–	–	–	–
	42SPT	P	7B23	4.0	2.0	500	250	-15	27.0	–	–	11.0

Table 2. Screen-grids and HF pentodes – *continued*

Make	Type	Description	Base	Fil. volts	Fil. amps	Anode volts	Screen volts	Bias volts	Anode current (mA)	Screen current (mA)	Bias res. ohms	Slope mA/V
Dario	PF462	P	7B4	2.0	0.18	150	150	-0.5	3.0	-	-	1.85
	PF472	VP	7B4	2.0	0.18	150	150	-0.5-16	2.5	-	-	1.7
	TB622	S	4B5	2.0	0.18	150	90	-0.5	2.0	-	-	1.4
	TB552	VS	4B5	2.0	0.15	150	75	0.9	1.8	-	-	1.5
	TE424	S	5B17	4.0	1.0	200	100	-1.3	1.5	-	-	0.9
	TE524	S	5B17	4.0	1.0	200	100	-2.0	3.0	-	-	2.0
	TE554	VS	5B17	4.0	1.0	200	100	-1.5-40	3.0	-	V	2.0
	TE464	P	5B19/7B23	4.0	1.1	200	100	-2.0	3.0	-	-	2.3
	TF44	P	7B26	4.0	0.65	250	250	-2.4	4.0	-	-	3.4
	TE474	VP	5B19/7B23	4.0	1.1	250	100	-1.5-30	4.5	-	V	2.3
	TE564	VP	5B19/7B23	4.0	1.2	200	100	-2.0-22	4.25	-	V	2.5
	TF64	VP	7B26	4.0	0.65	250	250	-3.0-45	11.5	-	V	-
	TF713	P	7B23	13.0	0.2	200	100	-2.0	3.0	-	-	2.1
	TF313	VP	7B26	13.0	0.2	200	100	-3.0-50	8.0	-	V	1.8
	TB5613	VP	7B26	13.0	0.2	200	100	-2.0-22	4.5	-	V	2.2
	TB4620	P	5B19	20.0	0.18	200	-	-2.0	3.0	-	-	2.2
	TB4720	VP	5B19	20.0	0.18	200	-	-2.0-50	4.0	-	V	2.0
Elco	VP41	VP	7B26	4.0	0.65	250	250	-3.0-40	12.0	4.5	180	3.5
	VPU1	VP	7B26	13.0	0.2	250	250	-3.0-40	12.0	4.5	180	3.5
Ever ready	K50M	VP	4B8/7B4	2.0	0.18	135	135	0.7	3.0	-	V	1.5
	K50N	VP	7B5	2.0	0.14	135	60	-1.5	2.0	-	V	1.4
	K40B	S	4B5	2.0	0.18	150	90	0	2.9	-	-	1.5
	K40N	VS	4B5/7B4	2.0	0.18	150	90	0.7	2.5	-	V	1.4
	K40M	VS	5B17/7B23	4.0	1.0	200	110	-1.5-40	6.0	-	V	2.5
	A50M	VP	5B19/7B23	4.0	1.0	200	100	-2-50	4.5	-	V	2.3
	A50N	VP	5B19/7B23	4.0	1.2	200	100	-2.0	4.25	-	V	2.5
	A50P	VP	7B26	4.0	0.65	250	250	-3.0	11.5	-	V	2.0
	A50A	P	5B19/7B23	4.0	1.0	200	100	-2.0	3.0	-	V	2.3
	A50B	P	7B26	4.0	1.65	250	250	-2.4	4.0	-	-	3.4
	EF9/39	VP	8S24/04	6.3	0.2	250	100	-2.5	6.0	-	-	2.2
	C50N	VP	7B26	13.0	0.2	250	200	-2.0	9.0	-	V	2.2
	C50B	P	7B26	13.0	0.2	200	200	-2.2	2.5	-	-	2.8
Ferranti	VS2	VS	4B5	2.9	9.1	150	70	-	-	-	-	1.0
	VPT2	VP	7B4	2.0	0.15	150	75	-	-	-	-	1.6
	SPT4A	P	7B23	4.0	1.0	250	100	-1.5	2.0	1.0	-	3.0
	VPT4	VP	5B19	4.0	1.0	250	100	-3.0	5.5	3.0	V	-
	VPT4B	VP	7B23	4.0	1.0	250	100	-2.0	6.0	3.0	V	3.6
	SPTS	P	7B23	13.0	0.3	250	100	-1.5	2.0	1.0	-	3.0
	VPTS	VP	7B23	13.0	0.3	250	100	-3.0	5.5	2.0	V	-
	VPTA	VP	7B23	13.0	0.2	250	100	-	4.2	2.0	V	-
	VPTSB	VP	7B23	13.0	0.3	250	100	-2.0	6.0	3.0	V	3.6
Hivac	XSG 1.5V	S	4D2	1.5	0.08	50	30	0	0.55	0.25	-	0.30
	XW 1.5V	P	5D1	1.5	0.08	50	45	0	0.75	0.2	-	0.52
	XSG 2.0V	S	4D2	2.0	0.08	50	30	0	0.6	0.3	-	0.4

Make	Model	Type	Base									
Lissen	XVS 2.0V	VS	4D2	2.0	0.08	50	30	0	0.4	0.15	—	0.33
	XW 2.0V	P	5D1	2.0	0.08	50	45	0	0.95	0.3	—	0.60
	SG215	S	4B5	2.0	0.15	150	75	-1.5	2.7	0.8	—	1.0
	SG220	S	4B5	2.0	0.2	150	70	-1.5	2.4	0.9	—	1.5
	SG220SW	S	4B10	2.0	0.2	150	70	-1.5	2.4	0.9	V	1.5
	VS215	VS	4B5	2.0	0.15	150	75	0.14	6.0	1.7	V	1.0
	HP215	P	4B5/7B4	2.0	0.15	150	70	-1.5	1.5	0.3	—	1.2
	VP215	VP	4B5/7B4	2.0	0.15	150	70	0.9	3.75	0.75	—	1.25
	AC/SL	S	5B17	4.0	1.0	200	80	-1	3.8	0.4	250	2.2
	AC/SH	S	5B17	4.0	1.0	200	80	-1.5	7.4	0.5	200	3.5
	AC/VS	VS	5B17	4.0	1.0	200	80	-1.5 –40	4.4	0.6	V	3.0
	AC/VH	VS	5B17/7B23	4.0	1.0	200	80	-1.5 –40	9.3	1.6	V	3.3
	AC/HP	P	5B17/7B23	4.0	1.0	200	100	-2	4.2	1.4	350	3.2
	AC/VP	VP	5B17/7B23	4.0	1.0	200	100	-1.5 –30	5.7	2.3	V	3.0
	VP13	VP	7B23	13.0	0.3	200	100	-1.5 –30	6.3	2.0	V	3.0
	SG215	S	4B5	2.0	0.15	150	80	—	—	—	—	1.1
	SG2V	VS	4B5	2.0	0.15	150	80	—	—	—	—	1.2
	SG410	S	4B5	4.0	0.1	150	80	—	—	—	—	1.25
	AC/SG	S	5B17	4.0	1.0	200	80	—	—	—	—	3.25
	AC/SGV	VS	5B17	4.0	1.0	250	90	—	—	—	—	3.5
Marconi	Z14	P	O7	1.4	0.05	90	70	0	1.2	0.25	—	0.75
	S23	S	4B5	2.0	0.1	150	70	-1.5	1.3	0.6	—	1.1
	S24	S	4B5	2.0	0.1	150	75	0	4.5	0.5	—	1.4
	VS2	VS	4B5	2.0	0.1	150	75	0.9	4.5	0.5	—	1.25
	VS24	VS	4B5	2.0	0.15	150	150	0.9	4.4	0.3	—	1.5
	VS24/K	P	4B8/7B4	2.0	0.15	150	60	0	1.7	0.6	—	1.5
	Z21	VP	7B4	2.0	0.1	150	150	0	2.8	0.7	—	1.7
	VP21	VP	4B8/7B4	2.0	0.1	150	70	-1.5	3.0	0.9	—	1.1
	W21	S	5B17	2.0	0.1	150	80	-1.5	2.4	0.3	—	1.4
	MS4	S	5B17	4.0	1.0	250	80	-2.0	2.5	1.2	550	1.1
	MS4B	S	5B17	4.0	1.0	250	80	—	—	—	440	3.2
	MS4B/K	S	5B17	4.0	1.0	250	80	-2	9	2	V	3.2
	VMS4	VS	5B17	4.0	1.0	250	80	-2	9	2	V	2.4
	VMS4/K	VS	5B17	4.0	1.0	250	80	-1	5	1.2	V	2.6
	VMS4B	VS	5B17	4.0	1.0	250	100	-1.75	3.3	1.0	400	2.9
	MSP4	P	5B17/7B23	4.0	1.0	250	240	-4	8.5	3.2	—	4.0
	MSP41	P	5B17/7B23	4.0	1.0	250	100	-2	3.0	1.0	V	3.2
	VMP4	VP	5B17/7B23	4.0	1.0	250	100	-2	7.0	3.5	V	3.5
	VMP4/K	VP	5B17	4.0	1.0	250	100	-2.0	8.0	5.0	V	2.5
	VMP4G	VP	7B23	4.0	1.0	250	125	-3.0	7.6	1.9	—	2.7
	W42	T	7B30	4.0	0.6	250	250	-2.5	8.0	2.25	V	1.5
	KTZ41	P	7B41	6.3	1.5	250	125	-3.0	7.6	0.5	V	7.5
	Z63	VP	O47	6.3	0.3	250	100	-3.0	8.0	1.9	V	1.225
	W63	VP	O47	6.3	0.3	250	100	-3.0	7.6	2.3	—	1.5
	KTW61	VP	O47	6.3	0.3	250	100	-3.0	8.0	1.9	V	2.9
	KTW63	T	O47	6.3	0.3	250	125	-3.0	7.6	0.5	V	1.5
	KTZ63	VP	7B23	13.0	0.3	250	250	—	2.0	—	V	1.225
	W30	VP	7B23	13.0	0.3	250	100	-1.0	—	—	—	4.5
	W31	S	5B17	16.0	0.25	200	—	—	—	—	V	4.0
	DS	S	5B17	16.0	0.25	200	70	—	—	—	V	1.1
	DSB						80	—	—	—	—	3.2

195

Table 2. Screen-grids and HF pentodes – *continued*

Make	Type	Description	Base	Fil. volts	Fil. amps	Anode volts	Screen volts	Bias volts	Anode current (mA)	Screen current (mA)	Bias res. ohms	Slope mA/V
Marconi – *continued*												
	VDS	VS	5B17	16.0	0.25	200	80	–	–	–	V	2.4
	VDSB	VS	5B17	16.0	0.25	200	–	–	–	–	V	2.2
	S12	S	4D2	2.0	0.06	100	30	0	2.5	0.4	–	0.7
	ZA1	–	Acorn	4.0	0.25	250	100	-3.0	2.0	0.7	1500	1.4
	Z62	P	O47	6.3	0.45	300	150	-2.0	10.0	2.0	–	7.5
	ZA2	P	Special	6.3	0.15	250	100	-3.0	2.0	0.7	–	1.4
Mazda	SP141	P	OM7	1.4	0.05	90	90	–	1.8	–	–	0.8
	SG215	S	4B8/7B4	2.0	0.15	150	80	-1.5	1.5	0.25	–	1.1
	S215A	S	4B8/7B4	2.0	0.15	150	80	–	1.9	0.3	–	1.1
	S215B	S	4B8/7B4	2.0	0.15	150	80	-1.5	1.5	0.3	–	1.7
	S215VM	VS	4B8/7B4	2.0	0.15	150	80	0.8	1.0	0.15	–	1.4
	SP210	P	7B4	2.0	0.1	150	150	-1	1.1	0.33	–	1.2
	SP215	P	7B4	2.0	0.15	150	150	-1.5	1.35	0.47	–	1.3
	VP210	VP	7B4	2.0	0.1	120	70	-1.5	1.8	0.63	–	1.03
	VP215	VP	7B4	2.0	0.15	150	150	-1.5	1.1	0.385	–	0.82
	SP22	P	OM3	2.0	0.1	150	150	-1.0	1.1	0.38	–	1.2
	VP22	VP	OM3	2.0	0.1	150	140	-1.5	1.2	0.32	–	0.02
	VP23	VP	OM3	2.0	0.05	150	150	-2.0	1.0	0.35	–	0.8
	AC/SG	S	7B23	4.0	1.0	200	80	-1.5	8.5	–	–	2.4
	SC/S2	S	5B17	4.0	1.0	200	80	-1.5	7.0	–	–	4.4
	AC/S1VM	VS	5B17	4.0	1.0	200	100	-1.5	5.7	–	–	1.1
	AC/SGVM	VS	5B19/7B23	4.0	1.0	200	80	-2	5.8	–	–	2.0
	AC/S2Pen	P	7B23	4.0	1.0	250	150	-4.25	5.25	1.75	–	5.5
	AC/SP1	P	7B23	4.0	1.0	250	250	-3.0	4.9	4.1	300	2.65
	AC/VP1	VP	7B23	4.0	0.65	250	250	-4	8.8	2.2	–	2.0
	AC/VP2	VP	7B26	4.0	0.65	250	250	-4	8.8	2.2	–	2.0
	VP41	VP	OM24	4.0	0.65	250	250	-4	8.6	2.3	–	2.0
	SP41	P	OM24	4.0	0.95	250	250	-2.1	11.1	2.8	150	8.4
	SP42	P	OM24	4.0	0.95	200	200	-1.25	27.0	6.75	37	9.0
	SP1320	P	7B23	13.0	0.2	250	250	-2.0	3.5	0.3	–	1.9
	VP1320	VP	7B23	13.0	0.2	250	250	-1.5	4.7	1.25	–	2.0
	VP1321	VP	7B23	13.0	0.2	250	250	-4	8.8	2.2	–	2.0
	VP1322	VP	7B26	13.0	0.2	250	250	-4	8.8	2.2	–	2.0
	VP133	VP	OM24	13.0	0.2	200	200	-0.7	7.2	2.0	–	2.35
	SP2220	P	7B23	22.0	0.2	250	250	-3.0	4.9	4.1	–	2.65
	DC2/SG	S	7B23	20.0	0.1	200	100	-1.5	11.0	–	–	2.4
	DC2/SGVM	VS	7B23	20.0	0.1	200	100	-4	8.0	–	–	1.6
Mullard	DF1	P	8S6	1.4	0.05	90	90	0	1.2	–	–	0.75
	DF5	P	4D23	1.5	0.067	45	13.5	0	0.125	–	–	0.17
	DAS1	T	4D2	2.0	0.06	120	60.0	-2.7	1.5	–	–	0.58
	SP2	P	7B4	2.0	0.18	135	135.0	0	3.0	1.0	–	1.8
	PM12	T	4B5	2.0	0.15	150	75	–	4.25	–	–	1.1
	PM12A	T	4B5	2.0	0.18	135	75	0	2.0	–	–	1.5
	PM12M	VT	4B5	2.0	0.18	150	90	0.7	2.5	–	–	1.4

Type	Class	Base	V_h	I_h	V_a	V_s	V_g	I_a	g_m	r_a	μ
VP2	VP	7B4	2.0	0.18	135	135	0.7	3.0	1.25	–	1.5
VP2B	VP	7B5	2.0	0.14	135	60	-1.5	2.0	–	–	1.4
AP4	P	ACORN	4.0	0.2	250	100	-3.0	2.0	0.7	600	1.4
S4V	S	5B17	4.0	1.0	200	75	-1.0	1.5	–	460	1.1
S4VA	T	5B17	4.0	1.0	200	110	-1.5	2.75	–	–	2.0
S4VB	VT	5B17	4.0	1.0	200	110	-1.5–40	4.6	–	V	2.5
MM4V	VT	5B17	4.0	1.0	200	100	-1.5–40	6.0	–	200	1.2
VM4V	VS	5B17	4.0	1.0	250	250	-3.0	8.5	–	250	4.7
TSP4	P	7B26	4.0	1.3	200	100	-2.0	10.5	2.0	–	2.3
SP4	P	5B19/7B23	4.0	0.65	250	250	-2.4	3.0	–	500	3.4
SP4B	P	7B26	4.0	1.0	200	100	-2–50	4.0	1.5	–	2.3
VP4	VP	5B19/7B23	4.0	1.2	200	100	-2.0	4.5	–	V	2.5
VP4A	VP	5B19/7B23	4.0	0.65	250	250	-3.0	4.25	1.8	V	2.0
VP4B	VP	7B26	4.0	0.15	250	100	-3.0	11.5	4.25	V	1.4
4762	P	ACORN	6.3	0.2	250	100	-3–50	2.0	–	–	1.7
EF5	VP	8S24	6.3	0.2	250	100	-2.0	8.0	–	V	1.8
EF6/36	P	8S24/047	6.3	0.2	250	250	-2.5	3.0	–	–	1.8
EF8/38	P	8S25/051	6.3	0.2	250	100	-2.5	8.0	–	–	2.2
EF9/39	P	8S24/047	6.3	0.2	250	100	-2.0	6.0	–	–	2.2
SP13	P	8S24	13.0	0.2	200	200	-2.2	3.3	–	400	2.8
SP13C	P	7B26	13.0	0.2	200	100	-2.0	2.5	0.9	600	2.2
VP13A	VP	8S24	13.0	0.2	200	200	-2.0	4.0	1.4	V	2.2
VP13C	VP	7B26	13.0	0.2	200			9.0	3.6	V	0.75
Z14	P	O7	1.4	0.05	90	90	-1.5	–	–	–	1.1
S23	S	4B5	2.0	0.1	150	70	0	1.3	0.6	–	1.4
S24	S	4B5	2.0	0.15	150	70	0.15	4.5	0.5	–	1.25
VS2	VS	4B5	2.0	0.15	150	75	0.9	5.0	2.0	–	1.5
VS24	VS	4B5	2.0	0.15	150	75	0.9	4.5	0.5	–	1.5
VS24K	VS	4B5	2.0	0.15	150	75	0.9	4.4	0.3	–	1.5
VS24K	VS	4B5	2.0	0.1	150	75	-0.5	4.4	0.3	–	1.7
Z21	P	4B8/7B4	2.0	0.1	150	150		1.7	0.6	–	1.4
Z22	P	7B4	2.0	0.1	150	150	0	–	–	–	1.1
VP21	VP	7B4	2.0	0.1	150	60	0	2.8	0.7	–	1.4
W21	VP	4B8/7B4	2.0	1.0	150	150	-1.5	3.6	1.2	–	1.1
MS4	S	5B17	4.0	1.0	250	70	-2.0	2.4	0.3	–	3.2
MS4B	S	7B23	4.0	1.0	250	80	-2–30	3.4	1.2	550	3.2
VMS4	VS	5B17	4.0	1.0	250	80	-1–15	7.5	2.0	250	2.4
VMS4B	VS	5B19/7B23	4.0	1.0	250	80		5.0	1.2	V	2.9
VMS4/B	VS	5B19/7B23	4.0	1.0	250	80	-1.75	–	–	V	2.0
MSP4	P	5B17/7B23	4.0	1.0	240	100	-4.0	3.3	1.0	400	4.0
MSP41	P	5B17/7B23	4.0	1.0	250	240		9.0	3.2	–	3.2
VMP4	VP	5B19/7B23	4.0	1.0	250	100	-2.0	–	–	V	3.5
VMP4G	VP	7B23	4.0	1.0	250	100	-3.0	8.0	5.0	V	2.8
W42	VP	7B30	4.0	0.6	250	125	-1.5	7.6	1.9	65	1.5
KTZ41	T	7B41	4.0	1.5	250	250	-3.0	18.0	5.25	1000	12.0
KTZ73	P	O30	6.0	0.16	250	100	-3.0	2.0	0.25	V	1.5
KTW73M	T	O30	6.0	0.17	250	100		6.5	1.3	V	1.7
KTW74M	T	O30	13.0	0.16	250	150	-2.0	–	–	–	1.5
Z62	P	O47	6.3	0.45	300	125	-3.0	10.0	2.3	160	7.5
Z63	P	O50	6.3	0.3	250	100	-4–40	2.0	0.5	1200	1.225
W63	VP	O50	6.3	0.3	250			7.6	1.9	V	1.5

197

Table 2. Screen-grids and HF pentodes – *continued*

Make	Type	Description	Base	Fil. volts	Fil. amps	Anode volts	Screen volts	Bias volts	Anode current (mA)	Screen current (mA)	Bias res. ohms	Slope mA/V
Ostram – *continued*	KTW61	VP	O47/O50	6.3	0.3	250	100	–3.0	.0	2.3	V	2.9
	KTW63	VP	O50	6.3	0.3	250	125	–3.0	7.6	1.5	V	1.5
	KTZ63	T	O50	6.3	0.3	250	125	–3.0	2.0	0.5	1200	1.23
	W30	VP	7B23	13.0	0.3	250	250	–	–	–	V	4.5
	W31	VP	7B23	13.0	0.3	250	100	–2.5	8.1	5.0	V	2.78
	DS	S	5B17	16.0	0.25	200	70	–	–	–	–	1.1
	DSB	S	7B23	16.0	0.25	200	80	–	–	–	–	3.2
	VDS	VS	5B17	16.0	0.25	200	80	–	–	–	–	2.4
	VDSB	VS	5B17	16.0	0.25	200	–	–	–	–	–	2.2
	S12	T	4D2	2.0	0.06	100	30	0	2.5	0.4	–	0.7
	ZA2	P	Acorn	6.3	0.15	250	100	–	–	–	–	1.4
Record	S2	S	4B5	2.0	0.12	150	75	–0.9	1.5	0.3	–	1.4
	VS2	VS	4B5	2.0	0.12	150	75	–0.5	1.0	0.1	–	1.5
	HFP2	P	4B5	2.0	0.12	150	150	–1.5	1.9	0.7	–	1.9
	VHP2	VP	7B23	2.0	0.12	150	150	–0.9–17	2.5	0.6	–	1.7
	AC/S	S	5B17	4.0	1.0	200	100	–2.0	3.0	0.8	500	3.0
	AC/VS	VS	5B19/7B23	4.0	1.2	200	100	–1.5–40	3.0	0.8	V	3.0
	AC/VFP	P	7B26	4.0	1.0	200	100	–2.0	3.5	0.6	600	3.5
	AC/HPB	P	5B19/B23	4.0	0.65	250	250	–2.0	2.9	0.8	500	4.0
	AC/VHFP	VP	7B26	4.0	1.0	200	100	–2.0–35	5.0	1.3	V	3.5
	AC/VHPB	VP	7B26	4.0	0.65	250	250	–1.0–50	10.0	2.5	V	4.0
	HFP/13	P	7B26	13.0	0.2	200	100	–2.0	3.0	1.5	450	2.4
	HFP/13L	P	8S24	13.0	0.2	200	100	–2.0	3.0	1.5	450	2.4
	HPB/13	P	7B26	13.0	0.2	200	200	–1.5	3.5	1.5	300	3.5
	VHFP/13	VP	7B26	13.0	0.2	200	100	–1.0–10	8.0	2.9	V	3.5
	VHFP/13L	VP	8S24	13.0	0.2	200	100	–1.0–10	8.0	2.9	V	3.5
	VHP/13	VP	7B26	13.0	0.2	200	100	–3.0–55	8.0	2.6	V	2.8
	VHP/13L	VP	8S24	13.0	0.2	200	100	–3.0–55	8.0	2.6	V	2.8
	NHPB/13	VP	7B26	13.0	0.2	200	200	–1.0–50	10.0	3.5	V	3.5
Triotron	S217	VP	7B4	2.0	0.2	150	150	–0.5–16	2.5	–	–	1.7
	S218	P	7B4	2.0	0.2	150	150	–0.5	3.0	–	–	1.85
	S215	S	4B5	2.0	0.18	150	90	–0.5	2.0	–	–	1.4
	S213	VS	4B5	2.0	0.15	150	75	0.9	4.0	–	–	1.5
	S434N	VP	5B19/7B23	4.0	1.1	200	100	–1.5–30	4.5	–	V	3.5
	S420	P	7B26	4.0	0.65	250	250	–3.0	11.5	–	V	–
	S440	P	7B26	4.0	0.65	250	250	–2.4	4.0	–	–	3.4
	S435N	P	5B19/7B23	4.0	1.1	200	100	–2.0	3.0	–	–	3.5
	S415N	VS	5B17	4.0	1.0	200	100	–1.5–40	3.0	–	V	–
	S410N	S	5B17	4.0	1.0	200	60	–1.3	1.5	–	–	1.0
	S430N	S	5B17	4.0	1.0	200	100	–2.0	3.0	–	V	2.0
	S1324	P	7B26	13.0	0.2	200	100	–2.0	3.0	–	–	2.4
	S1328	S	8S24	13.0	0.2	200	100	–2.0	3.0	–	–	2.4
	S1323	VP	7B26	13.0	0.2	200	100	–3–55	8.0	–	V	1.8
	S2034N	VP	5B19	20.0	0.18	200	100	–2–35	5.0	–	V	3.5
	S2035N	P	5B19	20.0	0.18	200	100	2.0	3.0	–	V	3.5

Tungsram

Type		Base									
SE211	VS	4B5	2.0	0.12	150	75	-9.5	1.0	0.1	—	1.3
SE211C	VS	4B5	2.0	0.12	150	75	-5	1.0	0.1	—	1.5
HP210	P	4B8/7B4	2.0	0.12	150	150	-1.5	1.9	0.7	—	1.9
HP210C	P	7B4	2.0	0.12	150	150	-1.5	1.9	0.7	—	1.9
HP210NC	P	4B8/7B4	2.0	0.12	150	150	-1.5	1.9	0.7	—	1.9
SP2B	P	7B3	2.0	0.05	135	135	-0.5	2.6	1.0	—	1.0
SP2D	P	7B3	2.0	0.1	150	150	-1.0	1.45	0.35	—	1.7
SS210	T	4B5	2.0	0.12	150	75	-0.9	1.5	0.3	—	1.4
VP2B	VP	7B3	2.0	0.05	135	135	0.15	2.5	0.8	—	0.65
VP2D	VP	7B3	2.0	0.1	150	150	-1.5–12	1.3	0.6	—	2.0
HP211C	VP	7B4	2.0	0.12	150	150	-0.17	2.6	0.6	V	1.7
AS4125	VS	5B17	4.0	1.2	200	100	-1.5–40	3.0	0.8	V	3.0
AS4120	T	5B17	4.0	1.0	200	100	-2.0	3.0	0.8	500	3.0
HP4101	P	5B19/7B23	4.0	1.0	250	100	-3.0	3.5	1.8	600	3.5
HP4115	P	7B23	4.0	1.0	200	100	-2.0	4.5	1.5	150	3.2
SP4B	P	7B26	4.0	0.65	250	250	-3.0	3.2	1.5	500	4.0
HP4106	VP	5B19/7B23	4.0	1.0	250	100	-1.5–35	5.0	1.3	V	3.5
VP4B	VP	7B26	4.0	0.65	250	250	-1.50	6.0	2.5	V	4.0
EF12	P	OF3	6.3	0.2	300	250	-2.0	3.0	1.0	500	3.0
SP6S	P	8S24	6.3	0.2	250	100	-2.0	3.0	1.0	500	2.0
VP6S	VP	OF3	6.3	0.2	250	100	-3.0–50	8.0	2.5	V	1.7
EF11	VP	8S24	6.3	0.2	300	125	-2.0	6.0	2.0	250	2.2
EF6	P	8S24	6.3	0.2	300	125	-2.0	3.0	1.1	—	2.0
EF5	VP	8S24	6.3	0.2	250	125	-3.0–50	8.0	2.6	V	1.7
EF9/39	VP	8S24/O47	6.3	0.2	300	300	-2.5–55	6.0	1.7	V	2.2
SP13	P	7B26	13.0	0.2	200	100	-2	3.0	1.5	450	2.4
SP13B	P	7B26	13.0	0.2	200	200	-1.5	3.5	1.5	—	3.5
HP13	VP	7B26	13.0	0.2	200	100	0.10	8.0	2.9	V	3.5
VP13	VP	7B26	13.0	0.2	200	100	-3.55	8.0	2.6	V	2.8
VP13B	VP	7B26	13.0	0.2	250	200	-1.50	10.0	2.0	V	3.5
EF8	HF HEX	8S25	6.3	0.2	250	250	-2.5	8.0	0.25	—	1.8
HP2118	VP	5B19	20	0.18	200	100	-2.0	5.0	1.1	—	3.5
HP2018	P	5B19	20	0.18	200	100	-2.0	4.0	1.2	—	3.5
HP1118	VP	7C3	10	0.18	250	100	-3.0	8.2	2.0	—	1.6
HP1018	P	7C3	10	0.18	250	100	-3.0	2.0	0.5	—	1.22
SS2018	S	5B17	20	0.18	200	100	-3.0	3.0	1.0	—	3.0
S2018	S	5B17	20	0.18	200	60	-3.0	4.0	1.2	—	1.2

Table 3. Diodes

Make	Type	Description	Base	Filament		Max. diode volts	Max. diode current
				volts	amps		
Brimar	10D1	DD	5B12	13.0	0.2	—	—
	6H6G	DD	—	6.3	0.3	—	—
Cossor	220DD	DD	5B12	2.0	0.2	—	—
	DDL4	DD	5B12	4.0	0.75	—	—
	DD4	DD	5B12	4.0	0.75	—	—
	OM3	DD	O38	6.3	0.2	—	—
Dario	TB24	DD	5B12	4.0	0.65	—	—
	TB213	DD	5B12	13.0	0.2	—	—
Ever Ready	A20B	DD	5B12	4.0	0.65	200	0.8
	EB34	DD	O38	6.3	0.2	200	0.8
	C20C	DD	5B12	13.0	0.2	200	0.8
Ferranti	ZD	DD	5B12	7.0	0.2	—	—
Hivac	Ac/DD	DD	5B12	4.0	1.0	—	—
Marconi	D41	D	5B12	4.0	0.3	—	—
	D42	DD	4B18	4.0	0.6	75	15.0
	D63	DD	O38	6.3	0.3	100	2.0
Mazda	DD207	DD	4B3	2.0	0.075	—	—
	DD41	DD	OM18	4.0	0.5	—	—
	V914	DD	5B12	4.0	0.3	—	1.0
	DD620	DD	5B12	6.0	0.2	—	1.0
	DD101	DD	OM18	10.0	0.2	—	—
Mullard	2D2	DD	5B12	2.0	0.9	125	0.5
	2D4A	DD	5S2	4.0	0.65	200	0.8
	2D4B	DD	7B16	4.0	0.35	200	0.8
	EB34	DD	8S18/O38	6.3	0.2	200	0.8
	EAB1	DDD	8S19	6.3	0.2	200	0.8
	2D13	DD	5B12	13.0	0.2	200	0.8
	2D13A	DD	5S1	13.0	0.2	200	0.8
	2D13C	DD	5B12	13.0	0.2	200	0.8
Osram	D41	D	5B12	4.0	0.3	—	—
	D42/43	D	4B18/4B19	4.0	0.6	75	15.0
	D63	DD	O38	6.3	0.3	100	2.0 each
Record	Ac/DD4A	DD	5B12	4.0	0.65	100	0.8
	DDA/13	DD	5B12	13.0	0.2	200	0.8
	DDA/13L	DD	8S16	13.0	0.2	200	0.8
Triotron	D400	DD	4B12	4.0	0.65	200	0.8
	D1300	DD	8S16	13.0	0.2	200	0.8
Tungsram	DD4	DD	5B12	4.0	0.65	200	0.8
	DD4D	DD	7B16	4.5	0.4	100	4.0
	D418	D	4B15	4.0	0.18	200	1.5
	DD6DS	DD	8S16	6.3	0.2	200	0.8
	EB4	DD	8S18	6.3	0.2	200	0.8
	EAB1	DDD	8S19	6.3	0.2	200	0.8
	DD13/13S	DD	5B12/8S16	13.0	0.2	200	0.8
	DD18	DD	5B11	8.0	0.18	100	1.5

Table 4. Diode combinations

Make	Type	Description	Base	Fil. volts	Fil. amps	Anode volts	Screen volts	Amp. factor	Slope (mA/V)	Bias volts	Bias res. (ohms)	Anode current (mA)	Output (mW)
Brimar	11A2	DDT	7B19	4.0	1.0	200	—	50	2.8	-2.0	—	3.0	—
	11D3	DDT	7B19	13.0	0.2	250	—	100	1.1	-2.0	5000	0.4	—
	11D5	DDT	7B19	13.0	0.15	250	—	40	1.5	-3.0	750	3.8	—
	6Q7G	DDT	—	6.3	0.3	250	—	70	1.2	-3	4000	1.1	—
	6R7G	DDT	—	6.3	0.3	250	—	16	1.9	-9	1000	9.5	—
Cossor	21ODDT	DDT	5B2	2.0	0.1	150	—	27.5	1.1	0	—	2.3	—
	2102	DDT	6UX2	2.0	0.12	150	—	30	1.3	0	—	2.5	—
	DDT	DDT	7B19	4.0	1.0	200	—	41	2.4	-3.0	—	3.4	—
	DD/Pen	DDP	7B34	4.0	1.0	250	200	—	2.7	-2.5	—	5.0	—
	420TDD	DDP	7B22	4.0	2.0	250	250	—	7.0	-5.5	—	34.0	—
	13DHA	DDT	7B19	13.0	0.2	250	—	125	1.5	-1.5	—	1.0	—
	DDT16	DDT	7B19	16.0	0.25	200	—	40	2.5	-3.0	—	5.0	—
	202DDT	DDT	7B19	20.0	0.2	200	—	41	2.4	-3.0	—	3.5	—
Dario	BBC12	DDT	5B2	2.0	0.1	130	—	16	1.5	-4.2	—	2.5	—
	TBC14	DDT	7B19	4.0	0.65	250	—	27	2.0	-7.0	—	4.0	—
	TE444	D Tetrode	7B50	4.0	1.1	200	33	1000	3.0	-2.3	—	0.35	—
	TBL14	DDP	7B32	4.0	2.25	250	250	—	9.5	—	—	36.0	—
	TBL44	DDP	7B33	4.0	2.25	250	250	—	9.5	—	—	36.0	—
	TBC113	DDT	7B19	13.0	0.2	200	—	27	2.0	-5.0	—	4.0	—
Ekco	DT41	DDT	7B19	4.0	0.65	200	—	29	3.0	-3.5	470	7.5	—
Ever Ready	K23A	DDT	5B2	2.0	0.1	150	—	16.5	1.4	-1.5	—	2.5	—
	K23B	DDT	5B2	2.0	0.12	135	—	30	1.2	-1.5	—	1.95	—
	A23A	DDT	7B19	4.0	0.65	250	—	27	2.0	-7.0	—	4.0	—
	A27D	DDP	7B33	4.0	2.25	250	250	—	10.0	-6.0	—	36.0	4300
	EBC3/EBC33	DDT	8S21/041	6.3	0.2	275	—	30	2.0	-6.25	—	5.0	—
	EBL1/31	DDP	8S27/053	6.3	1.5	250	250	—	9.5	-6.0	—	36.0	4300
	C23B	DDT	7B19	13.0	0.2	200	—	27	2.0	-5.0	—	4.0	—
Ferranti	H2D	DDT	SB2	2.0	0.1	150	—	39	—	—	—	4.5	—
	H4D	DDT	7B19	4.0	1.0	250	—	39	2.7	-3.0	—	4.5	—
	PT4D	DDP	7B32	4.0	2.0	250	250	—	7.5	-6.0	140	7.5	3600
	HSD	DDT	7B19	13.0	0.3	200	—	39	2.7	-3.0	—	4.5	—
	HAD	DDT	7B19	13.0	0.2	200	—	39	2.7	-3.0	—	3.3	—
	PTSD	DDP	7B32	13.0	0.3	250	250	—	7.5	-6.0	140	7.5	3600
Hivac	DDT215	DDT	5B2	2.0	0.15	150	—	20	1.6	-3.0	—	3.0	—
	AC/DDT	DDT	7B19	4.0	1.0	200	—	35	2.3	-4.0	800	5.0	—
	AC/DD	DDTetrode	7B32	4.0	2.0	250	250	—	8.0	-9.5	160	32.0	3000
	DDT213	DDT	7B19	13.0	0.3	200	—	35	2.3	-4.0	800	5.0	—
Marconi	HD14	DD	O4	1.4	0.05	90	—	65	0.275	0	—	0.14	—
	HD21	DDT	5B2	2.0	0.2	150	—	27	1.5	-1.5	—	—	—
	HD22	DDT	5B12	2.0	0.2	150	—	27	1.5	-1.5	—	—	—
	HD23	DDT	5B2	2.0	0.15	150	—	40	1.4	-1.5	—	2.0	—
	HD24	DDT	5B2	2.0	0.1	150	—	40	1.4	-1.7	—	1.7	—
	WD40	VPDD	9B6	4.0	1.0	250	100	—	3.5	—	—	—	—

Table 4. Diode combinations – *continued*

Make	Type	Description	Base	Fil. volts	Fil. amps	Anode volts	Screen volts	Amp. factor	Slope (mA/V)	Bias volts	Bias res. (ohms)	Anode current (mA)	Output (mW)
Marconi – *continued*	MHD	DDT	7B19	4.0	1.0	200	–	40	2.2	–	750	3.2	–
	DH42	DDT	7B19	4.0	0.6	250	–	70	1.2	–3.0	2000	1.5	–
	DL63	DDT	O41	6.3	0.3	250	–	37	1.65	–3	–	5.0	–
	DN41	DDP	7B32	4.0	2.3	250	250	–	10.0	–4.4	90	32.0	4400
	DH63	DDT	O41	6.3	0.3	250	–	70	1.2	–3.0	2000	1.1	–
	WD30	VPDD	9B6	13.0	0.3	250	100	–	2.6	–	–	–	–
	DH30	DDT	7B19	13.0	0.3	200	–	80	4.5	–2.0	1000	2.7	–
	DHD	DDT	7B24	16.0	0.25	200	–	40	2.2	–	–	–	–
Mazda	H141D	DT	OM6	1.4	0.05	90	–	65	0.48	–	–	0.065	–
	HL21/DD	DT	5B2	1.4	0.05	90	–	65	0.48	–	–	0.065	–
	HL21/DD	DDT	5B2	2.0	0.15	150	–	32	1.5	–2.0	–	2.0	–
	L21/DD	DDT	5B2	2.0	0.1	150	–	18.5	1.85	–5.0	–	2.8	–
	L22/DD	DDT	OM2	2.0	0.1	150	–	18.5	1.85	–5.0	–	2.3	–
	HL23/DD	DDT	OM2	2.0	0.05	150	–	25	1.2	–1.5	–	0.6	–
	AC/HLDD	DDT	7B19	4.0	1.0	250	–	36	2.6	–3.0	700	4.3	–
	AC/HLDDD	Triple DT	9B3	4.0	1.0	250	–	35	2.7	–3.0	700	4.3	–
	AC2/PENDD	DDP	7B32	4.0	2.0	250	250	–	8.0	–5.3	140	32.0	3500
	AC5/PENDD	DDP Tet	7B22	4.0	2.0	250	250	–	9.0	–8.5	175	40.2	5800
	PEN 45/DD	DDP Tet	OM25	4.0	2.0	250	250	–	9.0	–8.5	175	40.0	5800
	HL41/DD	DDT	OM21	4.0	0.65	250	–	30	2.5	–7.4	1400	2.2	–
	HL42/DD	DDT	OM21	4.0	0.65	250	–	23	2.9	–1.25	450	2.8	–
	HLDD1320	DDT	7B19	13.0	0.2	250	–	39	2.0	–3.0	700	4.3	–
	HL133/DD	DDT	OM21	13.0	0.2	250	–	32	2.5	–2.2	1750	1.25	–
	PENDD1360	DDP	7B32	13.0	0.6	250	250	–	8.0	–5.3	140	32.0	–
	DC2HLDD	DDT	7B19	25.0	0.1	200	–	30	2.0	–3.0	700	3.75	–
	PENDD4020	DDP	7B32	40.0	0.2	250	250	–	7.0	–7.75	150	43	3900
	PEN453/DD	DD Tet	OM25	45.0	0.2	200	200	–	12.0	–10.0	130	64.0	3750
Mullard	DAC1	DT	8S3	1.4	0.05	90	–	65.0	0.275	0	–	0.14	–
	TDD2	DDT	5B2	2.0	0.1	150	–	16.5	1.4	–5.5	–	2.5	–
	TDD2A	DDT	5B2	2.0	0.12	135	–	30.0	1.2	–1.5	–	1.95	–
	SD4	D Tetrode	7B50	4.0	1.0	250	100	–	3.0	–	–	–	–
	TDD4	DDT	7B19	4.0	0.65	250	–	27.0	2.0	–7.0	1500	4.0	–
	PEN4DD	DDP	7B33	4.0	2.25	250	250	–	9.5	–6.0	150	36.0	4300
	EBC3/33	DDT	8S21/041	6.3	0.2	250	–	30.0	2.0	–6.25	–	5.0	–
	EBF2	DDP	8S27	6.3	0.2	275	100	–	1.8	–2.0	–	5.0	–
	EBL1/31	DDP	8S27/053	6.3	1.5	250	250	–	9.5	–6.0	–	36.0	4300
	TDD13C	DDT	7B19	13.0	0.2	250	–	27.0	2.0	–6.0	1250	4.0	–
	PEN40DD	DDP	7B33	44.0	0.2	200	200	–	8.0	–5.0	–	45.0	4000
	CBL1/31	DDP	8S27/053	44.0	0.2	200	200	–	8.0	–8.5	–	45.0	4000
Osram	HD14	DT	O4	1.4	0.05	90	–	65	0.275	–	–	–	–
	HD21	DDT	5B2	2.0	0.2	150	–	27	1.5	–1.5	–	–	–
	HD22	DDT	5B2	2.0	0.2	150	–	27	1.5	–1.5	–	–	–
	HD23	DDT	5B2	2.0	0.15	150	–	40	1.4	–1.5	–	1.7	–

Make	Model	Type	Base										
	HD24	DDT	5B2	2.0	0.1	150	—	40	1.4	-1.7	—	1.7	—
	WD40	VPDD	9B6	4.0	1.0	100	—	—	2.6	—	—	—	—
	MHD4	DDT	7B19	4.0	1.0	250	—	40	2.2	-4.0	1000	4.0 each	—
	DH42	DDT	7B19	4.0	0.6	250	250	70	1.2	-3.0	1500	—	3500
	DH41	DDP	7B32	4.0	2.3	250	250	—	10.0	-3.5	90	32.0	—
	DN41	DDP	7B32	4.0	2.3	250	—	—	10.0	-5.0	120	32.0	—
	DH73M	DDT	O41	6.0	0.17	250	—	44	2.0	—	—	—	—
	DL74M	DDT	O41	13.0	0.16	250	—	36	1.6	—	—	1.1 each	—
	DH63	DDT	O41	6.3	0.3	250	—	70	1.2	-3.0	2000	4.2 each	—
	DL63	DDT	O41	6.3	0.3	250	—	36	1.6	-3.0	1500	—	—
	WD30	VPDD	9B6	13.0	0.3	250	100	—	2.6	—	—	2.7	—
	DH30	DDT	7B19	13.0	0.3	200	—	80	4.5	-2.0	1000	—	—
	DHD	DDT	7B19	16.0	0.25	200	—	40	2.2	—	—	1.0	—
Record	DDTR2	DDT	5B2	2.0	0.1	135	—	30	1.4	-3.0	—	4.0	—
	AC/DDTR	DDT	7B19	4.0	0.65	250	—	40	3.6	-5.0	1000	4.0	—
	DDTR/13	DDT	7B19	13.0	0.2	200	250	40	3.6	-5.0	1000	4.0	—
	DDTR/13L	DDT	8S21	13.0	0.2	200	—	40	3.6	-5.0	1000	2.5	—
Triotron	DT215	DDT	5B2	2.0	0.1	135	—	16	1.0	-4.5	—	4.0	—
	DT436	DDT	7B19	4.0	0.65	250	250	27	2.0	-7.0	—	36.0	—
	DP495/6	DDP	7B33/7B32	4.0	2.25	250	—	—	—	-6.0	—	4.0	—
	DT1336	DDT	7B19	13.0	0.2	200	200	27	2.0	-5.0	—	45.0	—
	DP4480	DDP	7B33	44.0	0.2	200	—	280	8.0	-8.5	—	1.0	—
Tungsram	DDT2	DDT	5B16	2.0	0.1	135	—	30	1.4	-1.5	—	2.5	—
	DDT2B	DDT	5B16	2.0	0.1	135	—	16	1.0	-4.5	—	4.0	—
	DDT4/S	DDP	7B19/8S21	4.0	0.65	250	250	40	3.6	-5.0	—	36.0	—
	DDPP4B/M	DDP	7B32/7B33	4.0	2.0	250	250	—	10.0	—	150	5.0	3600
	EBF11	DDP (HF)	OF4	6.3	0.2	300	125	—	1.8	—	—	5.0	—
	DDT6S	DDT	8S21	6.3	0.2	250	—	30	2.5	-5.5	1000	2.0	—
	EBC3/33	DDT	8S21/O41	6.3	0.2	300	—	—	2.0	-6.25	—	36.0	3600
	EBF2	DDP (HF)	8S27	6.3	0.2	300	300	—	1.8	-2.0	—	4.0	—
	EBL1/21	DDP	8S27/O53	6.3	1.4	250	250	40	10.0	-6.0	150	45.0	—
	DDT13/S	DDT	7B19/8S21	13.0	0.2	250	—	—	3.6	-5.0	1000	2.0	3200
	DDPP39/M/S	DDP	7B32/7B33/8S27	39.0	0.2	200	200	—	8.5	—	170		—
	DDPP6B	DDP	7B32	6.3	1.4	250	250	—	10.0	-6.0	—		—

203

Table 5. General purpose triodes

Make	Type	Base	Fil. volts	Fil. amps	Anode volts	Amp. factor	Impedance (ohms)	Slope (mA/V)	Bias volts	Anode current (mA)	Bias res. (ohms)
Brimar	4215A	–	1.0	0.25	45	6	25 000	0.4	–3.0	0.8	–
	HLA2	5B15	4.0	1.0	200	50	9 000	5.5	–2.0	8.0	400
	4D1	7B18	13.0	0.2	200	40	10 000	4.0	–3.0	5.0	800
	6C5G	–	6.3	0.3	250	20	10 000	2.0	–8	8.0	1000
	6J5G	–	6.3	0.3	250	20	7 700	2.6	–8	9.0	900
Cossor	210RC	4B4	2.0	0.1	150	40	50 000	0.8	–1.5	0.45	–
	210HL	4B4	2.0	0.1	150	24	22 000	1.1	–1.5	2.0	–
	210HF	4B4	2.0	0.1	150	24	15 000	1.5	–1.5	2.25	–
	210DET	4B4	2.0	0.1	150	15	13 000	1.15	–1.5	4.5	–
	210LF	4B4	2.0	0.1	150	14	10 000	1.4	–3.0	4.5	–
	41MRC	5B15	4.0	1.0	200	50	19 500	2.6	–1.0	2.5	–
	41MH	5B15	4.0	1.0	200	72	18 000	4.0	–1.5	1.5	–
	41MHF	5B15	4.0	1.0	200	41	14 500	2.8	–2.0	2.5	–
	41MHL	5B15	4.0	1.0	200	52	11 500	4.5	–3.0	4.0	–
	41MLF	5B15	4.0	1.0	180	15	7 900	1.9	–4.5	7.5	–
	DHL	5B15	16.0	0.25	200	58	13 000	4.5	–1.5	3.8	–
	41MTL	5B15	4.0	1.0	250	44	15 000	3.0	–3.0	4.0	–
	41MTB	5B15	4.0	1.0	250	104	40 000	2.6	–1.0	3.4	–
	41MTA	5B15	4.0	1.0	200				–		–
Dario	TB282	4B4	2.0	0.1	150	28	22 000	1.3	–2.0	2.0	–
	TB172	4B4	2.0	0.1	150	17	12 000	1.4	–3.0	4.5	–
	TB102	4B4	2.0	0.1	150	10	8 000	1.25	–6.0	5.0	–
	TB122	4B4	2.0	0.2	135	12	6 000	2.0	–6.0	5.0	–
	TE994	5B15	4.0	1.0	200	100	25 000	4.0	–1.6	1.0	–
	TE384	5B15	4.0	1.0	200	38	25 000	1.5	–2.5	1.5	–
	TE244	5B15	4.0	1.0	200	24	10 000	2.4	–3.5	6.0	–
	TE094	5B15	4.0	1.0	200	9	7 000	1.3	–16.0	12.0	–
	TC113	7B18	13.0	0.2	200			3.3	–3.7	5.0	–
	TB9920	5B15	20.0	0.18	200	100		3.0	–		–
Ever Ready	K30K	4B4	2.0	0.1	135	30	21 500	1.4	–1.5	2.2	–
	K30D	4B4	2.0	0.1	150	18	12 000	1.5	–4.5	4.0	–
	K30E	4B4	2.0	0.1	135	18	12 000	1.5	–4.5	2.0	–
	A30B	5B15	4.0	0.65	200	72	20 600	3.5	–2.0	2.2	–
	A30D	5B15	4.0	0.65	250	40	11 500	3.5	–4.5	6.5	–
	C30B	7B18	13.0	0.2	200	40	12 000	3.3	–3.7	5.0	–
Ferranti	D4	5B15	4.0	1.0	200	40	12 500	3.3	–3.0	4.0	650
	DA	7B18	13.0	0.2	200	40	12 500	3.3	–3.0	3.7	650
	DS	5B15	13.0	0.3	200	40	12 500	3.3	–3.0	4.0	650
Hivac	XH1.5V	4D1	1.5	0.08	50	25	50 000	0.5	0	0.45	–
	XD1.5V	4D1	1.5	0.08	50	20	50 000	0.4	0	0.45	–
	XH2.0V	4D1	2.0	0.08	50	28	50 000	0.56	0	0.45	–
	XD2.0V	4D1	2.0	0.08	50	21	38 000	0.56	0	0.65	–
	H210	4B4	2.0	0.1	150	25	22 000	1.15	–3	1.1	–

Lissen	D210	4B4	2.0	0.1	150	16	12000	1.35	−4.5	2.4	—
	D210SW	4B11	2.0	0.1	150	16	12000	1.35	−4.5	2.4	—
	L210	4B4	2.0	0.1	150	12	7500	1.6	−6.0	4.2	460
	AC/HL	5B15	4.0	1.0	200	35	10000	3.5	−2.75	6.0	460
	HL13	7B18	13.0	0.3	200	35	10000	3.5	−2.75	6.0	—
	H2	4B4	2.0	0.1	150	50	45000	1.1	—	—	—
	HL2	4B4	2.0	0.1	150	35	22000	1.6	—	—	—
	L2	4B4	2.0	0.1	150	20	10000	2.0	—	—	—
Marconi	AC/HL	5B15	4.0	1.0	200	40	10000	4.0	—	—	—
	H2	4B4	2.0	0.1	150	35	25000	1.1	−1.5	1.5	—
	HL21	4B4	2.0	0.1	150	27	18000	1.5	−1.5	2.0	—
	HL2	4B4	2.0	0.1	150	27	18000	1.5	−1.5	2.0	—
	HL2/K	4B4	2.0	0.1	150	27	18000	1.5	−1.5	2.0	—
	HL210	4B4	2.0	0.1	150	24	20000	1.2	−3	1.2	—
	L21	4B4	2.0	0.1	150	16	8900	1.8	−6.0	2.2	—
	L210	4B4	2.0	0.1	150	11	12000	0.9	−7.5	2.5	—
	H42	7B18	4.0	0.6	250	100	66000	1.5	−2	1.0	400
	MH41	5B15	4.0	1.0	200	80	13000	6.0	−2.0	—	700
	MH4	5B15	4.0	1.0	200	40	11100	3.6	−3.0	5.5	850
	MHL4	5B15	4.0	1.0	250	20	8000	2.5	−9	1.0	2000
	H63	O39	6.3	0.3	250	100	66000	1.5	−2.0	7.5	—
	L63	O34	6.3	0.3	250	20	7700	2.6	−9	3.0	—
	H30	7B18	13.0	0.3	200	80	13300	6.0	−2.5	20	—
	L30	7B17	13.0	0.3	200	12	2860	4.2	−10	—	—
	DH	5B15	16.0	0.25	200	40	10800	3.7	—	—	—
	ET1	4B1	1.0	0.1	4—10	—	—	0.08	—	—	—
	H11	4DS1	1.0	0.1	100	15	30000	0.5	0	—	—
	L11	4D51	1.0	0.1	100	4.4	7700	0.57	0	—	—
	H12	4D1	2.0	0.06	100	26	108000	0.24	0	2.5	—
	L12	4D1	2.0	0.06	100	4.8	6000	0.8	—	—	—
	A537	4DS	4.0	0.4	150	15.5	10000	1.55	—	—	—
	A577	5B14	4.0	1.0	250	6	3000	2.0	—	—	—
	MH40	5B15	4.0	1.0	200	45	18000	2.5	—	—	—
Mazda	HA1	Acorn	4.0	0.25	180	25	12500	2.0	−5.0	4.5	—
	HA2	Acorn	6.3	0.15	180	25	12500	2.0	−5.0	4.5	—
	H2	4B4	2.0	0.1	150	50	45000	1.1	0	2.5	—
	HL2	4B4	2.0	0.1	150	32	21000	1.5	−1.5	2.7	—
	L2	4B4	2.0	0.1	150	19	10000	1.9	−3.0	5.3	—
	AC/HL	5B15	4.0	1.0	200	35	11700	3.0	−3.5	5.0	700
	AC2/HL	5B15	4.0	1.0	200	75	11500	6.5	−1.75	4.5	390
	HL41	OM19	4.0	0.65	250	36	10300	3.5	−3.1	2.2	1400
	P41	OM19	4.0	0.95	250	17	—	8.0	−10	30.0	—
	AC/P4	5B13	4.0	1.0	700	20	2800	7.0	−35	5.0	—
Mullard	HL1320	7B18	13.0	0.2	250	30	10000	3.0	−4.5	7.5	600
	HL133	OM20	13.0	0.2	250	36	10600	3.4	−1.95	1.3	1500
	DC3HL	5B15	25.0	0.1	200	35	11700	3.0	−3.5	5.0	700
	DC51	4D1	1.5	0.067	45	25	66000	0.38	0	0.34	—
	DA1	4D1	2.0	0.05	40	32	80000	0.4	0.25	0.25	—
	PM1A	4B4	2.0	0.1	150	50	41600	1.2	−1.0	1.0	—
	PM1HF	4B4	2.0	0.1	150	18	22500	0.8	−3—−4.5	1.5	—

Table 5. General purpose triodes – continued

Make	Type	Base	Fil. volts	Fil. amps	Anode volts	Amp. factor	Impedance (ohms)	Slope (mA/V)	Bias volts	Anode current (mA)	Bias res. (ohms)
Mullard – continued	PM1HL	4B4	2.0	0.1	135	28	23 400	1.2	-1.5	2.3	–
	PM2HL	4B4	2.0	0.1	135	30	21 500	1.4	-1.5	2.2	–
	PM1LF	4B4	2.0	0.1	150	11	12 000	0.9	-7.5	4.0	–
	PM2DX	4B4	2.0	0.1	135	18	18 000	1.0	-4.5	2.0	–
	PM2DL	4B4	2.0	0.1	135	18	12 000	1.5	-4.5	2.0	–
	AT4	Acorn	4.0	0.25	200	25	12 500	2.0	-6.0	4.5	–
	994V	5B15	4.0	0.65	200	135	35 000	3.6	-1.5	1.35	1000
	904V	5B15	4.0	0.65	200	72	20 600	3.5	-2.0	2.2	900
	484V	5B15	4.0	1.0	200	48	21 800	2.2	-3.0	2.8	1000
	354V	5B15	4.0	0.65	250	40	11 500	3.5	-4.5	6.5	700
	244V	5B15	4.0	0.65	200	25	9 000	2.8	-5.5	5.5	1000
	154V	5B15	4.0	0.65	200	15	7 500	2.0	-7.5	9.0	800
	4761	Acorn	6.3	0.15	180	25	12 500	2.0	-5.0	4.5	–
	HL13	8S20	13.0	0.2	200	40	12 000	3.3	-3.7	5.0	740
	HL13C	7B18	13.0	0.2	200	40	12 000	3.3	-3.7	5.0	740
Osram	H2	4B4	2.0	0.1	150	35	35 000	1.0	-1.5	1.5	–
	HL2	4B4	2.0	0.1	150	27	18 000	1.5	-1.5	2.0	–
	HL2/K	4B4	2.0	0.1	150	27	18 000	1.5	-1.5	2.0	–
	HL210	4B4	2.0	0.1	150	24	20 000	1.2	–	–	–
	H210	4B4	2.0	0.1	150	35	50 000	0.7	–	–	–
	L21	4B4	2.0	0.1	150	16	8 900	1.8	-6.0	2.2	–
	H42	5B15	4.0	0.6	250	100	66 000	1.7	-2.0	1.0	2000
	MH41	5B15	4.0	1.0	250	80	13 300	6.0	-2.5	3.6	700
	MH4	5B15	4.0	0.1	250	40	11 000	3.6	-4.0	5.0	750
	MHL4	5B15	4.0	1.0	250	20	8 000	2.5	-8.0	8.0	1000
	H63	O39	6.3	0.3	250	100	66 000	1.5	-2.0	1.0	2000
	L63	O39	6.3	0.3	250	20	7 700	2.6	–	–	–
	H30	5B15	13.0	0.3	250	80	13 300	6.0	–	–	–
	DH	5B15	16.0	0.25	200	40	10 800	3.7	–	–	–
	H12	4D1	2.0	0.06	100	26	21 600	1.2	–	–	–
	A577	5B14	4.0	1.0	250	60	3 000	2.0	–	–	–
	MH40	5B15	4.0	1.0	200	45	18 750	2.4	–	–	–
	HA1	Acorn	4.0	0.25	180	20	11 700	1.7	-6.5	4.5	–
	HA2	Acorn	6.3	0.15	180	25	12 500	2.0	–	–	–
Record	H2	4B4	2.0	0.1	200	30	23 000	1.3	-3.0	1.0	1000
	L2	4B4	2.0	0.12	150	17	15 000	1.2	-5.0	3.2	1000
	DL2	4B4	2.0	0.1	150	18	14 000	1.3	-4.5	3.0	–
	AC/NHL	5B15	4.0	0.65	250	33	11 000	3.5	-4.5	5.0	1000
	NHL/13	7B18	13.0	0.2	200	30	12 000	3.5	-4.0	6.0	1000
	NHL/13L	8S20	13.0	0.2	200	30	12 000	3.5	-4.0	6.0	1000

Triotron	HD2	4B4	2.0	0.08	200	15	15 000	1.0	−5.0	5.0	—
	TD2	4B4	2.0	0.1	150	10	8 000	1.25	−6.0	5.0	—
	A214	4B4	2.0	0.1	150	17	12 000	1.4	−4.5	4.0	—
	W213	4B4	2.0	0.1	150	28	22 000	1.3	−2.5	1.0	—
	A440N	5B15	4.0	1.0	200	100	25 000	4.0	−1.6	1.0	—
Tungsram	A240N	5B15	20.0	0.18	200	100	25 000	4.0	−1.5	0.2	—
	HR2	4B4	2.0	0.06	135	25	40 000	0.6	−1.5	1.2	—
	HR210	4B4	2.0	0.1	200	30	23 000	1.3	−3.0	1.0	—
	HL2	4B4	2.0	0.13	135	30	21 000	1.5	−1.5	2.2	—
	LD210	4B4	2.0	0.1	150	18	14 000	1.3	−4.5	3.0	—
	LL2	4B4	2.0	0.2	135	30	11 500	2.6	−2.5	3.0	—
	HL4+	5B15	4.0	0.65	250	33	11 000	3.5	−4.5	5.0	1000
	HL4g	7B18	4.0	0.65	250	33	11 000	3.5	−4.5	5.0	1000
	LL4C	5B13	4.0	1.2	350	10	—	3.5	—	—	—
	HL13	7B18	13.0	0.2	200	30	12 000	3.5	−5.5	6.0	1000

Table 6. Power output triodes

Make	Type	Base	Fil. volts	Fil. amps	Anode volts	Impedance	Slope (mA/V)	Bias volts	Anode current (mA)	Bias res. (ohms)	Output (mW)	Optimum load (ohms)
Brimar	PA1	5B15	4.0	1.1	200	1500	12.0	−9.0	50.0	260	1250	4000
Cossor	215P	4B4	2.0	0.15	150	4000	2.25	−7.5	10.0	—	150	9000
	220P	4B4	2.0	0.2	150	4000	2.25	−7.5	11.0	—	190	9000
	220PA	4B4	2.0	0.2	150	4000	4.0	−4.5	10.0	—	180	9000
	230XP	4B4	2.0	0.3	150	1500	3.0	−18.0	22.0	—	450	3500
	2P	4B4	2.0	2.0	250	1150	7.0	−22.0	40.0	—	—	3000
	2XP	4B4	2.0	2.0	300	900	7.0	−36.0	50.0	—	—	4000
	41MP	5B15	4.0	1.0	200	2500	7.5	−7.5	24.0	320	1250	3000
	41MXP	SB15	4.0	1.0	200	1500	7.5	−12.5	40.0	300	2000	2000
	4XP	4B4	4.0	1.0	250	900	7.0	−28.5	48.0	600	3000	3000
	DP	7B18	16.0	0.25	200	2800	6.0	−7.5	25.0	300	—	3500
	402P	7B18	40.0	0.2	200	1330	7.5	−9.5	30.0	320	—	2500
Dario	TB052	4B4	2.0	0.15	150	4200	1.2	−18.0	7.0	—	150	11000
	TB062	4B4	2.0	0.33	150	3000	2.0	−10.5	13.0	—	5550	6000
	TB032	4B4	2.0	0.2	150	2000	1.5	−30.0	12.0	—	500	6000
	TF104	4B4	4.0	2.0	550	2500	4.0	−36.0	45.0	—	—	—
	TF364	4B4	4.0	2.0	400	3000	3.8	−92.0	63.0	—	—	—
	TD044	4B4	4.0	0.65	250	1300	2.7	−40.0	40.0	—	—	—
	TD4	4B4	4.0	1.0	300	1200	5.0	−38.0	48.0	—	—	—
Ever Ready	K30G	4B4	2.0	0.2	135	6000	2.0	−6.0	5.0	—	500	7000
	S30C	4B4	4.0	1.0	300	1200	5.0	−38.0	50.0	600	3500	2300
	S30D	4B4	2.0	2.0	300	1200	5.0	−38.0	50.0	600	3500	2300
Ferranti	L2	4B4	2.0	0.1	150	6800	1.6	—	5.6	—	—	—
	LP4	4B4	4.0	1.0	250	980	5.5	−35.0	48.0	730	2800	2500
Hivac	XL1.5V	4D1	1.5	0.08	50	20000	0.6	−1.0	0.7	—	—	—
	XLO1.5V	4D1	1.5	0.08	50	20000	0.65	−1.0	0.9	—	—	—
	XP1.5V	4D1	1.5	0.08	50	7250	0.72	−4.5	1.75	—	—	—
	XP2.0V	4D1	2.0	0.08	50	12500	0.84	−1.0	1.0	—	—	—
	XLO2.0V	4D1	2.0	0.08	50	12500	0.92	−1.1	1.1	—	—	—
	XP2.0V	4D1	2.0	0.08	50	6000	1.0	−3.0	2.0	—	—	—
	P215	4B4	2.0	0.15	150	3600	2.2	−12.0	8.0	—	150	10000
	P220	4B4	2.0	0.2	150	4700	3.0	−7.5	6.0	—	175	9000
	PP220	4B4	2.0	0.2	150	2300	3.0	−12.0	12.5	—	250	5000
	PX230	4B4	2.0	0.3	150	1850	3.5	−15.0	17.5	—	450	4000
	PX230SW	4B11	2.0	0.3	150	1850	3.5	−15.0	17.5	—	450	4000
	AC/L	5B15	4.0	1.0	200	2350	4.25	−13.5	17.0	760	675	6300
	PX41	4B4	4.0	1.0	250	830	6.0	−40.0	48.0	830	2500	3500
	PX5	4B4	4.0	2.0	400	1480	6.5	−34.0	62.5	530	5750	3000
Lissen	LP2	4B4	2.0	0.2	150	3500	3.5	—	—	—	200	—
	P220	4B4	2.0	0.2	150	4000	1.75	—	—	—	100	—
	PX240	4B4	2.0	0.4	200	1500	3.0	—	—	—	800	—
Marconi	LP2	4B4	2.0	0.2	150	3900	3.85	−6.0	7.0	—	—	9700
	P215	4B4	2.0	0.15	150	5000	1.4	−12.0	8.5	—	—	12000
	P2	4B4	2.0	0.2	150	2150	3.5	−12.0	14.0	—	200	6000

Type	Base										
Mazda											
ML4	5B15	4.0	1.0	200	2860	4.2	−8.0	25	400	—	6000
PX4	4B4	4.0	1.0	300	830	6.0	−50.0	50.0	1000	4500	3500
PX25	4B4	4.0	2.0	400	1265	7.5	−30.0	6.25	530	5500	4000
PX25A	4B4	4.0	2.0	400	580	6.9	−103.0	62.5	1630	8400	4500
DA30	4B4	4.0	2.0	500	910	3.85	−134.0	60.0	1100	—	—
DA60	4L1	6.0	4.0	500	835	3.0	−135.0	120.0	—	—	2800
DA100	4L1	6.0	2.7	1000	1410	3.9	−150.0	100.0	—	—	—
DA250	4M1	10.0	2.0	2500	2300	7.0	−130.0	80.0	—	—	—
DA41	4UX1	7.5	2.5	1000	17500	3.6	0	—	—	—	—
DL	5B15	16.0	0.25	200	2660	4.5	—	—	—	—	—
P220	4B4	2.0	0.2	150	3700	3.4	−7.0	5.5	—	180	10000
P220A	4B4	2.0	0.2	150	1850	3.5	−14.0	15.0	—	350	4100
PA20	4B4	2.0	2.0	300	1000	6.5	−29.0	42.0	690	2650	2750
AC/P	5B15	4.0	1.0	200	2650	3.75	−13.5	17.0	750	650	6000
AC/P1	5B15	4.0	1.0	200	1450	3.7	−28.0	24.0	1200	1000	5000
PP5/400	4B4	4.0	2.0	400	1500	6.0	−32.0	62.5	510	5900	2700
PP3/250	4B4	4.0	1.0	300	1000	6.5	−30.0	42.0	715	2650	2750
Per pair in push-pull											
PA40	4B4	4.0	2.0	400	425	4.5	−85.0	210.0	400	33500	3700
PP3521	7B17	35.0	0.2	250	600	10.0	−25.0	70.0	360	2300	2000
DC2/P	5B15	35.0	0.1	200	2650	3.75	−13.5	17.0	800	650	6000
Mullard											
DD51	4D1	1.5	0.67	45	10000	0.5	−3.0	1.7	—	—	—
DA2	4D1	2.0	0.05	40	13600	0.5	−2.15	1.25	—	—	7000
DA3	4D1	2.0	0.05	40	7600	0.62	−2.8	1.8	—	—	9000
PM2A	4B4	2.0	0.2	135	6000	2.0	−6.0	5.0	—	150	3700
PM2	4B4	2.0	0.2	150	4400	1.7	−12.0	6.6	—	—	3700
PM252	4B4	2.0	0.2	150	2000	3.5	−12.0	14.0	—	—	—
PM202	4B4	2.0	0.2	150	2000	3.5	−12.0	14.0	—	—	—
164V	5B15	4.0	0.65	200	3640	4.5	−8.5	13.0	—	—	—
104V	5B15	4.0	1.0	250	3300	3.2	−16.0	20.0	—	—	10000
TT4	5B15	4.0	1.0	250	3300	3.2	−16.0	20.0	—	500	10000
TT4A	5B15	4.0	1.0	250	4400	4.1	−9.0	20.0	1500	500	5000
AC104	4B4	4.0	1.0	200	2850	3.5	−14.0	11.0	1000	400	6000
AC064	4B4	4.0	1.0	200	2000	3.0	−21.0	20.0	—	400	5000
AC044	4B4	4.0	1.0	300	1200	5.0	−38.0	50.0	—	620	2300
C042	4B4	2.0	2.0	300	1200	5.0	−38.0	50.0	—	3500	2300
D024	4B4	4.0	1.85	400	1070	7.5	−40.0	63.0	—	3500	3200
D026	4B4	4.0	2.0	400	950	3.8	−92.0	63.0	1500	7100	3000
D030	4B4	4.0	1.85	500	890	3.5	−140.0	60.0	—	7500	—
D010	4B4	6.0	0.85	400	2850	0.85	−130.0	25.0	5500	2500	6000
D025	4B4	6.0	1.1	400	800	3.75	−112.0	63.0	1780	7000	4000
D020	4B4	7.5	1.1	425	2000	2.5	−66.0	40.0	1650	5000	5000
Osram											
EC31	O34	6.3	0.65	250	3300	3.2	−16.0	20.0	—	500	10000
L12	4D1	2.0	0.06	100	6000	0.8	−4.5	1.9	—	100	1.2
LP2	4B4	2.0	0.2	150	4170	3.6	−6.0	5.6	—	—	—
P215	4B4	2.0	0.15	150	5000	1.4	—	—	—	—	—
P2	4B4	2.0	0.2	150	2150	3.5	−12.0	14.0	—	200	—
ML4	5B15	4.0	1.0	250	2860	4.2	−16.0	14.0	1000	3500	7000
PX4	4B4	4.0	1.0	300	830	6.0	−42.0	50.0	900	5500	4000
PX25	4B4	4.0	2.0	400	1265	7.5	−31.0	62.5	530	8400	3200
PX25A	4B4	4.0	2.0	400	580	6.9	−103.0	62.5	1630	11000	4500
DA30	4B4	4.0	2.0	500	580	6.9	−134.0	60.0	—	—	6000

Table 6. Power output triodes – *continued*

Make	Type	Base	Fil. volts	Fil. amps	Anode volts	Impedance	Slope (mA/V)	Bias volts	Anode current (mA)	Bias res. (ohms)	Output (mW)	Optimum load (ohms)
Osram – *continued*	DET5	4B4	4.0	2.0	600	1265	7.0	–	–	–	35 000	–
	DA60	4L1	6.0	4.0	500	835	3.0	–135.0	120.0	1100	–	2 800
	DA100	4L1	6.0	2.7	1000	1410	3.9	–150.0	100.0	–	30 000	6 800
Double	DET19	7UX10	6.3	0.8	300	3340	2.1	–	–	–	16 000	–
	DA41	4UX3	7.5	2.5	1000	17 500	3.6	–	–	–	–	–
	DET12	4B4	7.5	3.2	1250	–	–	–	–	–	70 000	–
	DET14	4UX3	7.5	3.0	1500	–	–	–	–	–	80 000	–
	DA250	4M1	10.0	2.0	2500	2290	7.0	–130.0	80.0	–	800 000	12 000
Triotron	DL	5B15	16.0	0.25	200	2660	4.5	–	–	–	–	–
	ZD2	4B4	2.0	0.15	150	4200	1.2	–18.0	7.0	–	–	–
	UD2	4B4	2.0	0.33	150	2000	2.0	–15.0	12.0	–	–	–
	E235	4B4	2.0	0.2	150	3000	3.0	–7.5	13.0	–	–	–
	E430N	5B15	4.0	1.0	200	7000	1.3	–16.0	12.0	–	–	–
	K480	4B4	4.0	2.0	550	2500	4.0	–36.0	45.0	–	–	–
	K435/10	4B4	4.0	0.65	250	1300	2.7	–40.0	40.0	–	–	–
	T1325	7B49	13.0	0.2	200	–	3.3	–3.7	5.0	–	–	–
Tungsram	LP220	4B4	2.0	0.2	150	3500	3.4	–6.0	5.0	–	200	7 500
	P215	4B4	2.0	0.15	150	3300	1.5	–12.0	8.0	–	260	7 000
	SP220	4B4	2.0	0.2	150	2200	3.0	–18.0	14.0	–	360	6 700
	LL4	5B15	4.0	1.2	350	–	3.5	–	24.0	–	–	–
	P12/250	4B4	4.0	1.0	250	850	6.0	–33.0	48.0	700	2 800	2 400
	P15/250	4B4	4.0	1.0	250	660	6.0	–44.0	60.0	750	3 500	2 500
	015/400	4B4	4.0	1.0	400	1800	5.0	–38.0	30.0	1000	3 700	7 000
	P26/500	4B4	4.0	2.0	500	670	4.7	–100.0	62.5	1600	6 500	5 000
	P27/500	4B4	4.0	2.0	500	1100	8.0	–32.0	62.5	500	5 000	4 000
	P25/500	4B4	6.0	1.1	500	1000	3.0	–112.0	62.5	1950	4 000	7 000
	P60/500	4L1	6.0	4.0	600	1000	3.5	–125.0	116.0	1080	15 000	3 000
	P25/450	4B4/4UX3	7.5	1.25	450	2000	2.0	–82.0	55.0	1500	5 100	5 000
	PX2100	4B4	7.5	1.25	425	5000	1.6	–39.0	18.0	2000	1 600	10 200

Table 7. Output pentodes and tetrodes

Make	Type	Base	Fil. volts	Fil. amps	Anode volts	Screen volts	Slope (mA/V)	Bias volts	Bias res. (ohms)	Anode and screen current (mA)	Output (mW)	Optimum load (ohms)
Brimar	6F6G	IO36	6.3	0.7	250	250	2.35	-16.5	410	40.5	3 000	7 000
	6V6G	IO36	6.3	0.45	250	250	4.10	-12.5	240	49.5	4 250	5 000
	25A6G	IO36	25.0	0.3	180	135	2.5	-20	440	45	2 750	5 000
	PENB1	5B4	2.0	0.2	150	150	2.5	-4.5	—	9.6	—	18 000
	7A2	5B18/7B31	4.0	1.2	250	250	3.2	-17.5	330	38.5	3 500	8 000
	7A3	7B31	4.0	2.0	250	250	10.0	-6	150	38.0	3 750	—
	PENA1	5B4	4.0	1.0	250	250	3.6	-16.5	450	38.5	2 700	8 000
	7D5	7B31	13.0	0.315	250	250	2.35	-16.5	410	40.5	3 000	7 000
	7D8	7B31	13.0	0.65	250	250	10.0	-6	150	38.0	3 750	8 500
	7D3	7B31	40.0	0.2	180	135	2.5	-20	440	45.0	2 750	5 000
	7D6	7B31	40.0	0.2	250	250	10.0	-7.5	150	38.0	3 750	8 500
Cossor	210PT	4B6/5B4	2.0	0.2	150	150	2.5	-7.5	—	—	—	8 000
	220HPT	4B6/5B4	2.0	0.2	150	150	2.5	-3.0	—	—	—	20 000
(Tetrode)	220 OT	5B3	2.0	0.2	150	150	2.5	-4.5	—	23.0	—	20 000
	230PT	4B6/5B4	2.0	0.3	150	150	2.0	-15.0	350	36.0	—	10 000
	PT41	5B4	4.0	1.0	250	200	3.0	-12.5	—	—	—	8 000
	PT41B	5B4	4.0	1.0	400	300	2.25	-33.0	450	36.0	—	8 000
	MP Pen	5B18/7B27	4.0	1.0	250	250	3.5	-16.0	140	38.0	—	10 000
	42 MP Pen	7B27	4.0	2.0	250	250	7.0	-5.5	—	—	—	8 000
	41 MPT	7B23	4.0	1.0	250	200	4.8	—	—	—	—	—
	42 PT	7B23	4.0	2.0	250	250	7.0	—	—	—	—	—
	41MTS	7B43	4.0	1.0	250	100	1.6	—	—	—	—	—
	PT10	7B27	4.0	2.0	250	250	9.0	-7.5	—	—	—	5 000
(Tetrode)	42 OT	7B20	4.0	2.0	250	250	7.0	-5.5	130	—	—	6 500
	DP/Pen	7B27	16.0	0.25	250	250	3.5	-10.0	300	—	—	10 000
(Tetrode)	402 OT	7B21	40.0	0.2	250	250	7.0	-6.6	—	—	—	5 500
	40PPA	7B27	40.0	0.2	150	150	4.0	-25.0	600	42.0	—	4 000
	402 Pen	7B29	40.0	0.2	250	250	7.0	-6.7	140	—	—	5 500
	402 Pen/A	7B29	40.0	0.2	150	150	8.0	-9.0	—	—	—	2 500
Dario	TC432	4B6/5B4	2.0	0.2	150	150	2.4	-4.5	—	—	—	—
	TC434	5B4	4.0	0.25	300	200	1.7	-25.0	—	—	—	—
	TE534	7B27	4.0	1.1	250	250	2.5	-15.0	—	—	—	—
	TE434	5B4	4.0	1.1	250	250	2.8	-14.0	—	—	—	—
	TE634	5B18/7B27	4.0	1.35	250	250	2.7	-22.0	—	—	—	—
	TL44	7B27	4.0	1.75	250	250	9.5	—	—	—	—	—
	TL54	7B27	4.0	2.0	250	275	8.5	—	—	—	—	—
	TL413	7B27	33.0	0.2	200	200	8.0	-8.5	—	—	—	—
	TB4320	8S22	24.0	0.2	200	100	3.1	-19.0	—	—	—	—
	TBL226	5B4	24.0	0.18	200	—	8.0	-19.0	—	—	—	—
Ekco	OP41	7B27	4.0	1.8	300	250	9.0	-13.0	200	66.0	8 000	4 000
	OP42	7B27	4.0	1.8	250	250	11.0	-6.0	145	36.5	3 800	8 000

Table 7. Output pentodes and tetrodes – *continued*

Make	Type	Base	Fil. volts	Fil. amps	Anode volts	Screen volts	Slope (mA/V)	Bias volts	Bias res. (ohms)	Anode and screen current (mA)	Output (mW)	Optimum load (ohms)
Ever Ready	K70B	5B4	2.0	0.15	135	135	2.2	-4.5	—	—	340	19 000
	K70D	5B4	2.0	0.3	135	135	3.0	-2.4	—	—	300	24 000
	A70B	7B27	4.0	1.35	250	250	2.8	-22.0	—	—	3 800	6 000
	A70D	7B27	4.0	1.95	250	250	9.5	-5.8	—	—	3 800	8 000
	A70E	7B27	4.0	2.1	250	275	8.5	-14.0	—	—	8 800	3 500
	EL32	O30	6.3	0.2	250	250	2.8	-18.0	—	—	3 600	8 000
	EL3/33	8S23/048	6.3	0.9	250	250	9.0	-6.0	—	—	4 500	7 000
Ferranti	PT4	7B27	4.0	2.0	250	250	7.0	—	—	37.5	—	—
	PTA	7B27	13.0	0.3	250	250	—	—	—	37.5	—	—
	PTSA	—	26.0	0.3	250	250	—	-8.9	—	—	—	—
	PTZ	—	40.0	0.2	250	250	—	—	—	47.0	—	—
Hivac	XY1.5V	SD2	1.5	0.16	45	45	1.0	-1.5	—	2.1	—	—
	XY2.0V	5D2	2.0	0.16	50	50	1.4	-2.0	—	2.15	—	—
	Y220	4B12	2.0	0.2	150	150	2.5	-4.5	—	11.8	500	1 500
	Z220	4B12	2.0	0.2	150	150	2.5	-6.0	—	20.1	1 000	7 500
(Tetrode)	AC/Y	5B18/7B27	4.0	1.0	250	250	3.5	-10.0	300	36.3	3 000	6 500
(Tetrode)	AC/YY	7B27	4.0	2.0	250	250	7.5	-10.0	140	78.0	5 000	3 000
(Tetrode)	AC/Z	7B27	4.0	2.0	250	250	8.0	-5.5	160	36.3	3 000	6 500
(Tetrode)	AC/Q	7B27	4.0	1.35	375	250	6.0	-22	370	59.5	11 500	4 000
(Tetrode)	FY	4B12	4.0	1.0	250	250	5.0	-10	250	38.0	3 000	6 000
(Tetrode)	AC/QA	7B27	6.3	0.9	75	250	6.0	-22	370	59.5	11 500	4 000
(Tetrode)	Y13	7B27	13.0	0.3	250	250	4.0	-22	550	39.5	3 000	4 000
(Tetrode)	Z26	7B27	26.0	0.3	250	250	8.0	-11	250	44.0	3 000	4 000
Lissen	PT225	4B6/5B4	2.0	0.2	150	150	1.6	—	—	—	300	—
	PT240	4B6/5B4	2.0	0.4	200	150	2.3	—	—	—	1 000	—
	PT2A	4B6/5B4	2.0	0.2	150	150	2.5	—	—	—	1 100	—
	PT425	4B6/5B4	4.0	0.25	200	150	2.3	—	—	—	1 000	—
	PT611	4B6	6.0	0.1	150	150	1.4	—	—	—	300	—
	AC/PT	5B4/7B27	4.0	1.0	250	250	4.0	—	—	—	2 500	—
Marconi	N14	O8	1.4	0.1	90	90	1.55	-7.5	—	9.0	250	8 000
(Tetrode)	KT12	5B4	2.0	0.2	150	150	2.5	-4.5	—	9.2	500	17 000
(Tetrode)	PT2	5B4	2.0	0.2	150	150	2.5	-4.5	—	9.2	500	17 000
(Tetrode)	KT21	4B4	2.0	0.3	150	150	5.3	-2.5	—	12.3	750	10 000
(Tetrode)	KT24	4B4	2.0	0.2	150	150	3.2	-2.8	—	12.1	640	10 000
(Tetrode)	MKT4	7B27	4.0	1.0	250	225	3.0	-13.5	360	37.0	3 200	7 000
(Tetrode)	MPT4	7B27	4.0	1.0	250	200	3.0	-10.5	250	37.5	—	8 000
(Tetrode)	MPT4K	7B27	4.0	1.0	250	200	3.0	-10.5	250	37.0	—	8 000
(Tetrode)	KT41	7B27	4.0	1.0	250	225	10.5	-4.4	90	48.5	4 300	6 000
	N40	7B27	4.0	1.0	250	225	2.9	—	—	—	—	—
	N41	7B27	4.0	2.0	250	250	10.5	-4.4	90	48.5	4 300	6 000
(Tetrode)	KT42	7B27	4.0	1.0	250	250	2.5	-16.5	420	39.5	3 000	7 000
	N42	7B27	4.0	1.0	250	250	2.5	-16.5	420	39.5	3 000	7 000
	N43	7B29	4.0	2.0	250	250	10.0	-4.5	—	50.0	—	5 400
	PT4	5B4	4.0	1.0	250	250	2.85	-10.0	400	38.0	2 500	7 500

Notes	Type	Base	Vf	If	Va	Vg2	gm	Vg				
	PT25	5B4	4.0	2.0	400	400	4.0	−16.0	250	75.0	10 000	5 000
	PT25H	5B4	4.0	2.0	400	400	6.5	−15	270	63.0	−	5 000
	PT16	5B4	4.0	1.0	300	300	4.8	−4.4	90	47.5	4 300	6 000
(Tetrode)	KT61	O48	6.3	0.95	250	250	10.5	−16.5	420	39.5	3 000	7 000
(Tetrode)	KT63	O48	6.3	0.7	250	250	2.5	−15	170	92.0	7 250	2 200
(Tetrode)	KT66	O48	6.3	1.27	400	300	6.3	−14.0	375	37.0	2 600	7 500
	N30	7B27	13.0	0.3	250	250	3.9	−14.0	375	37.0	2 600	7 500
(Tetrode)	KT30	7B27	13.0	0.3	250	250	3.9	−	−	−	−	−
	DPT	7B27/5B4	16.0	0.25	200	200	3.0	−4.5	95	54.0	3 000	6 500
(Tetrode)	KT31	7B46	13.0	0.6	200	200	10.0	−	−	−	−	6 500
	N31	7B46	26.0	0.3	180	180	10.0	−4.5	95	−	3 000	1 300
(Tetrode)	KT32	7B46	26.0	0.3	135	135	9.0	−7.6	95	80.0	3 500	3 000
(Tetrode)	KT33C	O48	26.0	0.6	200	200	10.0	−13.2	188	70.0	5 000	−
(Tetrode)	KT35	O48	13.0	0.3	200	200	10.0	−11.5	200	58.5	4 300	4 000
Mazda	KT44	7B23	4.0	2.0	400	300	6.3	−	−	−	−	−
	PEN141	OM4	1.4	0.1	90	90	1.75	−8.1	−	5.0	210	10 000
	PEN231	5B4	2.0	0.3	150	150	5.3	−2.2	−	5.5	290	19 000
	PEN220	5B4	2.0	0.2	150	150	2.5	−4.5	−	6.0	350	17 000
	PEN220A	5B4	2.0	0.2	150	150	2.5	−9.0	−	22.2	1 000	7 500
	PEN24	OM4	2.0	0.3	150	150	5.7	−3.3	−	6.0	440	16 000
	PEN25	OM4	2.0	0.15	150	150	4.5	−3.6	−	6.0	400	14 000
	AC/PEN	7B27	4.0	1.0	250	250	2.5	−15.5	250	32.0	3 300	7 500
	AC2/PEN	7B27	4.0	1.75	250	250	8.0	−5.3	140	38.0	3 500	6 700
	AC4/PEN	7B20	4.0	1.75	250	250	11.0	−8.75	114	77.0	6 900	3 300
	AC5/PEN	7B20	4.0	1.75	250	250	9.0	−8.5	175	47.5	5 800	4 500
	AC6/PEN	7B47	4.0	1.75	330	220	8.5	−6.9	90	77.0	−	−
	PEN44	OM22	4.0	2.1	275	275	11.0	−11.1	135	82.0	9 250	2 650
	PEN45	OM22	4.0	1.75	250	250	9.0	−8.5	175	47.5	4 850	5 200
	PEN46	OM23	4.0	1.75	330	220	8.5	−6.9	90	77.0	−	−
	PEN1340	7B27	13.0	0.4	250	250	6.5	−8.6	175	49.0	4 000	5 500
	PEN3520	7B27	35.0	0.2	250	250	7.0	−8.0	165	48.0	2 300	4 400
	DC2/PEN	7B27	35.0	0.1	200	200	2.5	−10.0	300	30.0	2 300	10 000
	PEN3820	7B20	38.0	0.2	200	200	12.0	−8.7	145	60.0	2 650	2 800
	PEN383	OM22	38.0	0.2	200	200	12.0	−8.7	145	60.0	2 650	2 800
Mullard	DL1	8S8	1.4	0.05	90	90	1.25	−3.0	−	−	170	22 500
	DL2	8S8	1.4	0.1	90	90	1.55	−7.5	−	−	240	8 000
	DL51	4D3	1.5	0.134	45	45	1.5	−1.5	−	−	−	−
	PM22	4B6/5B4	2.0	0.2	150	150	1.	−10.0	−	19.0	600	8 000
	PM22A	4B6/5B4	2.0	0.15	135	135	2.2	−4.5	−	7.0	340	19 000
	PM22C	5B4	2.0	0.3	150	150	3.0	−20.0	−	27.0	1 450	8 000
	PM22D	5B4	2.0	0.3	135	135	3.0	−2.4	−	5.8	300	24 000
	PEN4VA	5B18/7B27	4.0	1.35	250	250	2.8	−22.0	500	39.0	3 800	6 000
	PEN4VB	7B27	4.0	1.95	250	250	9.5	−5.8	145	41.0	3 800	6 000
	PENA4	7B27	4.0	1.95	250	250	9.5	−5.8	145	41.0	3 800	8 000
	PENB4	7B27	4.0	2.1	250	275	8.5	−14.0	175	79.0	8 800	3 500
	PEN428	7B27	4.0	2.1	375	275	8.0	−20.5	165	71.0	8 000	6 500
	PM24	4B6/5B4	4.0	0.15	150	150	1.75	−11.0	650	25.0	−	8 000
	PM24A	5B4	4.0	0.275	300	200	2.0	−22.5	1 000	23.5	−	10 000

Table 7. Output pentodes and tetrodes – *continued*

Make	Type	Base	Fil. volts	Fil. amps	Anode volts	Screen volts	Slope (mA/V)	Bias volts	Bias res. (ohms)	Anode and screen current (mA)	Output (mW)	Optimum load (ohms)
Mullard – *continued*	PM24B	5B4	4.0	1.1	250	250	3.0	−17.0	500	35.6	2 800	7 000
	PM24B	5B4	4.0	1.0	400	300	2.1	−40.0	1 100	37.0	—	8 000
	PM24C	5B4	4.0	1.0	400	200	3.0	−28.0	850	34.5	—	12 000
	PM24E	5B4	4.0	2.0	500	200	4.0	−35.0	750	59.0	—	7 000
	L2/32	8S22/050	6.3	0.2	250	250	2.8	−18.0	—	—	3 600	8 000
	EL3/33	8S23/048	6.3	0.9	250	250	9.0	−6.0	—	—	4 500	7 000
	EL35	O48	6.3	1.35	250	250	5.0	−15.5	—	—	—	—
	EL6/36	8S23/048	6.3	1.3	250	250	15.0	−7.0	—	—	8 000	3 500
	PEN13C	7B27	13.0	0.5	250	250	6.0	−11.9	250	39.0	3 200	6 400
	PEN26	8S22	24.0	0.2	200	100	3.1	−19.9	420	45.0	3 000	5 000
	PEN36C/CL33	7B27/048	33.0	0.2	200	200	8.0	−8.5	—	—	4 000	4 500
	CL4	8S22	33.0	0.2	200	200	8.0	−8.5	—	—	4 000	4 500
	CL6	8S22	35.0	0.2	200	100	8.0	−9.5	—	—	4 000	4 500
Osram	N14	O8	1.4	0.1	90	90	1.55	—	—	—	—	—
	N15	O12	1.4/2.8	0.05/0.1	90	90	2.0	—	—	—	—	—
(Tetrode)	KT2	5B4	2.0	0.2	150	150	2.5	−4.5	—	9.5	—	17 000
(Tetrode)	KT21	5B4	2.0	0.3	150	150	5.3	−2.5	—	6.5	—	19 000
(Tetrode)	KT24	5B4	2.0	0.2	150	150	3.2	−3.2	—	12.0	800	10 000
(Tetrode)	PT7	7B27	2.0	0.3	240	150	—	—	—	—	1 500	—
	ZA1	Acorn	4.0	0.25	250	100	1.4	—	—	—	—	—
(Tetrode)	MKT4	5B18/7B27	4.0	1.0	250	225	3.0	−11.0	300	37.0	8 000	8 000
	MPT4	7B27	4.0	1.0	250	225	3.0	—	—	—	—	—
(Tetrode)	KT41	7B27	4.0	2.0	250	250	10.5	−4.4	90	50.0	—	5 400
	N41	7B27	4.0	2.0	250	250	10.0	—	—	—	—	—
(Tetrode)	KT42	7B27	4.0	1.0	250	250	2.5	−16.5	420	29.5	—	7 000
	N42	7B27	4.0	1.0	250	250	2.5	—	—	—	—	—
	N43	7B29	4.0	2.0	250	250	10.0	−4.5	—	40.0	—	5 400
	PT4	5B4	4.0	1.0	250	250	2.85	−16.0	400	40.0	2 500	7 500
	PT5	5B4	4.0	1.0	1250	300	4.0	—	—	—	80 000	—
	DET8	7B27	4.0	2.0	400	200	4.0	—	—	—	—	—
	PT10/14	7B23	4.0	1.25	500	250	—	—	—	—	20 000	—
	PT25	5B4	4.0	2.0	400	250	4.0	−16.0	240	75.0	10 000	4 000
	PT25H	5B4	4.0	2.0	400	400	6.5	−12.5	300	39.0	2 000	6 000
	KT73	O48	6.0	0.4	175	175	2.5	—	—	—	38 000	—
	KT8	5B17	6.3	1.27	600	300	—	—	—	—	—	—
(Tetrode)	KT61	O48	6.3	0.95	250	250	10.5	−4.1	90	47.5	4 300	6 000
(Tetrode)	KT63	O48	6.3	0.7	250	250	2.5	−16.5	420	39.5	3 000	7 000
(Tetrode)	KT66	O48	6.3	1.27	400	300	6.3	−30.0	—	—	50 000	2 800
	N30	7B27	13.0	0.3	250	250	3.9	—	—	—	—	—
	N30G	7B27	13.0	0.3	250	250	3.9	—	—	—	—	—
	KT30	7B27	13.0	0.3	250	250	3.9	−14.0	375	37.0	3 000	7 500
(Tetrode)	KT72	O48	16.0	0.17	175	175	2.5	−12.5	300	36.0	2 000	6 000
(Tetrode)	KT74	O48	15.0	0.16	175	175	2.5	—	—	—	—	—
	DPT	7B27/B4										

Make	Type	Base										
(Tetrode)	KT31	7B46	26.0	0.3	200	200	10.0	−4.4	90	50.0	2500	5500
	N31	7B46	26.0	0.3	200	200	10.0	–	–	–	–	–
(Tetrode)	KT32	O48	26.0	0.3	135	135	9.0	−7.6	95	80.0	3500	1300
(Tetrode)	KT35	O46	26.0	0.3	200	200	10.0	–	–	–	–	–
(Tetrode)	KT33C	O46	26.0	0.6	200	200	10.0	−13.2	188	70.0	5000	3000
Record	PT2	4B6/5B4	2.0	0.22	150	150	3.0	−6.0	–	8.0	1000	14000
	PT2C	5B4	2.0	0.26	150	150	2.0	−12.0	–	20.0	3000	6000
	AC/PT	5B18/7B27	4.0	1.2	350	250	3.5	−18.0	400	40.0	3000	7000
	AC/PTA	5B18/7B27	4.0	1.2	250	250	3.5	−16.5	400	41.0	3000	7000
	AC/PT4VB	7B27	4.0	2.0	250	250	10.0	−6.0	150	40.0	3600	7000
	PT/24M	5B4	4.0	1.1	250	100	4.0	−15.0	400	42.0	3100	7500
	PT/24DA	7B27	24.0	0.2	200	100	8.0	−19.0	400	45.0	3000	5000
	PT/24DAL	8S23	24.0	0.2	200	100	8.0	−19.0	400	45.0	3000	5000
	PT/35DA	7B27	35.0	0.2	200	200	8.5	−8.0	170	50.0	3200	4400
Triotron	P225	4B6/5B4	2.0	0.2	150	150	2.0	−4.5	–	10.0	500	15000
	P469	7B27	4.0	2.0	250	275	8.5	−14.0	–	–	–	–
	P441N	7B27	4.0	1.35	250	250	4.0	−22.0	500	37.0	3800	7000
	P44n	5B18/7B27	4.0	1.1	250	250	3.5	−15.0	650	28.0	2000	7500
	P496	7B27	4.0	1.5	250	250	9.5	−6.0	–	–	–	–
	P425	5B4	4.0	0.25	300	200	1.7	−25.0	–	–	–	–
	P435	5B4	4.0	1.1	250	250	3.5	−14.0	–	–	–	–
	P3580	7B27	33.0	0.2	200	200	8.0	−23.0	–	–	–	–
	P2060	8S23	24.0	0.2	200	100	3.1	−19.0	–	–	–	–
	P2460	5B4	24.0	0.18	200	200	8.0	−19.0	–	–	–	–
	P2020N	5B4	20.0	0.18	200	135	2.5	−18.0	1000	19.0	1350	9000
Tungsram	PP2/S	5B5/8S8	2.0	0.14	135	135	2.1	−5.0	–	8.0	440	19000
	PP222	4B6/5B4	2.0	0.22	150	150	3.0	−6.0	–	9.0	600	14000
	PP225	5B4	2.0	0.26	135	135	2.0	−12.0	–	18.0	1000	6000
	PP4/S	5B4/8S8	4.0	1.1	250	250	4.0	−15.0	400	42.0	2800	7500
	APP4A/S	7B31/8S22	4.0	1.2	250	250	3.5	−16.5	400	40.5	3000	7000
	APP4B/S	7B27/8S23	4.0	1.95	250	250	10.0	−6.0	140	40.0	3600	7000
	APP4C	7B24	4.0	2.0	250	250	10.0	−6.0	140	40.0	4000	7000
	APP4E	7B27	4.0	2.1	375	275	8.5	−13.5	175	80.0	8000	3500
	APP4	7B30	4.0	2.0	250	250	10.0	−6.0	150	40.0	4000	7000
	PP6AS	8S23	6.3	0.2	250	250	2.8	−18.0	500	37.0	2250	8000
	PP6BS	8S23	6.3	1.2	250	250	10.0	−5.5	140	40.0	3600	7000
	PP6B	6UX9	6.3	1.2	250	250	10.0	−5.5	140	40.0	3600	7000
	PP6C	7B27	6.3	1.2	250	250	10.0	−5.5	140	40.0	3600	7000
	PP6E	7B27	6.3	1.2	375	275	8.5	−17.0	200	80.0	8800	3500
	EL2	8S22	6.3	0.2	250	250	2.8	−18.0	480	37.0	3600	8000
	EL3/33	8S23/O48	6.3	1.0	250	275	10.0	−7.0	175	40.5	3600	7000
	EL5	8S23	6.3	1.35	250	275	8.5	−14.0	175	79.0	8800	3500
	EL6/36	8S23/O48	6.3	1.4	250	250	15.0	−7.0	85	80.5	8200	3500
	6M6G	O48	6.3	1.0	250	250	10.0	−7.0	175	40.5	3600	7000
Double P	ELL1	8S26	6.3	0.45	250	275	1.3	−21.5	600	44.6	5400	16000
	PP13A	7B27	13.0	0.3	250	250	2.5	−16.5	410	40.5	3000	7000
	PP24	7B29	24.0	0.2	200	100	8.0	−19.0	400	45.0	3000	5000
	CL6/PP37	8S22/7B29	35.0	0.2	200	100	8.5	−9.5	140	50.0	4000	22000
	PP34	7B29	35.0	0.2	200	200	8.5	−8.0	170	50.0	3200	4400
	PP35	7B27	35.0	0.2	200	200	8.5	−8.0	170	50.0	3200	4400
	PP36	7B24	35.0	0.2	200	200	8.5	−8.5	170	50.0	3200	5000
	CL33	O48	35.0	0.2	200	200	8.5	−8.0	170	50.0	3200	4400

Table 8. Double output valves

Make	Type	Circuit	Base	Fil. volts	Fil. amps	Anode volts	Screen volts	Quiescent current (mA)	Peak current (mA)	Bias volts	Output (mW)	Optimum load (ohms)
Cossor	220B	Class B	7B2	2.0	0.2	120	–	2.5	–	0	–	12 000
	240B	Class B	7B2	2.0	0.4	120	–	4.0	–	0	–	8 000
	2103	QPP	7UX5	2.0	0.26	150	150	4.0	–	-10.5	–	35 000
	240QP	QPP	7B6	2.0	0.4	120	120	3.5	–	-9.0	–	24 000
Dario	TB402	Class B	7B2	2.0	0.2	150	–	–	–	-10.5	–	–
	BLL32	QPP	9B1	2.0	0.45	135	135	–	–	–	–	–
Ever Ready	K33A	Class B	7B2	2.0	0.2	120	–	3.0	–	0	1250	14 000
	K33B	Class B	7B2	2.0	0.2	120	–	3.0	–	-4.5	1450	14 000
	K77A	QPP	9B1	2.0	0.45	150	150	4.0	–	-13.5	2000	16 000
Ferranti	HP2	Class B	7B2	2.0	0.4	150	–	3.0	–	–	–	–
Hivac	B230	Class B	7B2	2.0	0.3	150	–	2.5	32.0	0	1250	14 500
	DB240	Driver	7B12	2.0	0.4	120	–	3.0	3.0	-4.5	–	–
		Class B				150	–	2.5	32.0	0	–	–
Lissen	QP240	QPP	7B6	2.0	0.4	150	150	8.0	32.0	-18	1250	14 500
	BB240	Class B	7B2	2.0	0.4	150	–	–	–	–	1400	14 500
Marconi	QP21	QPP	7B6	2.0	0.4	150	150	3.5	–	-9	2400	30 000
	B21	Class B	7B2	2.0	0.2	150	–	2.2	–	-6	1200	12 000
	B30	Class B	7B2	13.0	0.3	180	–	–	–	–	1500	7 000
Mazda	QP30	QPP	7B6	2.0	0.3	110	110	5.3	–	-8.6	5000	17 000
	QP240	QPP	9B1	2.0	0.4	150	150	4.0	–	-11.5	700	15 000
	QP25	QPP	OM9	2.0	0.2	120	120	5.5	–	-9.75	1.2	15 000
	PD220	Class B	7B2	2.0	0.2	150	–	0.8	45.0	-1.15	–	–
	PD220A	Class B	7B2	2.0	0.2	150	–	2.5	50.0	-6.0	–	–
Mullard	PM2B	Class B	7B2	2.0	0.2	120	–	3.0	20.0	0	1250	14 000
	PM2BA	Class B	7B2	2.0	0.2	120	–	3.0	20.0	-4.5	1450	14 000
	QP22A	QPP	9B1	2.0	0.45	135	135	–	–	-12.0	1400	16 000
	QP22B	QPP	7B6	2.0	0.3	135	135	–	–	-11.7	1330	14 700
	ECC31	Double Triode	O42	6.3	0.95	250	–	–	–	-4.6	–	–
Osram	QP21	QPP	7B6	2.0	0.4	150	150	3.5	–	-9.0	1200	14 000
	B21	Class B	7B2	2.0	0.2	150	–	2.2	–	-6.0	1500	12 000
	B30	Class B	7B2	13.0	0.3	180	–	–	–	0	5000	7 000
Record	BB2A	Class B	7B6	2.0	0.25	150	–	2.5	–	-3.0	2000	10 000
	BB2B	Class B	7B6	2.0	0.25	135	–	–	–	0	1700	10 000
Triotron	E220B	Class B	7B2	2.0	0.2	150	–	–	–	–	–	–
Tungsram	CB215/S	Class B	7B2/8S5	2.0	0.22	135	–	3.0	21.0	0	1700	10 000
	CB220	Class B	7B2	2.0	0.25	150	–	3.0	26.7	-3.35	2000	10 000

Table 9. Metal rectifiers – Westectors

Make	Type	Class	Max. safe input voltage	Max. current output (mA)
Westinghouse	W.4	Half-wave	24 volts peak carrier	0.25
	W.6	Half-wave	36 volts peak carrier	0.28
	WX.6	Half-wave	36 volts peak carrier	0.12
	WM.142	Full-wave centre-tapped	24 volts each side of C.T.	0.5
	WM.162	Full-wave centre-tapped	36 volts each side of C.T.	0.5

(WM.142 and WM.162 are the new code numbers of the earlier WM.24 and WM.26 respectively.)

Table 9. Metal rectifiers – LT types

Make	Type	Output Volts	Output Amps	Nominal AC input (volts)	Replaces
Westing-house	LT.41	12	1	22	LT.5, LT.9, A.3
	LT.42	6	1	11	LT.1, LT.2, LT.4, LT.7, LT.8
	LT.44	12	2	22	LT.10, A.4
	LT.45	6	4	11	LT.11, A.6

Table 9. Metal rectifiers – high-tension types

Make	Type	Maximum smoothed DC output Volts	Maximum smoothed DC output mA	Max. current output (mA)	Half-wave Volts	Half-wave mA	Full-wave Volts	Full-wave mA	Capacity of each (V.D.) mFd	Working voltage (V.D.)	Remarks
Westinghouse	HT.14	130	20	30	135	30	80	60	6	200	Replaced by HT.41
	HT.15	200	30	40	250	80	140	120	4	200	Replaced by HT.41
	HT.16	300	60	60	400	90	240	200	4	400	Replaced by HT.42
	HT.17	200	100	150	250	150	150	300	8	250	Replaced by HT.41
(For Class B)	HT.17	150	25	–	150	40	–	–	8	350	Replaced by HT.41
(2 in series)	HT.17	500	120	150	–	–	300	550	6	500	Replaced by HT.41
Used voltage doubler only {	HT.41	250	60	100	300	90	150	180	8	250	–
	HT.42	450	100	100	540	150	270	300	8	400	–
	H.1	3.6	10	10	3.5	20	–	–	100	12	–
	H.10	36	10	10	35	20	–	–	10	50	–
	H.50	180	10	10	175	20	–	–	2	250	–
	H.75	270	10	10	260	20	–	–	2	400	–
	H.100	360	10	10	350	20	–	–	1	500	–
	H.120	432	10	10	420	20	–	–	0.85	600	–
	H.176	650	10	10	620	20	–	–	0.5	1100	–
	J.10	80	2	2	74–80	4	–	–	10	250	–
	J.50	400	2	2	370–400	4	–	–	2	650	–
	J.100	800	2	2	740–800	4	–	–	1	1250	–
	J.125	1000	2	2	920–1000	4	–	–	1	1500	–
	J.176	1400	2	2	1300–1400	4	–	–	0.5	2000	–
2 units in series	H.120	870	10	10	–	–	480	30	0.5	700	–
2 units in series	H.176	1300	10	10	–	–	720	30	0.25	1000	–
10 units in series	H.176	6500	10	10	–	–	3600	30	0.35	5000	–
2 units in series	J.10	170	2	2	–	–	74–80	6	10	250	–
2 units in series	J.50	850	2	2	–	–	370–400	6	2	650	–
2 units in series	J.100	1700	2	2	–	–	740–800	6	1	1250	–
4 units in series	J.125	4000	2	2	–	–	1600–1700	6	0.5	3000	–
2 units in series	J.176	3000	2	2	–	–	1300–1400	6	0.5	2000	–
10 units in series	J.176	15000	2	2	–	–	6500–7000	6	0.1	12000	–

Table 10. HT rectifying valves

Make	Type	Base	Filament		Anode volts Max. (RMS)	Output (mA)
			Volts	Amps		
Brimar	25Z4G	–	25.0	0.3	250	75
	5Z4G	–	5.0	2.0	350 + 350	125
	R1	4B17	4.0	1.0	250 + 250	60
	R2	4B17	4.0	2.5	350 + 350	120
	R3	4B17	4.0	2.5	500 + 500	120
	1D5	5B10	40.0	0.2	250	75
	(Mercury)	4B2	2.0	1.2	6000	3
Cossor	44SU	4B1	4.0	0.4	200	20
	412SU	4B1	4.0	1.0	250	70
	506BU	4B3	4.0	1.0	250 + 250	60
	408BU	4B3	4.0	1.0	250 + 250	30
	412BU	4B3	4.0	1.0	350 + 350	70
	442BU	4B3	4.0	2.5	350 + 350	120
	460BU	4B3	4.0	2.5	500 + 500	120
	431U	4B17	4.0	2.5	350 + 350	120
	441U	4B17	4.0	2.5	500 + 500	120
	4/100BU	4B3	4.0	2.5	500 + 500	200
	451U	4B17	4.0	3.5	1500 + 1500	250
	405BU	4B3	4.0	0.5	1500 + 1500	20
	SU2150A	4B16	2.0	1.5	5000	10
	SU2150	4B16	2.0	1.15	8000	2
	612BU	4B3	6.0	0.4	250 + 250	50
	825BU	4B3	7.5	2.0	500 + 500	120
	40SUA	5B10	40.0	0.2	250	75
	225DU	7B1	2 + 2	0.5 + 0.5	750 + 750	20
Dario	TW1	SB10	20.0	0.2	250	80
	TW2	7B15	30.0	0.2	250	120
	TBY233	7B15	33.0	0.18	250	120
	SW1	4B3	4.0	1.0	400	60
	FW1	4B3	4.0	1.0	300 + 300	75
	FW2	4B3	4.0	2.0	350 + 350	120
	TZ34	4B17	4.0	2.0	350 + 350	120
	FW3	4B3	4.0	2.0	500 + 500	120
	IFW1	4B17	4.0	2.5	500 + 500	120
Ekco	R41	4B3	4.0	2.0	350 + 350	120
Ever Ready	S11A	4B3	4.0	1.0	250 + 250	60
	A11B	4B17	4.0	2.4	350 + 350	120
	S11D	4B3	4.0	2.0	350 + 350	120
	A11D	4B17	4.0	2.0	350 + 350	120
	A11C	4B17	4.0	2.4	500 + 500	120
	AZ1/31	8S1	4.0	1.1	300 + 300	100
	CY31	O35	20.0	0.2	250	75
	C10B	5B10	20.0	0.2	250	75
Ferranti	R4	4B3	4.0	2.5	350 + 350	120
	R4A	4B3	4.0	2.5	500 + 500	120
(Mercury)	IR4	–	4.0	1.0	5000	3
(Mercury)	GR4	4B3	4.0	3.0	350 + 350	350
	RS	5B10	13.0	0.3	250	75
	RA	5B12	13.0	0.3	250 + 250	50
	RZ	5B10	20.0	0.2	250	75
Hivac	UU60/250	4B3	4.0	1.25	300 + 300	75
	UU120/350	4B3	4.0	2.5	350 + 350	120
	UU120/500	4B3	4.0	2.5	500 + 500	120
	U26	7B42	13 or 26	0.6 or .03	250	120
	MR1	4B1	4.0	3.0	1000	250
Lissen	UU41	4B3	4.0	1.0	300 + 300	80
	U650	4B1	5.6	0.5	300	40
Marconi	MU12	4B17	4.0	2.5	350 + 350	120
	MU14	4B17	4.0	2.5	500 + 500	120
	U5	4B3	5.0	1.6	400 + 400	45

Table 10. HT rectifying valves – *continued*

Make	Type	Base	Filament		Anode volts Max. (RMS)	Output (mA)
			Volts	*Amps*		
Marconi–	U8	4B3	7.5	2.4	500 + 500	120
continued	U9	4B3	4.0	1.0	250 + 250	75
	U10	4B3	4.0	1.0	250 + 250	75
	U12	4B3	4.0	2.5	350 + 350	120
	U14	4B3	4.0	2.5	500 + 500	120
	U16	4B2	2.0	1.0	5000	5
	U17	4B2	4.0	1.0	2500	30
	U18	4B3	4.0	3.75	500 + 500	250
	U20	4B3	4.0	3.75	850 + 850	125
	U30	7B42	26.0 / 13.0	0.3 / 0.6	250	120
	U31	O35	26.0	0.3	250	120
	U50	O2	5.0	2.0	350 + 350	125
	U52	O2	5.0	3.0	500 + 500	250
(Mercury)	GU1	4B1	4.0	3.0	1000	250
(Mercury)	GU5	4B2	4.0	3.0	1500	250
(Mercury)	GU50	4B2	4.0	3.0	1500	250
	A831	4B3	1.8	2.8	30 + 30	1.3 amp
Mazda	UU4	4B17	4.0	2.2	350 + 350	120
	UU5	4B17	4.0	2.3	500 + 500	120
	UU120/500	4B17	4.0	2.5	500 + 500	120
	UU6	OM17	4.0	1.4	350 + 350	120
	UU7	OM17	4.0	2.3	350 + 350	180
	UU8	OM17	4.0	2.8	350 + 350	250
	U4020	5B10	40.0	0.2	250	120
	U403	OM15	40.0	0.2	250	120
	UD41	7B48	4.0	1.15	550	35
	U21	4B16	2.0	1.85	4500	5
	U22	OM16	2.0	2.0	4500	5
(Mercury)	MU2	4B2	2.0	3.1	12500	5
Mullard	DW2	4B3	4.0	1.0	250 + 250	60
	DW3	4B3	4.0	2.0	350 + 350	120
	DW4/350	4B3	4.0	2.0	350 + 350	120
	DW4	4B3	4.0	2.0	500 + 500	120
	DW4/500	4B3	4.0	2.0	500 + 500	120
	IW2	4B17	4.0	1.2	250 + 250	60
	IW3	4B17	4.0	2.4	350 + 350	120
	IW4	4B17	4.0	2.0	350 + 350	120
	IW4	4B17	4.0	2.4	500 + 500	120
	IW4/500	4B17	4.0	2.4	500 + 500	120
	FW4/500	4B3	4.0	3.0	500 + 500	250
	CY1/31	8S15/035	20.0	0.2	250	75
	UR1	8S15	20.0	0.2	250	75
	UR1C	5B10	20.0	0.2	250	75
	CY/2/32	8S17/038	30.0	0.2	250 + 250	120
	UR3	8S17	30.0	0.2	250 + 250	120
	UR3C	7B15	30.0	0.2	250 + 250	120
	UY31	O35	50.0	0.1	250	125
	HVR1	4B2	2.0	0.3	6000	5
	HVR2	4B2	4.0	0.65	6000	3
Osram	MU12/14	4B17	4.0	2.5	500 + 500	120
	MU14	4B13	4.0	2.5	500 + 500	120
	U5	4B3	5.0	1.6	400 + 400	45
	U8	4B3	7.5	2.5	500 + 500	120
	U10	4B13	4.0	1.0	250 + 250	60
	U12/14	4B13	4.0	2.5	500 + 500	120
	U14	4B13	4.0	2.5	500 + 500	120
	U16	4B2	2.0	1.0	5000	5
	U17	4B2	4.0	1.0	2500	30
	U18/20	4B3	4.0	3.75	500 + 500 / 250 + 850	250 / 125

Table 10. HT rectifying valves – *continued*

Make	Type	Base	Filament		Anode volts Max. (RMS)	Output (mA)
			Volts	Amps		
Osram –	U23	4B2	4.0	3.3	1750	250
continued			25.0	0.3	180	120
	U30	7B42	26.0	0.3	220	75
			13.0	0.6	250	120
	U31	O35	26.0	0.3	250	120
	U50	O2	5.0	2.0	400 + 400	110
	U52	O2	5.0	3.0	500 + 500	250
	U71	O35	30.0	0.17	250	100
	U74	O35	30.0	0.16	250	75
(Mercury)	GU1	4B1	4.0	3.0	1000	250
(Mercury)	GU5	4B2	4.0	3.0	1500	250
	GU50	4B2	4.0	3.0	1500	250
Philips	373	4B1	4.0	1.0	220	40
(Miniwatt)	505	4B1	4.0	1.0	400	60
	506	4B3	4.0	1.0	300 + 300	75
	506K	4B17	4.0	1.2	250 + 250	60
	1560	4B3	5.0	2.0	300 + 300	125
	1561	4B3	4.0	2.0	500 + 500	120
	1801	4B3	4.0	0.6	250 + 250	30
	1802	4B3	4.0	0.5	250	30
	1803	4B1	4.0	–	500	30
	1805	4B3	4.0	1.0	250 + 250	60
	1807	4B3	4.0	2.0	350 + 350	120
	1815	4B3	4.0	2.3	500 + 500	180
	1817	4B3	4.0	4.0	350 + 350	300
	1821	4B3	4.0	1.0	250 + 250	60
	1831	4B3	4.0	1.0	700 + 700	60
	1832	4B1	4.0	1.2	700	120
	1861	4B17	4.0	2.4	500 + 500	120
	1867	4B17	4.0	2.4	350 + 350	120
	1876	8S12	4.0	0.3	850	5
	1877	4B16	4.0	0.,65	6000	5
	1881	4B17	4.0	1.0	250 + 250	60
	1881A	4B17	4.0	2.4	250 + 250	60
	AZ/31	8S1/02	4.0	1.1	300 + 300	100
	EZ2	8S16	6.3	0.4	350 + 350	60
	CY1/31/C	8S15/035/5B10	20.0	0.2	250	75
	CY2	5B10	30.0	0.2	250 + 250	120
Record	1FW4A	4B17	4.0	2.0	400 + 400	120
	FW350	4B13	4.0	1.0	300 + 300	80
	FW3	4B17	4.0	2.0	350 + 350	120
	FW5	4B13	4.0	2.0	500 + 500	120
	FW6	4B13	4.0	2.0	600 + 600	180
	UFW/30	5B10	30.0	0.2	275	120
	UFW/30L	8S17	30.0	0.2	275	120
	HW/20	5B10	20.0	0.2	250	80
	HW/20L	8S15	20.0	0.2	250	80
	HW/30	5B10	30.0	0.2	275	120
Triotron	G429	4B1	4.0	0.3	250	30
	G470	4B13	4.0	1.0	300 × 300	70
	G4120	4B13	4.0	2.0	500 × 500	120
	G4120N	4B17	4.0	2.0	500 × 500	120
	G2080	5B10	20.0	0.2	250	80
	G3060	8S17	30.0	0.2	125 × 125	120
	G3120	7B15	30.0	0.2	250	120
Tungsram	PV4	4B3	4.0	2.0	350 + 350	120
	PV4200	4B3	4.0	2.0	500 + 500	120
	PV4201	4B3	4.0	2.0	600 + 600	180
	AP4V	4B17	4.0	2.0	350 + 350	120
	RV120/350/S	4B3/8S1	4.0	2.4	350 + 350	120
	RV120/500/S	4B3/8S1	4.0	2.4	500 + 500	120

Table 10. HT rectifying valves – *continued*

Make	Type	Base	Filament		Anode volts Max. (RMS)	Output (mA)
			Volts	Amps		
Tungsram – *continued*	RV200/600	4B3	4.0	2.8	600 + 600	200
	PV75/1000	4B3	2.2	4.0	1000 + 1000	75
	PV100/2000	4B3	4.0	2.2	2000 + 2000	100
	PVA6S	8S16	6.3	0.25	350 + 350	60
	PVB6S	8S16	6.3	0.65	350 + 350	100
	PVC6S	8S16	6.3	0.9	350 + 350	175
	EZ2	8S16	6.3	0.4	350 + 350	60
	EZ3	8S16	6.3	0.65	400 + 400	100
	EZ4	8S16	6.3	0.9	400 + 400	175
	V20/S	5B10/8S15	20.0	0.2	250	80
	PV25	7B15	25.0	0.3	275 and 275	120
	V30	5B10	30.0	0.2	275	120
	PV29/S	7B15/8S17	30.0	0.2	125 and 125	60
	PV30	7B15	30.0	0.2	275	120
	PV30S	8S17	30.0	0.2	275	120
	V2118	5B10	20.0	0.18	250	80
	PV308	7C1	30.0	0.18	250	100

Table 11. Barretters

Make	Type	Base	Current (amps)	Voltage range
Atlas	150A/4	4B20	0.2	100–200
	150A/C	8S27	0.2	100–200
	150B/UX4	4-pin US	0.3	100–200
	130B	6-pin US	0.3	85–170
	110B	6-pin US	0.3	75–145
	150C	4B20	0.18	100–200
Dario	T1	ES cap	0.2	100–200
Marconi	171	ES cap	0.16	100–200
	202	ES cap	0.2	120–200
	251	ES cap	0.25	100–180
	301	ES cap	0.3	138–221
	302	ES cap	0.3	112–195
	303	ES cap	0.3	86–129
	304	ES cap	0.3	95–165
Osram	301	ES cap	0.3	138–221
	302	ES cap	0.3	112–195
	303	ES cap	0.3	86–129
	304	ES cap	0.3	95–165
	251	4B20	0.25	100–180
	202	4B20	0.2	120–200
Philips (Miniwatt)	C1/C	8S33/4B20	0.2	90–230
	C2	8S33	0.2	60–120
	C3	8S33	0.2	100–200
	C9	8S33	0.2	35–100
	C13	8S33	Special low-voltage resistance lamp	
	1941	4B20/ES	0.3	100–240
	1933	4B20	0.1	50–160
	1934	4B20	0.25	85–195
	1927	4B20	0.18	60–120
	1928	4B20	0.18	100–210
	1920	4B20	0.25	40–70
	1904	4B20/and bayonet cap	0.1	40–70
Tungsram	BR201	4B20	0.2	100–200
	BR201/S	8S33	0.2	100–200

Table 12. Gas-filled relays

Make	ztype	Base	Filament		Anode volts	Anode current
			volts	amps		
Brimar	4039A	5B13	4.0	1.0	500	100 mA
Cossor	GDT4B	5B13	4.0	1.75	350	100 mA
	GDT4	5B13	4.0	1.5	500	20 mA
Marconi	GT1	5B15	4.0	1.3	1000	1.0 amp
	GT1A	5B15	4.0	1.3	300	0.6 amp
	GT1B	5B15	4.0	1.35	120	2.0 mA
	GT1C	5B15	4.0	1.3	500	1.0 amp
Mazda	T11	5B13	4.0	1.2	700	300 m/A
	T21	5B13	4.0	1.2	200	300 m/A
	T31	5B13	4.0	1.5	400	500 m/A
	T41	OM19	4.0	1.5	400	500m/A
Osram	GT1	5B15	4.0	1.3	1000	0.3 amp
	GT1A	5B15	4.0	1.3	300	0.2 amp
	GT1B	5B15	4.0	1.35	120	2.0 mA
	GT1C	5B15	4.0	1.3	500	0.3 amp

Table 13. Tuning indicators

Make	Name	Base	Type	Operation characteristics
Brimar	6G5/6U5	–	Cathode Ray	Fil. 6.3 volts, 0.3 amps; max. anode 250 volts
Cossor	3180	NE1	Neon	145–160 volts
	3184	NE2	Neon	145–160 volts
	41ME	8S31	Cathode Ray	Fil. 4.0 volts, 0.3 amps; max. anode 250 volts
Dario	TM14	8S31	Cathode Ray	Fil. 4.0 volts, 0.3 amps; max. anode 250 volts
Ever Ready	A39A	8S31	–	Fil. 4.0 volts, 0.3 amps; max. anode 250 volts
Marconi	Y61/62 Y63/64	O59	Cathode Ray	Fil. 6.3 volts, 0.3 amps; max. anode 250 volts
Mazda	AC/ME	7B40	Cathode Ray	Fil. 4.0 volts, 0.5 amps; max. anode 250 volts
	ME41	OM27	Cathode Ray	Fil. 4.0 volts, 0.5 amps; max. anode 250 volts
	ME91	OM27	Cathode Ray	Fil. 9.0 volts, 0.2 amps; max. anode 200 volts
	ME920	7B40	Cathode Ray	Fil. 9.0 volts, 0.2 amps; max. anode 250 volts
Mullard	TV4	8S31	Cathode Ray	Fil. 4.0 volts, 0.3 amps; max. anode 250 volts
	TV4A	8S31	Cathode Ray	Fil. 4.0 volts, 0.3 amps; max. anode 250 volts
	*TV6	8S31	Cathode Ray	Fil. 6.3 volts, 0.2 amps; max. anode 250 volts
	EM1	8S31	Cathode Ray	Fil. 6.3 volts, 0.2 amps; max. anode 250 volts
	EM3	8S31	Cathode Ray	Fil. 6.3 volts, 0.2 amps; max. anode 250 volts
	EM4	8S32	Cathode Ray	Fil. 6.3 volts, 0.2 amps; max. anode 250 volts
	EFM1	8S30	Cathode Ray	Fil. 6.3 volts, 0.2 amps; max. anode 250 volts
Osram	Y61/62/63/64	O59	Cathode Ray	Fil. 6.3 volts, 0.3 amps; max. anode 250 volts
	Y73	O59	Cathode Ray	Fil. 6.0 volts, 0.16 amps; max. anode 180 volts
Tungsram	ME4S	8S31	Cathode Ray	Fil. 4.0 volts, 0.3 amps; max. anode 250 volts
	VME4	7B40	Cathode Ray	Fil. 4.0 volts, 0.5 amps; max. anode 250 volts
	ME6S	8S31	Cathode Ray	Fil. 6.3 volts, 0.2 amps; max. anode 250 volts
	EM1	8S31	Cathode Ray	Fil. 6.3 volts, 0.2 amps; max. anode 250 volts
	EM4	8S32	Cathode Ray	Fil. 6.3 volts, 0.2 amps; max. anode 250 volts
	EMFI	8S30	Cathode Ray	Fil. 6.3 volts, 0.2 amps; max. anode 300 volts

Table 14. American types

Type	Description	Base	Fil. or beater (volts)	Fil. or Heater current (amps)	Anode volts	Screen volts	Bias volts	Bias. res. (ohms)	Anode current (mA)	Screen current (mA)	Slope mA/V (*=conv. condt. μA/V)	Impedance (ohms)	Amp. factor	Output (watts)	Optimum load (ohms)
00A	Triode (Gas)	4UX3	5.0	0.25	45	—	—	—	1.5	—	0.666	30 000	20	—	—
0Z4	Rectifier (Gas)	078	—	—	RMS 300	—	—	—	—	—	—	—	—	75 mA	—
01-A	Triode	4UX3	5.0	0.25	45	—	—	—	1.5	—	0.666	30 000	20	—	—
1A4	HF Pentode	4UX5	5.0	0.06	180	67.5	-3–15	—	2.3	0.8	0.75	1 meg.	750	—	—
1A5	LF Pentode	09	1.4	0.05	90	90	-4.5	—	4.0	0.8	0.85	300 000	255	0.115	25 000
1A6	Heptode	6UX3	2.0	0.06	180	67.5	-3–22.5	—	1.3	2.4	*300	500 000	—	—	—
1A7	Pentagrid	—	1.4	0.05	45	0–3	0	—	0.6	—	*250	600 000	—	—	—
1A7V	VMu. Pentagrid	010	2.0	0.05	90	90	-3.0	1.2	2.3	0.6	—	—	—	—	—
1B4	HF Pentode	4UX5	2.0	0.06	180	67	-3.0	—	1.7	0.6	0.65	35 000	20	—	—
1B5/255	DD Triode	6UX2	2.0	0.06	135	—	—	—	0.8	—	0.575	1.5 meg.	—	—	—
1C5	LF Pentode	09	1.4	0.1	90	90	-7.5	—	7.5	1.6	1.55	115 000	180	0.24	8000
1C6	Heptode	6UX3	2.0	0.12	180	67.5	-3–14	—	1.5	2.0	*325	750 000	—	—	—
1C7	Pentagrid	010	2.0	0.12	180	67.5	-3.0	—	1.5	0.8	*325	750 000	—	—	—
1D5	HF Pentode	07	2.0	0.06	180	67.5	-3.0	—	2.3	2.4	0.75	1 meg.	—	—	—
1D7	Pentagrid	010	2.0	0.06	180	67.5	-3.0	—	1.3	0.6	0.3	500 000	—	—	—
1E5	HF Pentode	07	2.0	0.06	180	67	-3.0	—	1.7	0.6	0.65	1.5 meg.	—	—	—
1E7	Twin LF Pentode	014	2.0	0.24	135	135	-7.5	—	7.5	2.2	1.425	260 000	—	0.575	24 000
1F4	LF Pentode	5UX1	2.0	0.12	180	180	-4.5	432	8.0	2.4	1.7	200 000	—	0.31	16 000
1F5	LF Pentode	08	2.0	0.12	180	180	-4.5	432	8.0	2.4	1.7	200 000	—	0.31	16 000
1F6	DD LF Pentode	6UX13	2.0	0.06	180	67.5	-1.5	—	2.2	0.7	0.65	1 meg.	—	—	—
1F7	DD LF Pentode	015	2.0	0.06	180	67.5	-1.5	—	2.2	0.7	0.65	1.5 meg.	—	—	—
1H4	Triode	03	2.0	0.06	180	—	-13.5	—	3.1	—	0.9	10 300	—	—	—
1H5	Diode Triode	04	1.4	0.05	90	—	0	—	0.14	—	0.275	140 000	9.3	—	—
1H6	DD Triode	013	2.0	0.06	135	—	-3.0	—	0.8	—	0.575	35 000	65	—	—
1J6	Class B	016	2.0	0.24	135	—	—	—	5.0	—	—	—	20	2.1	10 000
1LA4	LF Pentode	0L2	1.4	0.05	90	90	-4.5	—	4.0	0.8	0.85	300 000	255	0.115	25 000
1LA6	Frequency Changer	0L4	1.4	0.05	90	45	0–3.0	—	1.2	0.6	*250	600 000	—	—	—
1LH4	Diode Triode	0L1	1.4	0.05	90	—	0	—	0.14	—	0.275	240 000	65	—	—
1LN5	HF Pentode	0L3	1.4	0.05	90	90	0	—	1.6	0.35	0.8	1.1 meg.	880	—	—
1N5	HF Pentode	06	1.4	0.05	90	90	-4.0	—	1.2	0.3	0.75	1.5 meg.	1160	—	—
1N5V	HF Pentode	060	1.4	0.05	90	90	—	—	1.6	—	0.65	1 meg.	—	—	—
1Q5	LF Pentode	BB1	1.4	0.1	90	90	-4.5	—	9.5	1.6	2.1	1 meg.	—	0.27	8000
1R5	Pentagrid	BB2	1.4	0.05	90	67.5	—	—	1.7	3.0	—	—	—	—	—
1S4	LF Pentode	BB3	1.4	0.1	67.5	67.5	-7.0	—	7.2	1.5	—	—	—	—	—
1S5	LF Pentode	BB3	1.4	0.05	90	90	—	—	3.0	7.0	—	—	—	—	—
1T4	HF Pentode	BB4	1.4	0.05	67.5	67.5	—	—	3.7	1.25	—	—	—	—	—
1-V	Rectifier	4UX10	6.3	0.3	RMS 325	—	—	—	—	—	—	—	—	45 mA	—
2A3	Power Triode	4UX3	2.5	2.5	250	—	-45.0	750	60.0	—	5.25	800	4.2	3.5	2000
2A5	LF Pentode	6UX10	2.5	1.75	250	250	-16.5	410	34.0	6.5	2.65	30 000	190	3.0	7000
2A6	DD Triode	6UX6	2.5	0.8	250	—	-2.0	5000	0.8	—	1.1	90 000	100	—	—
2A7	Pentagrid	7UX5	2.5	0.8	250	100	-3–40	300	3.5	2.2	*520	—	—	—	—
2B7	DD HF Pentode	7UX6	2.5	0.8	250	100	-3.0	—	10.0	2.3	1.2	360 000	—	—	—
2E5	Tuning Indicator	6UX12	2.5	0.8	250	—	—	—	—	—	—	600 000	—	—	—
2G5	Tuning Indicator	—	2.5	0.8	250	—	—	—	—	—	—	—	—	—	—
3Q5	LF Pentode	011	2.8	0.05	90	90	-4.5	—	9.5	1.6	2.1	—	—	0.27	8000
5T4	Rectifier	02	5.0	2.0	RMS 450	—	—	—	—	—	—	—	—	250 mA	—
5U4	Rectifier	02	5.0	3.0	RMS 500	—	—	—	—	—	—	—	—	250 mA	—
5V4	Rectifier	036	5.0	2.0	RMS 400	—	—	—	—	—	—	—	—	200 mA	—
5X4	Rectifier	01	5.0	3.0	RMS 500	—	—	—	—	—	—	—	—	200 mA	—
5Y3	Rectifier	02	5.0	2.0	RMS 350	—	—	—	—	—	—	—	—	125 mA	—
5Y4	Rectifier	01	5.0	2.0	RMS 350	—	—	—	—	—	—	—	—	125 mA	—
5Z3	Rectifier	4UX2	5.0	3.0	RMS 350	—	—	—	—	—	—	—	—	250 mA	—
5Z4	Rectifier	036	5.0	2.0	RMS 350	—	—	—	—	—	—	—	—	125 mA	—
6A3	Power Triode	4UX3	6.3	1.0	250	—	-4.5	—	60.2	—	5.25	800	4.2	3.2	2500
6A4	LF Pentode	5UX1	6.3	0.3	180	180	-12.0	465	22.0	3.9	2.2	45 500	—	1.4	8000

223

Table 14. American types – continued

Type	Description	Base	Fil. or heater (volts)	Fil. or Heater current (amps)	Anode volts	Screen volts	Bias volts	Bias res. (ohms)	Anode current (mA)	Screen current (mA)	Slope mA/V (*=conv. condc. µA/V)	Impedance (ohms)	Amp. factor	Output (watts)	Optimum load (ohms)
6A6	Double Triode	042	6.3	0.8	300	—	0	0	—	—	—	—	35	10.0	—
6AG6	LF Pentode	049	6.3	1.2	250	250	-6.0	150	32.0	6.0	10.0	60 000	600	3.75	8500
6A7	Frequency Changer	7UX6	6.3	0.3	250	100	-3–40	300	3.5	2.2	*550	360 000	—	—	—
6A8	Frequency Changer	054	6.3	0.3	250	100	-3–40	300	3.5	2.2	*550	360 000	—	—	—
6AB5	Tuning Indicator	6UX12	6.3	0.15	180	Target 180	—	—	—	—	—	—	—	—	—
6AD6	Tuning Indicator	061	6.3	0.15	—	Target 150	—	—	—	—	—	—	—	—	—
6AE6	Twin Anode Control	063	6.3	0.15	250	Target 135	—	—	6.5	—	—	—	—	—	—
6AF6	Tuning Indicator	061	6.3	0.15	—	—	—	—	—	—	—	—	—	—	—
6B4	Power Triode	03	6.3	1.0	250	—	-45.0	—	60.0	—	5.25	800	4.2	3.2	2500
6B5	Double Triode	6UX8	6.3	0.8	300	—	—	—	43.0	—	2.25	24 000	54	5.0	7000
6B6	DD Triode	041	6.3	0.3	250	—	-2.0	5000	0.4	—	1.1	90 000	100	—	—
6B7	DD HF Pentode	7UX5	6.3	0.3	250	125	-3.0	250	7.5	2.1	1.1	650 000	700	—	—
6B8	DD HF Pentode	053	6.3	0.3	250	125	-3.0	—	9.0	2.3	1.1	600 000	800	—	—
6B8S	DD HF Pentode	053	6.3	0.3	250	100	-3–30	—	6.5	1.4	1.0	800 000	800	—	—
6C5	Triode	034	6.3	0.3	250	—	-8.0	1000	8.0	—	2.0	10 000	20	—	—
6C6	HF Pentode	6UX11	6.3	0.3	250	100	-3.0	600	2.0	0.5	1.25	1.5 meg.	1900	—	—
6C7	DD Triode	6UX11	6.3	0.3	250	—	-9.0	—	5.5	—	1.25	16 000	20	—	—
6D6	HF Pentode	6UX11	6.3	0.3	250	100	-3–40	300	8.2	2.0	1.6	800 000	1280	—	—
6D8	Pentagrid	054	6.3	0.15	250	100	-3.0	—	3.5	2.6	*550	400 000	—	—	—
6E5	Tuning Indicator	6UX12	6.3	0.3	250	—	—	—	—	—	—	—	—	—	—
6E6	Double Triode	7UX9	6.3	0.6	250	—	-27.5	—	36.0	—	3.4	7000	6.0	1.6	14 000
6E7	HF Pentode	7UX11	6.3	0.3	250	100	-3.0	—	7.5	1.75	1.5	770 000	20	—	—
6E8	Triode Hexode	058	6.3	0.3	250	100	—	—	3.3	—	*650	—	—	—	—
6F5	Triode	039	6.3	0.3	250	—	-2.0	2000	0.9	—	1.5	66 000	100	—	—
6F6	LF Pentode	049	6.3	0.7	250	250	-16.5	410	34.0	6.5	3.25	80 000	190	3.5	7000
6F7	Triode Pentode	7UX8	6.3	0.3	250	100	-3–35	500	6.5	1.5	1.1	850 000	900	—	—
6G5	Tuning Indicator	6UX12	6.3	0.3	250	—	0–22	—	—	—	—	—	—	—	—
6G6	LF Pentode	048	6.3	0.15	180	—	-9.0	—	15.0	2.5	2.3	—	—	—	—
6H4	Diode	064	6.3	0.15	100	—	—	—	4.0	—	—	—	—	—	—
6H6	Double Diode	037	6.3	0.15	—	—	—	—	—	—	—	—	—	—	—
6J5	Triode	034	6.3	0.3	250	—	-8.0	600	9.0	—	2.6	7700	20	—	—
6J7	HF Pentode	050	6.3	0.3	250	125	-3.0	—	2.0	0.5	1.25	1.5 meg.	1900	—	—
6K5	Triode	040	6.3	0.3	250	—	-3.0	3000	1.1	—	1.4	50 000	70	—	—
6K6	LF Pentode	048	6.3	0.4	250	250	-18.0	—	32.0	—	2.2	68 000	—	3.4	7600
6K7	HF Pentode	047	6.3	0.3	250	125	-3.0	200	10.5	2.6	1.65	600 000	1000	—	—
6K8	Triode Hexode	057	6.3	0.15	250	100	-3.0	300	2.5	4.5	*350	1 meg.	—	—	—
6L5	Triode	034	6.3	0.3	250	—	-9.0	—	8.0	—	1.9	9000	17	—	—
6L6	LF Pentode	060	6.3	0.9	250	250	-14.0	170	72.0	5.0	6.0	22 500	—	6.5	2500
6L7	Frequency Changer	055	6.3	0.3	250	150	-3.0	260	3.3	8.3	*350	1 meg.	—	—	—
6M6	LF Pentode	048	6.3	1.2	180	250	—	140	36.0	4.0	10.0	—	—	4.4	7000
6N5	Tuning Indicator	6UX12	6.3	0.15	—	—	0	—	—	—	—	—	—	—	—
6N6	Double Triode	044	6.3	0.8	300	—	-1.5–30	—	43.0	—	8.0	24 000	54	5.0	7000
6N7	Double Triode	042	6.3	0.8	300	—	-3.0	—	—	—	—	—	35	10.0	—
6P8	Triode Hexode	058	6.3	0.8	250	80	0	—	2.2	3.0	*650	750 000	—	—	—
6Q6	Diode Triode	065	6.3	0.15	250	—	-1.5–30	—	1.2	—	1.9	—	—	—	—
6Q7	DD Triode	041	6.3	0.3	250	—	-2.0	4000	1.1	—	1.2	58 000	70	—	—
6R7	DD Triode	041	6.3	0.3	250	—	-9.0	1000	9.5	—	1.9	8500	16	—	—
6S7	HF Pentode	047	6.3	0.15	300	100	-3.0	—	8.5	2.0	1.75	1 meg.	—	—	—
6T7	DD Triode	041	6.3	0.15	250	—	-3.0	—	1.2	—	1.05	62 000	65	—	—
6U5	Tuning Indicator	6UX12	6.3	0.3	250	—	0–22	—	—	—	—	—	—	—	—
6U7	HF Pentode	047	6.3	0.3	250	100	-3–40	300	8.2	2.0	1.6	800 000	1280	—	—
6V6	LF Pentode	060	6.3	0.45	250	250	-12.5	240	45.0	4.5	4.1	52 000	218	4.25	5000
6W7	HF Pentode	047	6.3	0.15	300	100	-3.0	—	2.0	0.5	1.225	1.5 meg.	—	—	—

Type	Description	Base	Heater V	Heater A	Anode V	Screen V	Grid V	Anode mA	Screen mA	Mutual cond. mA/V	Impedance Ω	Amp. factor	Output	Load Ω
6X5	Rectifier	037	6.3	0.6	RMS 350	—	—	—	—	—	—	—	75 mA	—
6Y5	Rectifier	—	6.3	0.8	RMS 350	—	—	—	—	—	—	—	50 mA	—
6Z4	Rectifier	037	6.3	0.5	RMS 350	—	—	—	—	—	—	—	60 mA	—
6ZY5	Rectifier	037	6.3	0.15	RMS 350	—	—	—	—	—	—	—	35 mA	—
7A6	Double Diode	OL9	6.3	—	150	—	—	8.0 (each anode)	—	—	—	—	—	—
7A7	HF Pentode	OL17	6.3	0.3	250	100	-3.5	8.6	2.0	2.0	800 000	1600	—	—
7A8	Frequency Changer	OL12	6.3	0.15	250	100	-3.5	3.0	2.8	*600	700 000	—	—	—
7B5	LF Pentode	OL11	6.3	0.4	250	250	-18.0	32.0	5.5	2.2	68 000	150	3.4	7600
7B6	DD Triode	OL12	6.3	0.3	250	—	-2.0	1.0	—	1.1	91 000	100	—	—
7B7	HF Pentode	OL13	6.3	0.15	250	100	-3.0	8.5	2.0	1.7	700 000	1200	—	—
7B8	Frequency Changer	OL13	6.3	0.3	250	100	-3.0	3.5	2.7	*550	360 000	—	—	—
7C5	LF Pentode	OL16	6.3	0.45	250	250	-12.5	45.0	4.5	4.1	52 000	218	4.25	5000
7C6	DD Triode	OL11	6.3	0.15	250	—	-1.0	1.3	—	1.0	100 000	100	—	—
7C7	HF Pentode	OL12	6.3	0.3	250	100	-3.0	2.0	0.5	1.2	1.5 meg.	1850	—	—
7Y4	Rectifier	OL10	6.3	0.5	RMS 350	—	—	—	—	—	—	—	60 mA	—
10	Power Triode	4UX3	7.5	1.25	450	—	-32.0	18.0	—	1.6	5000	—	1.6	10 000
12	Triode	4UX3	1.1	0.25	135	—	-10.5	3.0	—	0.44	15 000	6.6	—	—
12A6	Beam Power Output	048	12.6	0.15	250	250	-12.5	30.0	3.5	3.0	50 000	—	—	—
12A7	Diode Pentode	7UX7	12.6	0.3	135	135	-13.5	9.0	2.5	0.975	102 000	100	0.55	13 500
12A8	Pentagrid	054	12.6	0.15	300	100	-3.0	3.5	2.7	*550	360 000	—	—	—
12B6	Diode Triode	065	12.6	0.15	250	—	-2.0	0.9	—	1.1	91 000	—	—	—
12B7	HF Pentode	OL12	12.6	0.15	300	125	-3.0	9.2	2.3	2.0	800 000	—	—	—
12C8	DD HF Pentode	053	12.6	0.15	250	—	-3.0	10.0	—	1.325	600 000	—	—	—
12E5	Triode	034	12.6	0.15	250	—	-13.5	5.0	—	1.45	9500	—	—	—
12F5	Triode	039	12.6	0.15	250	—	-2.0	0.9	—	1.5	66 000	—	—	—
12G7	DD Triode	041	12.6	0.15	250	—	-3.0	9.0	—	1.2	58 000	—	—	—
12J5	Triode	034	12.6	0.15	250	—	-8.0	0.5	—	2.6	7700	—	—	—
12J7	HF Pentode	047	12.6	0.15	250	100	-3.0	10.5	—	1.225	2 meg.	—	—	—
12K7	HF Pentode	047	12.6	0.15	300	125	-3.0	2.5	2.6	1.65	600 000	—	—	—
12K8	Triode Hexode	057	12.6	0.15	250	100	-3.0	1.1	6.0	0.35	600 000	—	—	—
12Q7	DD Triode	041	12.6	0.15	250	—	—	0.9	—	1.2	58 000	—	—	—
12SA7	Pentagrid	066	12.6	0.15	300	100	-2.0	3.5	8.5	*450	1 meg.	—	—	—
12SC7	Double Triode	067	12.6	0.15	250	—	-2.0	2.0	—	1.325 each	53 000 each	—	—	—
12SF5	Triode	068	12.6	0.15	250	—	-2.5	0.9	—	1.5	66 000	—	—	—
12SG7	HF Pentode	069	12.6	0.15	250	150	-3.0	9.2	3.4	4.0	1 meg.	—	—	—
12SJ7	HF Pentode	070	12.6	0.15	250	100	-3.0	3.0	0.8	1.65	1.5 meg.	—	—	—
12SK7	HF Pentode	070	12.6	0.15	250	100	-3.0	9.2	3.4	1.65	1.5 meg.	—	—	—
12SQ7	DD Triode	071	12.6	0.15	250	—	-2.0	0.8	—	1.1	91 000	—	—	—
12SR7	DD Triode	071	12.6	0.15	250	—	-9.0	9.5	—	1.9	8500	—	—	—
12Z3	Rectifier	4UX10	12.6	0.3	RMS 250	—	—	—	—	—	—	—	60 mA	—
14A4	Triode	OL19	12.6	0.15	250	—	-8.0	—	—	2.6	7700	—	—	7000
14A5	Beam Power Output	OL16	12.6	0.15	250	250	-12.5	30.0	3.5	3.0	50 000	—	3.5	6500
14A7/12B7	HF Pentode	OL12	12.6	0.15	250	100	-3.0	9.2	2.6	2.0	800 000	—	—	—
14B8	Pentagrid	OL13	12.6	0.15	250	100	-3.0	3.5	2.7	*550	360 000	—	—	—
14C7	HF Pentode	OL12	12.6	0.15	250	100	-3.0	2.2	0.7	1.575	1 meg.	—	—	—
14F7	Double Triode	OL18	12.6	0.15	250	—	-2.0	2.3	—	1.6	44 000	—	—	—
15	LF Pentode	5UX8	2.0	0.22	135	67.5	-1.5	1.85	0.3	1.6	800 000	—	—	—
18	LF Pentode	6UX10	14.0	0.3	250	250	-16.5	34.0	6.5	2.35	80 000	190	2.1	—
19	Class B	6UX1	2.0	0.26	135	—	0	—	—	—	—	—	—	—
20	Power Triode	4UX3	3.3	0.132	135	—	-22.5	6.5	—	0.525	6300	3.3	0.11	—
22	Screen Grid	4UX6	3.3	0.132	135	67.5	-1.5	3.7	1.3	0.5	325 000	—	—	—
24	Screen Grid	5UX8	2.5	1.75	250	90	-3.0	4.0	1.7	1.0	400 000	400	—	—
24A	Screened Tetrode	5UX8	2.5	1.75	180	90	-3.0	4.0	1.7	1.05	600 000	630	—	—
25A6	LF Pentode	049	25.0	0.3	180	135	-20.0	38.0	7	2.5	40 000	100	2.75	5500
25A7	Diode Pentode	062	25.0	0.3	100	100	-15.0	20.5	4.0	1.8	50 000	—	0.77	4500
25B5	Double Triode	6UX15	25.0	0.3	180	100	—	46.0	5.8	2.3	15 200	35	3.8	3800
25B8	Triode Pentode	072	25.0	0.3	100	100	0	7.6	2.0	1.9	75 000	—	—	—
25D8	Diode-Triode-Pentode	073	25.0	0.15	100	—	-3.0	—	—	1.9 / 1.1 (Pentode / Triode)	200 000 / 91 000	Pentode / Triode	—	—
25L6	LF Pentode	060	25.0	0.3	110	110	-7.5	49.0	4.0	8.2	10 000	82	2.2	2000
25N6	Double Triode	044	25.0	0.3	180	110	0	45.0	7.0	11.4	11 400	25	2.0	2000
25R	Rectifier	6UX5	25.0	0.3	RMS 250	—	—	—	—	—	—	—	80 mA	—

225

Table 14. American types – continued

Type	Description	Base	Fil. or heater (volts)	Fil. or Heater current (amps)	Anode volts	Screen volts	Bias volts	Bias res. (ohms)	Anode current (mA)	Screen current (mA)	Slope mA/V (*=conv. condc. µA/V)	Impedance (ohms)	Amp. factor	Output (watts)	Optimum load (ohms)
25X6	Rectifier	–	25.0	0.15	RMS 125	–	–	–	–	–	–	–	–	60 mA	–
25Y4	Rectifier	035	25.0	0.15	RMS 125	–	–	–	–	–	–	–	–	75 mA	–
25Y5/25Z5	Rectifier	6UX5	25.0	0.3	RMS 250	–	–	–	–	–	–	–	–	85 mA	–
25Z6	Rectifier	037	25.0	0.3	RMS 250	–	–	–	–	–	–	–	–	85 mA	–
26	Triode	4UX3	1.5	1.05	180	–	-14.5	–	6.2	–	1.15	7300	8.3	–	–
27	Triode	5UX6	2.5	1.75	250	–	-21.0	–	5.2	–	0.97	9250	9.0	–	–
30	Triode	4UX3	2.0	0.06	180	–	-13.5	–	3.1	–	0.9	10300	9.3	–	–
31	Triode	4UX3	2.0	0.13	180	–	-30.0	–	12.3	–	1.05	3600	3.8	0.375	5700
32	HF Tetrode	4UX5	2.0	0.06	180	67.5	-12.0	–	1.7	0.4	0.65	1.2 meg.	780	–	–
33	LF Pentode	5UX1	2.0	0.26	135	135	–	–	2.8	1.0	2.0	–	–	1.0	6000
34	HF Pentode	4UX5	2.0	0.06	180	67.5	-3 to -22.5	–	6.5	2.5	0.62	1.2 meg.	620	–	–
35	HF Tetrode	5UX9	2.5	1.75	250	90	–	–	–	–	1.05	400 000	–	–	–
35A5	Beam Power Output	OL16	32.0	0.15	110	110	-7.5	–	41.0	7.0	5.8	14 000	–	–	–
35L6	Beam Power Output	048	35.0	0.15	110	110	-7.5	–	41.0	7.0	5.8	13 800	–	–	–
35R	Rectifier	6UX5	35.0	0.3	RMS 250	–	–	–	–	–	–	–	–	120 mA	–
35Z3	Rectifier	0L8	35.0	0.15	RMS 170	–	–	–	–	–	–	–	–	100 mA	–
35Z4	Rectifier	015	35.0	0.15	RMS 125	–	–	–	–	–	–	–	–	100 mA	–
35Z5	Rectifier	075	35.0	0.15	RMS 125	–	–	–	–	–	–	–	–	100 mA	–
36	Screened Tetrode	5UX8	6.3	0.3	250	90	-3.0	850	3.2	1.0	1.08	550 000	595	–	–
37	Triode	5UX6	6.3	0.3	250	–	-18.0	–	7.5	–	1.1	8400	9.2	–	–
38	LF Pentode	5UX7	6.3	0.3	250	250	-25.0	970	22.0	3.8	1.2	100 000	1.2	2.5	10 000
39/44	HF Pentode	5UX8	6.3	0.3	250	90	-3.0	400	5.8	1.4	1.05	1 meg.	1050	–	–
40	Triode	4UX3	5.0	0.25	180	–	-3.0	–	0.2	–	0.2	150 000	30	–	–
40Z5	Rectifier	075	45.0	0.15	RMS 125	–	–	–	–	–	–	–	–	100 mA	–
41	LF Pentode	6UX10	6.3	0.4	250	250	-18.0	480	32.0	5.5	2.2	68 000	150	3.4	7600
42	LF Pentode	6UX10	6.3	0.7	250	250	-16.5	410	34.0	6.5	2.35	80 000	190	3.5	7000
43	LF Pentode	6UX10	25.0	0.3	180	135	-20.0	440	38.0	7.5	2.5	40 000	100	2.75	5000
45	LF Triode	4UX3	2.5	1.5	250	–	-50.0	–	34.0	–	2.17	1600	3.5	1.6	3900
45Z5	Rectifier	075	45.0	0.15	RMS 125	–	–	–	–	–	–	–	–	100 mA	–
46	Dual Grid LF	5UX4	2.5	1.75	250	–	-33.0	450	22.0	6.0	2.35	2380	5.6	1.25	6400
47	LF Pentode	5UX1	2.5	1.75	250	250	-16.5	310	31.0	9.5	2.5	60 000	150	2.7	7000
48	Dual Grid LF	6UX14	30.0	0.4	125	100	-20.0	–	56.0	–	3.9	–	–	2.5	1500
49	Power Triode	5UX4	2.0	0.12	135	–	-20.0	1530	6.0	–	1.125	4175	4.7	0.17	11000
50	Power Triode	4UX3	7.5	1.25	450	–	-84.0	–	55.0	–	2.1	1800	3.8	4.6	4350
50C6	Beam Power Output	048	50.0	0.15	200	135	-14.0	–	16.0	2.2	7.1	18 300	–	–	–
50L6	Beam Power Output	048	50.0	0.15	110	110	-7.5	–	49.0	4.0	8.2	10 000	–	–	–
50Y6	Rectifier	038	50.0	0.15	RMS 117	–	–	–	–	–	–	–	–	75 mA	–
50Z7	Rectifier	074	50.0	0.15	RMS 117	–	–	–	–	–	–	–	–	65	–
51	HF Tetrode	5UX9	2.5	1.75	250	90	-3.0	–	6.5	2.5	1.05	400 000	–	–	–
53	Class B	7UX9	2.5	2.0	300	–	0	0	8.0	–	–	–	35	10.0	–
55	DD Triode	6UX6	2.5	1.0	250	–	-20.0	2500	8.0	–	1.1	7500	8.3	0.35	20 000
56	Triode	5UX6	2.5	1.0	250	–	-13.5	2500	5.0	–	1.45	9500	13.8	–	–
57	HF Pentode	6UX11	2.5	1.0	250	100	-3.0	600	2.0	0.5	1.22	1 meg.	–	–	–
58	HF Pentode	6UX11	2.5	1.0	250	100	-3 to -40	300	8.2	2.0	1.6	800 000	1280	–	–
59	Triple Grid Output	7UX4	2.5	2.0	250	250	-18.0	410	35.0	9.0	2.5	40 000	–	3.0	6000
70A7	Rectifier and Beam Power Output	062	70.0	0.15	RMS 117 / 110	–	–	–	40.0	–	5.8	–	–	60 mA	–
70L7	Rectifier and Beam Power Output	076	70.0	0.15	RMS 117 / 110	110	-7.5	–	43.0	6.0	7.5	15 000	–	70 mA	–
71A	Power Triode	4UX3	5.0	0.25	180	–	-40.5	–	20.0	–	1.7	1750	3	0.79	4800
75	DD Triode	6UX6	6.3	0.3	250	–	-2.0	5000	0.4	–	1.1	90 000	100	–	–
76	Triode	5UX6	6.3	0.3	250	–	-13.5	2500	5.0	–	1.45	9500	13.8	–	–
77	HF Pentode	6UX11	6.3	0.3	250	100	-3.0	1000	2.3	0.5	1.25	1.5 meg.	1500	–	–
78	HF Pentode	6UX11	6.3	0.3	250	125	-3 to -40	200	10.5	2.6	1.65	600 000	1000	–	–
79	Class B	6UX7	6.3	0.6	250	–	0	–	–	–	–	–	–	8.0	14 000

Type	Description	Base												
80	Rectifier	4UX11	5.0	2.0	RMS 350	—	—	—	—	—	—	—	—	125 mA	—
80A	Rectifier	4UX10	7.5	1.25	RMS 700	—	—	—	—	—	—	—	—	85 mA	—
81	Rectifier	4UX10	7.5	1.25	RMS 700	—	—	—	—	—	—	—	—	85 mA	—
82	Rectifier (mercury)	4UX7	2.5	3.0	RMS 450	—	—	—	—	—	—	—	—	115 mA	—
83	Rectifier (mercury)	4UX7	5.0	3.0	RMS 450	—	—	—	—	—	—	—	—	225 mA	—
83V	Rectifier	4UX8	5.0	2.0	RMS 400	—	—	—	—	—	—	—	—	200 mA	—
84	Rectifier	5UX5	6.3	0.5	RMS 350	—	—	—	—	—	—	—	—	50 mA	—
85	DD Triode	6UX6	6.3	0.3	250	—	−20.0	2500	8.0	—	1.1	7500	8.3	0.35	20 000
89	Triple Grid Output	6UX11	6.3	0.4	250	—	−31.0	970	32.0	—	1.8	2600	4.7	0.9	5500
V99	Triode	4UX9	3.0–3.3	0.06–0.063	90	—	−4.5	—	—	—	—	—	—	—	—
X99	Triode	4UX3	3.0–3.3	0.06–0.063	90	—	−4.5	—	—	—	—	—	—	—	—
112A	Triode	4UX3	5.0	0.25	180	—	−13.5	—	—	—	—	—	—	0.285	10 650
117Z6	Rectifier	077	117.0	0.15	RMS 117	—	—	—	—	—	—	—	—	60 mA	—
183	Triode	4UX3	5.0	1.25	250	—	−60.0	—	25.0	—	1.8	1800	3.2	2.0	4500
484	Triode	–	3.0	1.3	10	—	−9.0	—	6.0	—	1.35	9300	12.5	—	—
950	LF Pentode	5UX2	2.0	0.12	135	135	−16.5	—	7.0	2.0	0.95	105 300	100	0.45	13 500
2101	LF Pentode	5UX1	2.0	0.12	135	135	−4.5	—	8.0	2.6	1.7	200 000	340	0.45	16 000
2102	DD Triode	6UX2	2.0	0.12	135	—	−1.5	—	2.1	—	1.3	23 000	30	—	—
2103	Double LF Pentode	7UX1	2.0	0.26	135	135	−7.5	—	4.0	1.2	1.6	—	350	0.6	—
2151	LF Pentode	6UX10	14.0	0.3	250	250	−31.0	—	47.0	11.6	2.4	50 000	120	6.0	—

Table 15. Valve base diagrams

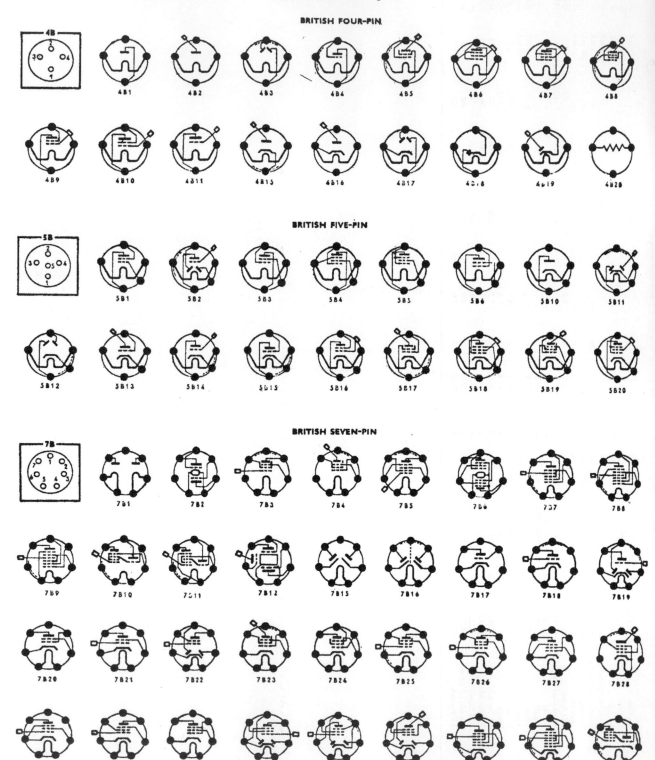

BRITISH FOUR-PIN

BRITISH FIVE-PIN

BRITISH SEVEN-PIN

Table 15. Valve base diagrams – *continued*

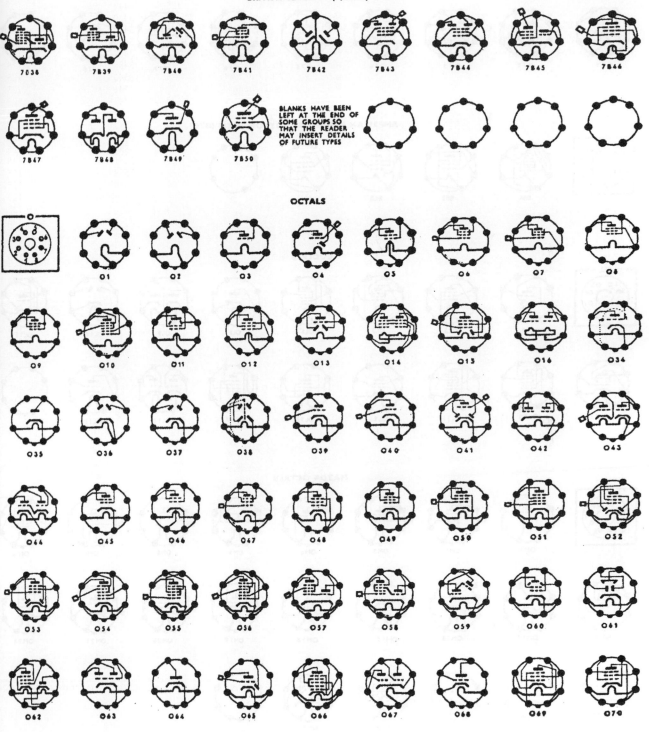

BRITISH SEVEN-PIN (contd.)

7B38 7B39 7B40 7B41 7B42 7B43 7B44 7B45 7B46

7B47 7B48 7B49 7B50

BLANKS HAVE BEEN LEFT AT THE END OF SOME GROUPS SO THAT THE READER MAY INSERT DETAILS OF FUTURE TYPES

OCTALS

O1 O2 O3 O4 O5 O6 O7 O8

O9 O10 O11 O12 O13 O14 O15 O16 O34

O35 O36 O37 O38 O39 O40 O41 O42 O43

O44 O45 O46 O47 O48 O49 O50 O51 O52

O53 O54 O55 O56 O57 O58 O59 O60 O61

O62 O63 O64 O65 O66 O67 O68 O69 O70

Table 15. Valve base diagrams – *continued*

OCTALS (contd.)

O71 O72 O73 O74 O75 O76 O77 O78

AMERICAN MINIATURE BUTTON BASES (B7G)

BB   BB1 BB2 BB3 BB4

LOCTALS

OL   OL1 OL2 OL3 OL4 OL8 OL9 OL10 OL11

OL12 OL13 OL14 OL15 OL16 OL17 OL18 OL19

MAZDA OCTALS

OM   OM1 OM2 OM3 OM4 OM5 OM6 OM7 OM8

OM9 OM15 OM16 OM17 OM18 OM19 OM20 OM21 OM22

OM23 OM24 OM25 OM26 OM27

Table 15. Valve base diagrams – *continued*

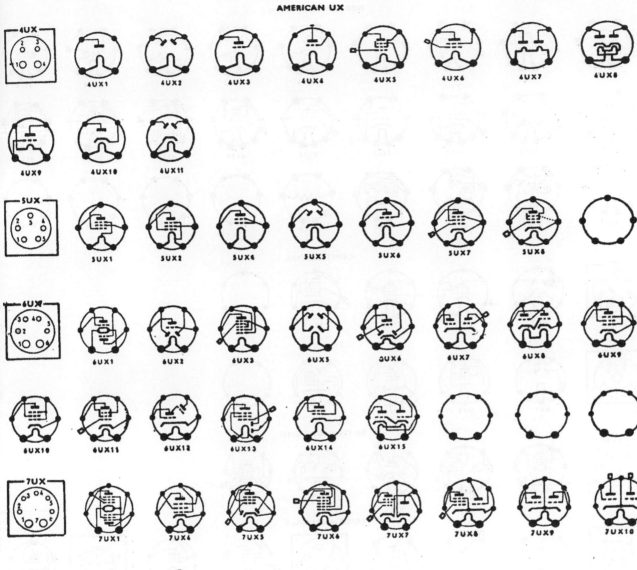

AMERICAN UX

SIDE CONTACT

Table 15. Valve base diagrams – *continued*

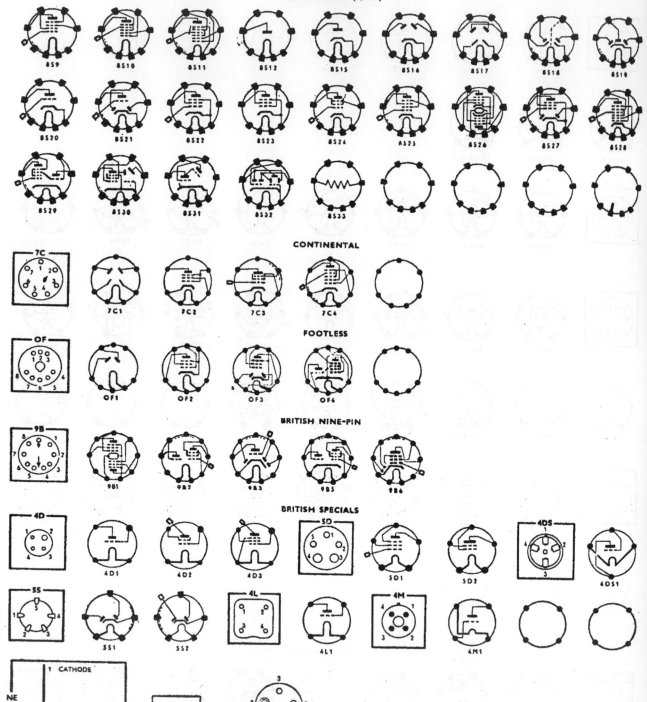

Table 16. American ballast tubes

Type	Volts droppe at 117.5 V	No. of pilot lamps	Rating lamps (amps)	Base code	Base type	Equivalent	Equivalent with base changed
42A	42.3	0	–	A	Octal	K42A, 42AG, K42AG, K43A	140R
42A1	42.3	0	–	AY	Octal	KY42A	140R
42A2	42.3	1	0.15	BY	Octal	KY42B	140R4
42B2	42.3	2	0.15	CY	Octal	KY42C	140R8
K42A	42.3	0	–	A	Octal	42A	140R
K42B	42.3	1	0.15	B	Octal	K42BG, K43B, 135K1	140R4
K42C	42.3	2	0.15	C	Octal	K42CG, BK42C, 95K2, K40C, 5516, 5530	140R8
K42D	42.3	2	0.15	D	Octal	K42DG, BK42D, 3326	140R44
KX42B	42.3	1	0.15	BX	4-pin	140R4	K42B
KX42C	42.3	2	0.15	CX	4-pin	140R8	K42C
KY42D	42.3	2	0.15	DY	Octal	2LR212	–
L42B2	42.3	1	0.25	B	Octal	BL42B, L42BG, 5547	104L4
L42BX	42.3	1	0.25	BX	4-pin	140L4, LX42B	L42B
L42C	42.3	2	0.25	C	Octal	BL42C, L42CG, L40C, 69.2037, 5548, 16035	140L8
L42D	42.3	2	0.25	D	Octal	BL42D, L42DG, 5549	104L44
L42DX	42.3	2	0.25	DX	4-pin	140L44	L42D
L42F	42.3	1	0.25	F	Octal	–	–
L42S1	42.3	1	0.25	S1	Octal	L0S1	–
L42S2	42.3	2	0.24	S2	Octal	L40S2	–
M42C	42.3	2	0.2	C	Octal	K42C or L42C and alter pilot lamps	–
49A	48.6	0	–	A	Octal	K49A, 49KA, K50A	165R
49A1	48.6	0	–	AY	Octal	KY49A	165R
49A2	48.6	1	0.15	BY	Octal	KY49B	165R4
49B2	48.6	2	0.15	CY	Octal	KY49C	165R8
K49A	48.6	0	–	A	Octal	49A	165R
K49B	48.6	1	0.15	B	Octal	BK49B, 49KB, K43B2, W43357, 115.41, 5533, 8593, 5693	165R4
K49C	48.6	2	0.15	C	Octal	49KC, BK49C, K50C, K49CB, A16040, 81966–2, 5534	165R8
K49D	48.6	2	0.15	D	Octal	49KD, BK49D, BK49D-10, 5633, 5518, 69116, 115.28, 3334, 3334A	165R44
KX49A	48.6	0	–	AX	4-pin	165R, 340	49A
KX49C	48.6	2	0.15	CX	4-pin	165R8, 50A2	K49C
KZ49B	48.6	1	0.15	B2	Octal	50B2MG	165R4
KZ49C	48.6	2	0.15	CZ	Octal	50A2MG	165R8
L49B	48.6	1	0.25	B	Octal	49LB, BL49B, 2UR224, 69.2033, 5511, 5550	165L4
L49C	48.6	2	0.25	C	Octal	49LC, L49–5.5C, BL49C, 2905, 5552, 16036	165L8
L49D	48.6	2	0.25	D	Octal	49LD, BL49D, 3CR-241, 5567	165L44
L49F	48.6	1	0.25	F	Octal	–	–
M49B	48.6	1	0.2	B	Octal	BM49B, 38710	–
M49C	48.6	2	0.2	C	Octal	BM49C	–
M49H	48.6	2	0.2	H	Octal	M49HG	–
55A	54.9	0	–	A	Octal	K55A	185R
55A1	54.9	0	–	AY	Octal	KY55A	185R
55A2	54.9	1	0.15	BY	Octal	KY55B	185R4
55B2	54.9	2	0.15	CY	Octal	KY55C	185R8
K55A	54.9	0	–	A	Octal	55A	185R
K55B	54.9	1	0.15	B	Octal	55KB, K55BG, K54B, BK55B, 3613, 5519, 7-TU-9, 5535, 16039	185R4

Table 16. American ballast tubes – *continued*

Type	Volts droppe at 117.5 V	No. of pilot lamps	Rating lamps (amps)	Base code	Base type	Equivalent	Equivalent with base changed
K55C	54.9	2	0.15	C	Octal	BK55C, 5536	185R8
K55D	54.9	2	0.15	D	Octal	BK55D, 115.22	185R44
K55H	54.9	2	0.15	H	Octal	K52H	–
L55B	54.9	1	0.25	B	Octal	2V4215, 2903, 5555, 8598, 2VR215	185L4
L55C	54.9	2	0.25	C	Octal	85LC, L55–5.5C, 2904	185L8
L55D	54.9	2	0.25	D	Octal	85LD	185L44
L55F	54.9	1	0.25	F	Octal	BL55F	–
M55F	54.9	1	0.2	F	Octal	–	–
M55H	54.9	2	0.2	H	Octal	M55HG, M52H	–
C9266	54.9	–	–	L	Octal	–	–
100R8	29.7	2	0.15	CX	4-pin	KX30C	K30C
120R8	36.0	2	0.15	CX	4-pin	KX36C	K36C
140L4	42.3	1	0.25	BX	4-pin	L42BX, LX42B	L42B
140L8	42.3	2	0.25	CX	4-pin	L42CX, LX42C	L42C
140L44	42.3	2	0.25	DX	4-pin	L42DX, LX42D	L42D
140R	42.3	0	–	AX	4-pin	–	42A
140R4	42.3	1	0.15	BX	4-pin	40B2, KX42B	K42B
140R8	42.3	2	0.15	CX	4-pin	40A2, KX42C	K42C
165L4	48.6	1	0.25	BX	4-pin	L49BX, LX49B	L49B
165L8	48.6	2	0.25	CX	4-pin	LX49C	L49C
165R	48.6	0	–	AX	4-pin	–	49A
165R4	48.6	1	0.15	BX	4-pin	50B2, KX49B	K49B
165R8	48.6	2	0.15	CX	4-pin	50A2, KX49C	K49C
185L4	54.9	1	0.25	BX	4-pin	LX55B	L55B
185L8	54.9	2	0.25	CX	4-pin	–	L55C
185R	54.9	0	–	AX	4-pin	50X3, KX55A	K55A
185R4	54.9	1	0.15	BX	4-pin	KX55B	K55B
185R8	54.9	2	0.15	CX	4-pin	50X3T, KX55C	K55C
200R	60.0	0	–	AX	4-pin	–	–
290L4	–	1	0.25	BX	4-pin	Special Type	–
300R4	79.5	1	0.15	BX	4-pin	KX80B	K80B

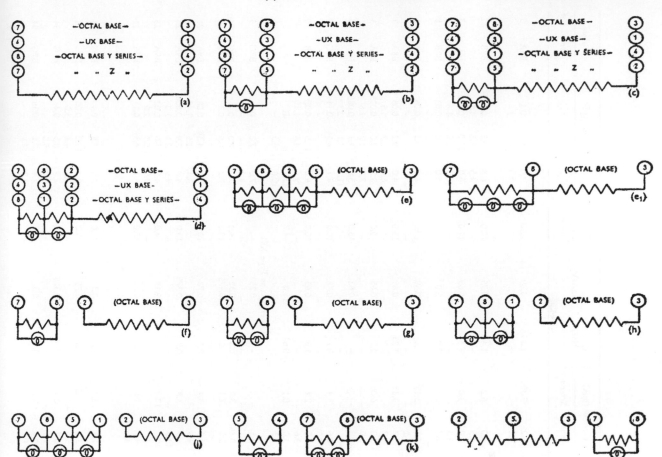

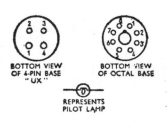

BOTTOM VIEW
OF 4-PIN BASE
" UX "

BOTTOM VIEW
OF OCTAL BASE

REPRESENTS
PILOT LAMP

Table 17. Further valve data
Frequency-changers

Type			Heater Volts	Heater Amps	Anode	Screen	Grid	Anode (mA)	Screen (mA)	r_a (MΩ)	g_m (mA/V)	Osc. volts (peak)	C_{gk}	C_{ak}	C_{ga}	Base Type	Ref.
Cossor																	
141TH	(TH$_x$)	mix	14.0	0.1	200	85	−2.0	3.2	3.35	1.25	0.69	12.0	4.0	9.2	0.1	B8A	3
		osc			110			4.2					5.5	2.3	1.2		
Marconi																	
X143	(TH)	mix	6.3	0.3	250	100	−2.0	3.0	6.2	1.4	0.75	10.0	6.8	9.5	0.002	B8B	42
		osc			160			4.5					4.5	3.5	1.1		
X147	(TH$_x$)	mix	6.3	0.3	250	100	−2.0	3.0	3.0	1.3	0.65	11.0	5.0	10.0	0.0003	IO	3
		osc			100			3.3					9.0	3.0	1.6		
X148	(TH)	mix	6.3	0.3	250	100	−2.0	1.7	2.2	2.0	2.0		5.0	8.0	0.03	B8B	8
		osc			250												
X150	(TH$_x$)	mix	6.3	0.225	250	83	−2.0	3.15	3.15	1.0	0.69	10.0	4.0	9.2	0.05	B8A	3
		osc			100			3.2									
X142	(TH$_x$)	mix	14.0	0.1	200	84	−2.0	3.2	3.35	1.25	0.69	13.0	4.0	9.2	0.05	B8A	3
		osc			110			4.2					6.4	2.7	1.5		
X145	(TH)	mix	28.0	0.1	175	100	−2.5	3.0	6.0	2.2	0.65	9.0	8.3	3.0	0.003	B8A	3
		osc			80			5.0					7.7	1.7	1.8		
X71M	(TH$_x$)	mix	13.0	0.16	250	100	−3.0				0.62	15.0	5.0	14.1	0.085	IO	3
		osc			100								11.0	7.1	2.3		
X78	(TH$_x$)	mix	6.3	0.3	250	75	0	4.5	3.4	0.7	0.78	10.0	4.1	4.34	0.11	B7G	48
		osc			100			4.5									
X81	(TH$_x$)	mix	6.3	0.3	250	100	−2.0	3.0	2.4	1.0	0.65	10.0	6.0	11.5	0.07	B8B	8
		osc			100			3.6					9.6	4.8	1.15		
X76M	(TH$_x$)	mix	13.0	0.16	250	100	−3.0	3.0		0.62	15.0	4.7	13.1		IO	3	—
		osc			100			3.3	0.1				10.6	6.3			
X101	(TH$_x$)	mix	19.0	0.1							Other data as Type X81						
X17	(H)	osc	1.4*	0.05	90	67.5	0	5.5		0.25	0.25		7.0	7.0	0.4	B7G	3
													3.8				
X18	(H)	mix	1.4*	0.05	90	67.5	0	0.86	3.0	0.6	0.32	15.0	7.0	7.0	0.4	B7G	54
X727/6BE6	(H)	osc	6.3	0.3	250	100	−1.5	3.0	7.1	1.0	0.475	10.0	7.2	8.6	0.3	B7G	29
X79	(TH$_x$)	mix	6.3	0.3	250	75	0	4.5	3.4	0.7	0.78	10.0	4.1	4.34	0.08	B9A	21
		osc			100			4.5					5.47	1.5	1.48		
X719/ECH81	(TH)	mix	6.3	0.3	250	100	−2.0	6.5	3.8	0.7	0.775	13.0	4.8	7.9	0.006	B9A	24
		osc			100		0	13.5					2.6	2.1	1.0		
LZ319/PCF80	(TP)	mix	9.0	0.3	170	170	−2.0	10	10		2.18	4.0	4.5	4.0	0.02	B9A	25
		osc			100		−2.0	14					3.0	0.5	2.0		
X109	(TH$_x$)	mix	19.0	0.1	175	75	−2.0	4.3	3.6	0.25	0.71	10.0	4.1	4.34	0.11	B9A	21
		osc			100		0	4.5									
Mazda																	
ICI IR5 }	(H)	mix	1.4*	0.05	90	76.5	0	1.6	3.2	0.6	0.3	37.0	7.0	7.5	0.4	B7G	3
6C31	(TH)	mix	6.3	0.85	250	100	−3.0	3.0	6.05	1.6	0.75	9.0	9.5	13.0	0.001	IO	3
		osc			80			5.0					11.5	4.4	3.0		
IC2	(H)	mix	1.4*	0.05	85	60	0	0.7	0.15	0.65	0.325	7.0	7.5	8.5	0.4	B7G	54
		osc			30			1.6					4.0	5.0			
IC3	(H)	mix	1.4*	0.025	85	68	0	0.6	0.14	0.8	0.3	5.7	7.4	8.1	0.36	B7G	54
		osc			35			1.5					3.9	4.8			

Type	El.	Section	V_h	I_h	V_a	V_{g2}	V_g									Base	Ref
6C9	(TH)	mix	6.3	0.45	250	100	−2.5	3.0	6.0	3.0	0.65	9.0	8.3	3.0	0.003	B8A	3
		osc			80	—	−2.5	5.0					7.7	1.7	1.8		
6C10	(TH$_x$)	mix	6.3	0.225	250	100	−2.5	3.6	3.5	1.03	0.71	17.0	4.0	9.2	0.05	B8A	3
		osc			115	—	−2.5	5.0					6.4	2.7	1.5		
30C1	(TP)	mix	9.0	0.3	170	170	−2.8	5.0	2.0	0.8	2.2	5.0	4.0	5.0	0.025	B9A	25
		osc			100	—		10.0					7.5	2.7	1.5		
10C1	(TH)	mix	28.0	0.1	175	100	−2.5	3.0	6.0	2.2	0.65	9.0	8.3	3.0	0.003	B8A	3
		osc			80	—		5.0					7.7	1.7	1.8		
10C2	(TP)	mix	28.0	0.1	150	150	0	4.7	1.3	—	2.1	3.25	7.5	2.6	0.012	B8A	19
		osc			80	—		5.0					4.1	1.6	1.7		
Mullard																	
DF97	(H)		1.4*	0.025	85	45	0	0.57	0.84	0.5	0.27	17.0	3.7	7.5	0.01	B7G	59
									0.7	0.6	0.25	10.0	7.0	10.0	0.5	IO	76
DK32	(H)	mix	1.4*	0.05	90	45	0	0.6	0.7	0.6			4.0	4.4	0.9	B7G	3
		osc						1.2					7.0	7.5	0.4		
DK91	(H)	mix	1.4*	0.05	90	45	0	0.8	1.9	0.8	0.25	15.0	7.0	8.5	0.4	B7G	54
DK92	(H)	mix	1.4*	0.05	—	60	0	0.7	0.15	0.65	0.325	6.0	7.5		0.36	B7G	54
		osc			85										0.01	IO	98
DK96	(H)	mix	1.4*	0.025	85	68	0	0.6	0.14	0.8	0.3	6.0	7.6	8.4	2.0		
KCF30	(TP)	mix	2.0*	0.2	120	60	−1.5	0.5	1.0	1.6	0.23	8.0	6.5	8.0	0.07		
		osc			100								9.0	4.0			
KK32	(O)	osc	2.0*	0.13	135	45	0	0.7	0.7	2.5	0.27	12.0	9.0	11.0	0.025	IO	76
		mix			135			2.1					6.3	8.5	1.5	B9A	44
ECF80	(TP)	mix	6.3	0.43	170	170	−5.5	5.2	1.5	0.87	2.1	5.0	5.2	3.8	0.01	B9A	44
		osc			100		−2.0	14.0					2.3	0.3	1.8		
ECF82	(TP)	mix	6.3	0.45	100	100	0	5.2	1.9	0.4	1.9	4.0	5.0	2.6	0.002	B8B	42
		osc			250			5.7					2.5	0.4	1.1		
ECH21	(TH)	mix	6.3	0.33	250	100	−2.0	3.0	6.2	1.4	0.75	14.0	6.8	9.5	0.003	IO	3
		osc			160			4.5					4.5	3.5	1.6		
ECH35	(TH$_2$)	mix	6.3	0.225	250	100	−2.0	3.0	3.0	1.3	0.65	11.0	5.0	10.0	0.1	B8A	3
		osc			100			3.3					9.0	3.0	1.2		
ECH42	(TH$_2$)	mix	6.3	0.23	250	85	2.0	3.0	3.0	1.0	0.75	11.0	4.0	9.2	0.006	B9A	24
		osc			115			4.8					5.5	2.3	1.0		
ECH81	(TH)	mix	6.3	0.3	250	100	2.0	6.5	3.8	0.7	0.775	13.0	4.8	7.9	0.01	B9A	24
		osc						13.5					2.6	2.1	1.0		
ECH83	(TH)	mix	6.3	0.3	12.6	12.6	0	0.1	0.35	3.8	0.16	2.5	4.8	7.9	0.006	IO	1
		osc			12.6			0.4					2.6	2.1	0.1	Cr8	2
EK32	(O)	mix	6.3	0.2	250	50	−2.0	1.0	0.8	2.0	0.55	21.0	9.0	10.5	0.1	B7G	29
EK2		osc			200			2.5					6.0	5.0		B9A	25
EK90	(H)		6.3	0.3	250	100	−1.5	3.0	7.1	1.0	0.475	10.0	7.2	8.6	0.3		
PCF80	(TP)		9.0	0.3	170	170	−5.5	5.2	1.5	0.87	2.1	5.0	5.5	3.8	0.025	B9A	44
							−2.0						2.3	0.3	1.5		
PCF82	(TP)	mix	9.5	0.30	250	100	0	5.2	1.9	0.4	1.9	4.0	5.0	2.6	0.01	B9A	44
		osc			250			5.7					2.5	0.4	1.8		
HK90	(H)	mix	12.6	0.15	200	85	−2.0	3.0	3.0	1.0	0.75	13.0	3.8	9.2	0.1	B8A	3
UCH42	(TH$_2$)	osc	14.0	0.1	100	—		3.1					5.5	2.3	1.2		
UCH21	(TH)	mix	20.0	0.1	200	100	−2.0	3.5	6.5	1.0	0.75	13.0	6.8	9.5	0.002	B8B	42
		osc			120			4.1					4.5	3.5	1.1		
UCH81	(TH)	mix	19.0	0.1	200	119	−2.6	3.7	8.1	1.0	0.78		4.8	7.9	0.006	B9A	24
		osc			100		0	13.5					2.6	2.1	1.0		
UCF80	(TP)	mix	27.0	0.1	170	170	−5.5	5.2	1.5	0.7	2.1	5.0	5.5	3.8	0.025	B9A	25
		osc			100		−2.0	14.0					2.3	0.3	1.5		

Other data as Type EK90 (HK90 row)

*Directly heated cathode (filament)
†Centre-tapped heater or filament, parallel connected

237

Table 17. Further valve data – continued
Screened tetrodes and pentodes

Type		Heater — Volts	Heater — Amps	Volts — Anode	Volts — Screen	Volts — Grid	Current (mA) — Anode	Current (mA) — Screen	r_a (MΩ)	g_m (mA/V)	C_{gk}	C_{ak}	C_{ga}	Base — Type	Base — Ref.
Cossor															
41SA		4.0	1.0	250	100	0	5.0	—	—	1.6	—	—	—	B7	38
OM5B		6.3	0.2	250	100	-2.0	3.0	0.8	2.5	1.8	—	—	—	IO	8
OM5C		characteristics as OM5B but suitable for use in D.C. amplifiers													
OM6	(VM)	6.3	0.2	250	100	-2.5	6.0	1.8	1.0	2.0	6.3	7.8	0.003	IO	8
6VP		6.3	0.2	250	100	-2.5	6.0	1.7	1.0	2.2	4.7	8.0	0.002	B8A	7
12DDP	(DD, VM)	17.0	0.1	170	85	-2.0	5.0	1.75	0.9	2.2	4.0	4.6	0.0025	B9A	12
Ferranti															
6G7	(VM)	6.3	0.3	250	150	-2.5	9.2	3.4	1.0	4.0	8.5	7.0	0.003	IO	14
6H7		6.3	0.3	250	150	-1.5	10.8	4.1	0.9	4.9	8.5	7.0	0.003	IO	14
6J7		6.3	0.3	250	100	-3.0	3.0	0.8	1.0	1.65	6.0	7.0	0.005	IO	10
6K7	(VM)	6.3	0.3	250	100	-3.0	9.2	2.6	0.8	2.0	6.5	7.5	0.005	IO	10
6S7	(VM)	6.3	0.15	250	100	-3.0	9.0	2.0	1.0	1.85	5.5	7.0	0.004	IO	10
Marconi and GEC															
V61	(VM)	6.3	0.3	250	100	-3.0	10.0	2.3	0.45	2.9	7.8	10.0	0.002	IO	8
V81	(VM)	6.3	0.3	250	100	-3.6	9.6	3.6		2.8	7.25	6.0	0.006	B8B	3
V101	(VM)	19.0	0.1	Other data as Type W81											
Z90		6.3	0.3	250	250	-2.0	10.0	3.0	—	6.3	8.2	5.4	0.007	B9G	1
W76	(VM)	13.0	0.16	250	100	-3.0	7.6	1.9	0.5	1.5	4.2	12.8	0.007	IO	8
D17	(SD)	1.4*	0.05	90	90	0	2.7	0.5	0.6	0.63	2.2	2.4	0.2	B7G	5
W17	(VM)	1.4*	0.05	90	67.5	0	3.5	1.4	0.5	0.9	4.5	7.5	0.006	B7G	2
Z77		6.3	0.3	250	250	-2.0	10.0	2.5	—	7.5	7.4	3.0	0.009	B7G	21
ZA2403	(SQ)														
W77		6.3	0.2	200	200	-2.5	8.0	2.0	0.5	2.5	4.6	6.5	0.009	B7G	21
ZA2400	(SQ)														
Z319	(SE)	6.3	0.3	350	250*	-1.7	15.0	1.2	0.5	19.0	8.0	3.0	0.003	B8A	46
WD709/EBF80	(VM, DD)	6.3	0.3	250	85	-2.0	5.0	1.75	1.4	2.2	4.2	4.9	0.0025	B9A	12
W719/EF85	(VM)	6.3	0.3	250	100	-2.0	10.0	2.5	0.5	6.0	7.2	3.7	0.007	B9A	10
W729	(VM)	6.3	0.3	170	170	-2.0	10.0	2.5	0.4	6.4	7.5	3.3	0.007	B9A	10
W727/6BA6	(VM)	6.3	0.6†	250	100	-1.0	11.0	4.2	1.0	4.4	5.5	5.0	0.0035	B7G	16
Z309		6.3	0.3	250	250	-2.0	20.0	5.25	0.5	15.0	13.0	2.5	0.007	B9A	22
Z719		6.3	0.2	170	170	-2.0	10.0	2.5	0.4	7.4	7.5	3.3	0.006	B9A	10
Z729		6.3	0.3	250	140	-2.0	3.0	—	2.0	1.85	4.0	5.5	0.025	B9A	23
Z579		6.3	0.6	250	250	-2.5	20.0	5.25	0.5	15.0	13.0	2.5	0.007	B9A	48
W107	(VM)	12.6	0.1	250	250	-2.0	8.0	2.0	0.5	2.5	4.2	7.0	0.006	B7G	22
Z359		12.6	0.3	250	250	-2.0	20.0	5.25	0.5	15.0	13.0	2.5	0.007	B9A	47
Mazda															
IF2		1.4*	0.05	90	67.5	0	2.9	1.2	0.6	0.92	3.6	7.5	0.008	B7G	2
IL4															
SP181		18.0	0.2	200	200	-1.5	10.9	2.7	0.7	8.5	10.75	5.25	0.005	MO	11
10F3		22.0	0.1	200	200	-2.35	6.0	1.6	—	6.5	9.0	4.6	0.0065	B8A	8
IF3	(VM)	1.4*	0.05	90	45	0	1.8	0.65	0.8	0.75	3.6	7.5	0.01	B7G	2
IT4															
IFI	(VM)	1.4*	0.025	85	64	0	1.65	0.55	1.0	0.85	3.3	7.8	0.01	B7G	64
IFDI	(SD)	1.4*	0.025	67.5	67.5	-1.5	0.17	0.055	—	0.17	1.8	2.7	0.3	B7G	65
IFD9															
IS5	(SD)	1.4*	0.05	67.5	67.5	0	1.6	0.4	0.6	0.63	2.2	3.3	0.4	B7G	5

Type															Base	
6F1			6.3	0.35	200	200	−1.8	10.0	2.6	0.9	9.0	9.0	4.6	0.0065	B8A	17
6F11			6.3	0.2	250	100	−1.8	4.4	1.35	2.8	2.2	5.3	6.7	0.004	B8A	8
6F12		(SQ)	6.3	0.3	250	250	−2.0	10.0	2.5	0.9	7.5	7.6	3.2	0.0045	B7G	21
6F12ANR																
6F14	(VM)		6.3	0.35	135	135	−1.3	27.0	6.5	—	10.6	8.8	4.6	0.007	B8A	8
6F15	(VM)		6.3	0.2	250	100	−2.5	7.0	2.0	1.7	2.3	5.1	6.8	0.0035	B8A	8
6F18			6.3	0.2	175	200	−1.3	12.0	3.5	—	4.5	5.2	5.0	0.0017	B9A	10
6F33			6.3	0.35	200	200	−4.0	5.75	3.1	—	3.55	7.3	4.5	0.01	B7G	21
30FS			7.3	0.3	170	170	−1.85	10.0	2.6	—	8.8	9.0	4.4	0.0073	B9A	10
30FL1	(T, BT)		9.4	0.3	170	170	−2.1	10.0	2.5	—	7.5	7.9	3.2	0.03	B9A	49
20F2			11.0	0.2	135	135	−1.3	27.0	6.5	—	10.6	8.8	4.6	0.007	B8A	8
10F9	(VM)		13.0	0.1	175	100	−2.5	7.0	2.0	1.0	2.3	5.1	6.8	0.0035	B8A	8
10F18	(VM)		13.0	0.1	175	100	−1.3	12.0	3.5	—	4.5	5.2	5.0	0.0017	B9A	10
10F1			22.0	0.1	200	200	−1.8	10.0	2.6	0.9	9.0	9.0	4.6	0.0065	B8A	17
Mullard																
DF64			0.62*	0.01	15	15	−0.75	0.05	0.017	1.2	0.09	1.8	2.0	0.2	B5A	3
DF61			1.25*	0.025	67.5	67.5	0	1.7	0.45	1.6	0.95	3.1	3.6	0.01	B5A	3
DF62			1.25*	0.1	45	45	0	3.0	0.8	0.5	2.0	4.0	4.0	0.01	B5A	2
DAF70	(SD)		1.25*	0.025	67.5	67.5	0	1.0	0.25	0.4	0.44	1.8	3.0	0.15	B8D	1
DF73	(VM)		1.25*	0.025	67.5	67.5	0	1.7	0.5	0.8	0.8	2.9	5.0	0.015	B8D	2
DF91	(VM)		1.4*	0.05	90	67.5	0	3.5	1.4	0.5	0.9	3.6	7.5	0.01	B7G	2
DF92			1.4*	0.05	90.0	64.0	0	3.7	1.4	0.5	1.0	3.6	7.5	0.01	B7G	2
DF96			1.4*	0.025	85.0	61.0	0	1.65	0.55	1.0	0.75	3.3	7.8	0.01	B7G	2
DF97	(SD)		1.4*	0.025	85.0	67.5	−1.5	1.75	0.65	0.4	1.0	3.7	7.5	0.01	B7G	59
DAF96	(SD)		1.4*	0.025	67.5	67.5	−1.5	0.7	0.055	—	0.17	1.8	2.7	0.3	B7G	5
DAF91	(VM)		1.4*	0.05	90	90	−1.5	2.7	0.63	0.5	0.72	2.0	2.8	0.4	B7G	5
KF35	(VM, SD)		2.0*	0.05	120	60	−1.5	1.45	0.5	1.5	1.08	8.0	10.0	0.01	IO	85
EAF41	(VM, DD)		6.3	0.2	250	110	−2.0	5.0	1.5	1.4	2.0	4.0	6.5	0.002	B8A	11
EBF89			6.3	0.3	250	100	−2.0	9.0	3.0	0.9	3.6	5.0	5.0	0.002	B9A	12
EF91		(SQ)	6.3	0.3	250	250	−2.0	10.0	2.5	1.0	7.6	7.0	2.0	0.008	B7G	21
MB083																
EF92	(VM)	(SQ)	6.3	0.2	250	150	−0.65	8.0	2.0	0.5	2.5	4.5	7.0	0.004	B7G	21
M8161																
EF93	(VM)	(SQ)	6.3	0.3	250	100	−1.0	11.0	4.2	1.5	4.4	5.5	5.0	0.0035	B7G	16
M8101																
EF95	(VM)	(SQ)	6.3	0.175	180	120	−2.0	7.7	2.4	0.69	5.1	4.0	2.8	0.02	B7G	14
M8100																
6AS6			6.3	0.175	120	120	−2.0	5.2	3.5	0.11	3.2	4.0	3.0	0.02	B7G	32
E180F			6.3	0.3	190	160	−1.0	13.0	3.0	0.035	16.5	7.9	2.9	0.02	B9A	45
EF22	(VM)		6.3	0.2	250	100	−2.5	6.0	1.7	1.2	5.5	5.5	6.4	0.002	B8B	3
EF37A			6.3	0.2	250	100	−2.0	3.0	0.8	2.5	1.8	5.5	8.5	0.02	IO	8
EF40	(VM)		6.3	0.2	250	140	−2.0	3.0	0.55	2.5	1.85	4.0	5.5	0.025	B8A	15
EF41			6.3	0.2	250	100	−2.5	6.0	1.7	1.0	2.2	4.7	8.0	0.002	B8A	7
EF42	(VM, SD)		6.3	0.33	250	100	−2.0	10.0	2.3	0.44	9.5	9.5	4.5	0.005	B8A	8
EAF42			6.3	0.2	250	250	−2.0	5.0	1.5	1.4	2.0	4.5	5.1	0.002	B8A	12
EF50			6.3	0.3	250	250	−2.0	10.0	3.0	1.0	6.5	8.3	5.2	0.007	B9G	1
EF54			6.3	0.3	250	250	−1.7	10.0	1.45	0.5	7.7	6.2	4.9	0.02	B9G	2
EF55			6.3	1.0	250	250	−4.5	40.0	5.5	0.055	12.0	15.0	12.0	0.15	B9G	1
EF70		(SQ)	6.3	0.2	100	100	−2.0	3.0	2.5	0.1	2.5	4.5	4.7	0.02	B8D	3
M8125																
EF71	(VM)		6.3	0.15	100	100	−1.2	7.2	2.2	0.26	4.5	4.4	4.0	0.015	B8D	4
EF72		(SQ)	6.3	0.15	100	100	−1.4	7.0	2.2	0.25	5.0	4.1	2.0	0.02	B8D	4
M8121																
EF73		(SQ)	6.3	0.2	100	100	−2.0	7.5	2.5	0.25	5.25	5.0	3.0	0.2	B8D	5
M8122																
EF74			6.3	0.2	100	100	−1.4	7.0	2.2	0.18	3.0	4.0	5.0	0.15	Special	

Table 17. Further valve data – *continued*

Screened tetrodes and pentodes – *continued*

Type		Heater		Volts			Current (mA)		r_a (MΩ)	g_m (mA/V)	Capacitances (pF)			Base	
		Volts	Amps	Anode	Screen	Grid	Anode	Screen			C_{gk}	C_{ak}	C_{ga}	Type	Ref.
Mullard – *continued*															
EF80		6.3	0.3	170	170	-2.0	10.0	2.5	0.4	7.4	7.5	3.3	0.007	B9A	10
EBF80	(VM, DD)	6.3	0.3	250	85	-2.0	5.0	1.75	1.4	2.2	4.2	4.9	0.0025	B9A	12
EF85	(VM)	6.3	0.3	250	100	-2.0	10.0	2.5	0.5	6.0	7.2	3.7	0.007	B9A	10
EF86		6.3	0.2	250	140	-2.0	3.0	0.6	2.5	1.8	4.0	5.5	0.025	B9A	23
EF89	(VM)	6.3	0.2	250	100	-2.0	9.0	3.0	1.0	3.6	5.5	5.1	0.002	B9A	36
EF97	(VM)	6.3	0.3	12.6	6.3	–	3.0	1.1	0.15	1.9	6.5	4.0	0.015	B7G	68
EF41	(VM)	12.6	0.1	170	100	-2.5	6.0	1.75	1.0	2.2	5.0	7.0	0.002	B8A	7
UAF41	(VM, SD)	12.6	0.1	200	85	-2.0	5.0	1.5	1.0	2.0	4.5	5.0	0.002	B8A	11
UAF42	(VM, SD)	12.6	0.1	200	85	-2.0	5.0	1.5	1.0	2.0	4.5	5.1	0.002	B8A	12
UF93	(VM)	12.6	0.15	Other data as Type EF93											
UF89	(VM)	12.6	0.1	170	110	-2.0	12.0	3.9	0.525	3.85	5.5	5.1	0.002	B9A	36
UBF80	(VM, DD)	17.0	0.1	170	85	-2.0	5.0	1.75	0.9	2.2	4.2	4.9	0.0025	B9A	12
UF80		19.0	0.1	170	170	-2.0	10.0	2.5	0.4	7.4	7.5	3.3	0.007	B9A	10
UF85		19.0	0.1	200	116	-2.3	11.4	3.1	0.35	6.1	6.9	3.2	0.007	B9A	10
UBF89	(VM, DD)	19.0	0.1	200	100	-1.5	11.0	3.3	0.6	4.5	5.0	5.2	0.002	B9A	12
UF42		21.0	0.1	170	170	-2.0	10.0	2.8	0.2	8.5	9.5	4.5	0.005	B8A	8

§ Screen and secondary cathode voltage

Table 17. Further valve data – continued

Output valves (triodes, tetrodes and pentodes, class A operation)

Type		Heater		Volts			Current (mA)		r_a (Ω)	g_m (mA/V)	R_K (Ω)	R_L (Ω)	Power output (W)	D (%)	Base	
		Volts	Amps	Anode	Screen	Grid	Anode	Screen							Type	Ref.
Cossor																
142BT	(BT)	14.0	0.2	180	180	−8.5	29.0	3.0	58 000	3.7	265	5 500	2.0	8	IO	36
16AS	(P)	16.5	0.3	170	170	−10.4	53.0	10.0	20 000	9.5	–	3 000	4.2	10	B9A	16
35AS	(BT)	35.0	0.15	200	110	−8.0	44.0	7.0	40 000	5.9	157	4 500	3.3	10	B8B	10
45IPT	(P)	45.0	0.1	170	170	−10.4	53.0	10.0	20 000	9.5	140	3 000	4.2	10	B8A	7
Marconi and GEC																
KT71	(BT)	48.0	0.16	175	175	−9.8	70.0	12.0	–	10.0	120	2 500	5.0	9.0	IO	36
KT81	(BT)	6.3	0.95	250	250	−4.4	40.0	7.5	–	10.8	90	6 000	4.3	8	B8B	10
KT76	(BT)	15.0	0.16	175	175	−13.0	35.0	6.0	–	2.5	300	5 000	2.0	4.5	IO	36
KT32	(BT)	26.0	0.3	135	135	−7.6	75.0	5.0	–	9.0	95	1 300	3.5	11	IO	36
KT101	(BT)	80.0	0.1	200	200	−12.5	60.0	10.0	–	10.0	180	3 000	5.0	12.0	B8B	10
KT101	(T)	80.0	0.1	175	–	−7.5	120.0		–	11.5	–	–	–		B8B	10
N19	(P)	1.4*	0.1†	90		−4.5	9.5	2.1	100 000	2.15	–	10 000	0.27	7	B7G	58
N18	(P)	1.4*	0.1†	90	90	−4.5	9.5	2.1	100 000	2.15	–	10 000	0.27	7	B7G	6
PX4	(T)	4.0*	1.0	300	–	−50.0	50.0		830	6.0	1 000	3 500	4.5	4	B4	1
KT61	(BT)	6.3	0.95	250	250	−4.4	40.0	8.5	–	10.5	90	6 000	4.3	8	IO	36
N77 QA2402 }	(P)	6.3	0.2	250	250	−12.0	16.0	3.0	130 000	2.6	680	16 000	1.4	10	B7G	25
A2134	(P)	6.3	0.635	165	165	−9.3	53.0	9.0	23 200	9.5	150	3 000	4.1	10	B7G	33
N78	(P)	6.3	0.64	250	250	−5.5	36.0	5.0	40 000	10.0	120	7 000	4.0	10	B7G	25
KT66	(BT)	6.3	1.27	250	250	−15.0	85.0	6.3	–	6.3	160	2 200	7.25	9	IO	36
N709	(P)	6.3	0.76	250	250	−7.5	48.0	–	38 000	11.3	120	5 000	6.0	10	B9A	16
N727/6AQ5	(BP)	6.3	0.45	250	250	−12.5	45.0	4.5	52 000	4.1	240	5 000	4.5	8	B7G	27
A1834	(DT)	6.3	2.5	135	–	−31.5	125.0	7.5	280	7.5	250	–			IO	26
LN309	(P)	12.6	0.3	165	165	−8.4	32.0	6.5	45 000	4.7	220	6 000	2.1	10	B9A	27
HN309	(TP)	12.6	0.3	165	165	–	32.0	6.0	45 000	4.7	220	6 000	2.1	10	B9A	27
KT33C	(BT)	13.0	0.6†	200	200	−13.3	60.0	10.0	–	10.0	190	3 000	5.0	8	IO	73
N37	(BT)	13.0	0.3	165	165	−11.4	29.0	5.4	23 200	9.5	330	6 000	2.3	10	B7G	25
N309/PL83	(P)	15.0	0.3	170	170	−2.5	32.0	4.2	41 000	10.0	68	5 000	1.65	7.8	B9A	14
N329	(P)	16.5	0.3	170	170	−10.6	50.0	9.0	20 000	9.0	180	3 000	4.0	10	B9A	16
N108	(P)	40.0	0.1	165	165	−9.3	53.0	9.0	23 200	9.5	150	3 000	4.1	10	B7G	25
Mazda																
1P1		1.4*	0.05†	85	85	−5.2	5.0	0.9	150 000	1.4	–	13 000	0.2	10	B7G	9
1P10 3S4 }	(P)	1.4*	0.1†	90	67.5	−7.0	7.4	1.4	–	1.57	–	8 000	0.27	12	B7G	6
1P11 3V4 }	(P)	1.4*	0.1†	90	90	−4.5	9.5	2.1	100 000	2.15	–	10 000	0.27	7	B7G	9
6P1	(BT)	6.3	0.8	250	250	−8.5	40.0	7.5	40 000	8.8	180	5 000	4.2	7	IO	36
6P25	(BT)	6.3	1.1	250	250	−8.5	40.0	8.0	40 000	8.8	180	5 000	4.5	7	IO	36
12E1	(BT)	6.3	1.6	800***	300***	–	–	–	–	–	–	–	–	–	IO	38
30P12	(BT)	12.6	0.3	170	180	−10.3	31.0	7.3	–	6.7	270	5 000	2.25	7	B9A	16
30PL1	(T,BT)	13.0	0.3	180	180	−9.6	28.0	6.5	–	6.0	270	6 000	2.0	7	B9A	27
13E1	(BT)	13.0	2.6†	800‡	300‡	–	–	–	–	–	–	–	–	–	B7A	2
20P3	(BT)	20.0	0.2	195	210	−11.5	51.0	12.7	–	7.4	180	3 700	4.5	7	IO	36
20P5	(BT)	20.0	0.2	180	150	−6.3	29.0	5.8	–	7.5	180	5 400	2.6	10	B8A	7
10P13	(BT)	40.0	0.1	180	150	−6.3	29.0	5.8	–	7.5	180	5 400	2.6	10	B8A	7
10P14	(BT)	40.0	0.1	195	210	−11.5	51.0	12.7	–	7.4	180	3 700	4.5	7	IO	36

Table 17. Further valve data – continued

Output valves (triodes, tetrodes and pentodes, class A operation) – continued

Type		Heater		Volts			Current (mA)		r_a (Ω)	g_m (mA/V)	R_K (Ω)	R_L (Ω)	Power output (W)	D (%)	Base Type	Ref.
		Volts	Amps	Anode	Screen	Grid	Anode	Screen								
Mullard																
DL66	(P)	1.25*	0.015	22.5	22.5	−1.4	0.3	0.075	300 000	0.35	—	75 000	0.0027	10	B5A	1
DL68	(P)	1.25*	0.025	22.5	22.5	−2.2	0.6	0.15	100 000	0.43	—	37 000	0.005	10	B5A	1
DL71	(P)	1.25*	0.025	45	45	−1.25	0.6	0.15	350 000	0.55	—	100 000	0.0063	10	Wires	6
DL75	(P)	1.25*	0.025	90	90	−2.5	1.75	0.4	450 000	0.85	—	60 000	0.05	10	B8D	3
DL64	(P)	1.25*	0.01	15.0	15.0	−1.5	0.16	0.04	400 000	0.18	—	100 000	0.00095	10	B5A	3
DL69	(P)	1.25*	0.025	90.0	90.0	−3.0	1.75	0.04	600 000	0.85	—	60 000	0.05	10	B5A	3
DL70	(P)	1.25*	0.11	150	90	−7.5	6.5	1.4	—	1.5	—	—	—	—	B8D	6
DL73	(P)	1.25*	0.2	100	100	−9.0	15.0	3.8	—	2.3	—	—	—	—	B8D	—
DL620	(P)	1.25*	0.05	67.5	67.5	−6.5	3.3	1.1	100 000	0.65	—	20 000	0.065	10	B5A	1
DL92	(P)	1.4*	0.1†	90	67.5	−7.0	7.4	1.4	100 000	1.57	—	8 000	0.27	12	B7G	6
DL93	(P)	1.4*	0.2†	150	90	−8.4	13.3	2.2	100 000	1.9	—	8 000	0.7	6	B7G	7
DL94	(P)	1.4*	0.1†	90	90	−4.5	9.5	2.1	100 000	2.15	—	10 000	0.27	7	B7G	9
DL96	(P)	1.4*	0.05	85	85	−5.2	5.0	0.9	150 000	1.4	—	13 000	0.2	10	B7G	9
KL35	(P)	2.0*	0.15	135	135	−4.5	5.6	—	150 000	2.2	—	19 000	0.34	10	IO	78
EBL21	(P,DD)	6.3	0.8	250	275	−6.2	44.0	5.8	50 000	9.5	125	5 700	5.5	10	B8B	6
EBL31	(P,DD)	6.3	1.2	250	250	−6.0	36.0	5.0	50 000	9.5	146	7 000	4.3	10	IO	15
ECL80	(TP)	6.3	0.3	200	200	−8.0	17.5	3.3	150 000	3.3	—	11 000	1.4	10	B9A	13
ECL82	(TP)	6.3	0.78	170	170	−11.5	41.0	7.5	16 000	7.5	—	3 900	3.3	10	B9A	37
EL6	(P)	6.3	1.2	250	250	−7.0	72.0	8.0	20 000	14.5	90	3 500	8.0	10	C8	12
EL36	(P)	6.3	1.2	250	250	−7.0	44.0	5.2	45 000	9.5	140	5 750	5.2	10	IO	36
EL22	(P)	6.3	0.7	250	250	−13.5	100.0	14.9	15 000	11.0	120	2 000	11.0	10	B8B	10
EL34	(P)	6.3	1.5	250	250	−13.5	100.0	13.5	13 500	11.0	120	2 500	10.5	10	IO	133
EL37	(P)	6.3	1.4	250	250	−7.0	36.0	5.2	40 000	10.0	170	7 000	4.2	10	IO	36
EL41	(P)	6.3	0.7	250	250	−12.5	33.0	4.1	90 000	3.2	250	3 500	1.55	8	B8A	4
EL41	(T)	6.3	0.7	250	—	−7.3	26.0	—	38 000	11.3	360	9 000	2.5	10	B8A	4
EL42	(P)	6.3	0.2	225	225	−10.8	48.0	5.5	90 000	3.2	360	5 200	—	10	B8A	7
EL84	(P)	6.3	0.76	250	250	−12.5	26.0	4.1	38 000	11.3	135	9 000	5.7	12	B9A	16
EL85	(P)	6.3	0.2	225	225	−12.5	45.0	4.5	90 000	3.2	360	5 000	2.8	8	B9A	26
EL90	(P)	6.3	0.45	250	250	−12.5	16.0	2.4	52 000	4.1	250	16 000	4.5	10	B7G	27
EL91	(P) [SQ]	6.3	0.2	250	250	−9.5	30.0	5.0	—	—	—	5 500	—	10	B7G	25
MB082	(P)	6.3	0.2	250	250	−2.3	16.0	5.0	130 000	2.6	680	16 000	1.4	10	B7G	27
PCL83	(TP)	12.6	0.3	170	170	−11.5	36.0	7.5	53 000	5.5	—	5 500	2.2	10	B9A	14
PL83	(TP)	15.0	0.3	170	170	−10.4	41.0	10.0	100 000	10.0	—	—	—	—	B9A	37
PCL82	(TP)	16.0	0.3	170	170	−6.0	53.0	4.0	16 000	7.5	—	3 900	3.3	10	B9A	16
PLB2	(P)	16.5	0.3	170	250	−8.5	36.0	—	20 000	9.0	165	3 000	4.0	10	B9A	36
PL33	(P)	19.0	0.3	250	250	−9.5	20.0	5.0	50 000	9.0	150	7 000	4.5	5	IO	36
PL33	(T)	19.0	0.3	250	—	−10.4	30.0	—	3 000	6.5	425	7 000	1.1	10	IO	36
UCL83	(TP)	40.0	0.1	170	170	−10.4	30.0	10.0	53 000	5.5	—	5 500	2.2	10	B9A	27
UL41	(P)	45.0	0.1	170	170	−12.0	53.0	10.0	20 000	9.5	140	3 000	4.2	10	B8A	7
UL46	(P)	45.0	0.1	170	170	−7.5	53.0	4.5	20 000	9.5	—	3 000	4.2	10	B9A	16
UL84	(P)	45.0	0.1	165	165	−13.0	73.0	4.0	20 000	10.5	—	2 400	5.6	10	B7G	42
HL92	(P)	50.0	0.15	110	110	−7.5	49.0	—	10 000	7.5	—	2 500	1.9	9	B7G	—
UBL21	(P,DD)	55.0	0.1	200	200	−13.0	55.0	9.5	25 000	8.0	200	3 500	4.8	10	B8B	6

* Maximum values for use in stabilised supply circuits. $I_f(max) = 300$ mA, $P_{A(max)} = 35$ W

† Maximum values for use in stabilised supply circuits. $I_f(max) = 800$ mA, $P_{A(max)} = 90$ W

Thermionic diodes

Table 17. Further valve data – continued

| Type | Heater | | Max. Input volts (R.M.S.) | Max. Rect. current (mA) | No. of diodes | Capacitances (pF) | | | Base | |
	Volts	Amps				a'-k	a"-k	a'-a"	Type	Ref.
Marconi										
D152	6.3	0.3	150	9.0	2	3.0	3.0	0.03	B7G	18
Marconi and GEC										
D77 (SQ)	6.3	0.3	120	50.0	2	3.5	3.5	0.025	B7G	18
QA2404										
Mazda										
ID13	1.4	0.15	130	0.5	1	0.6	–	–	B7G	13
DD41	4.0	0.5	175	5.0	2	4.0	4.25	0.06	MO	13
D1	4.0	0.2	125	5.0	1	2.1	–	–	B3G	–
6D1	6.3	0.15	125	5.0	1	2.1	–	–	B3G	–
6D3*	6.3	0.3	–	5.0	1	–	–	–	B7G	50
6D2	6.3	0.3	175	9.0	2	3.4	3.4	0.018	B7G	18
20D1	9.5	0.2	175	9.0	2	3.4	3.4	0.018	B7G	18
10D2	19.0	0.1	175	9.0	2	3.4	3.4	0.018	B7G	18
Mullard										
EA50	6.3	0.15	50	5.0	1	2.1	–	–	B3G	–
EB34	6.3	0.2	140	0.8	2	4.5	4.5	0.5	IO	53
EB41	6.3	0.3	150	9.0	2	<0.01	<0.01	<0.03	B8A	10
UB41	19.0	0.1	150	9.0	2	<0.01	<0.01	<0.03	B8A	10
DA90	1.4	0.15	117	0.5	1	0.4	–	–	B7G	13
EA76 (SQ)	6.3	0.15	150	9.0	1	2.5	–	–	B5B	1
M8123										
EB91 (SQ)	6.3	0.3	150	9.0	2	3.0	3.0	<0.025	B7G	18
M8079										

Amplifier triodes

Table 17. Further valve data – continued

| Type | | Heater | | Volts | | Anode current (mA) | r_a (Ω) | g_m (mA/V) | Capacitances (pF) | | | Base | |
		Volts	Amps	Anode	Grid				C_{fk}	C_{ak}	C_{fa}	Type	Ref.
Marconi													
DH147	(DD)	6.3	0.2	250	-5.5	5.0	15 000	2.0	2.4	-	-	IO	29
DH149	(DD)	6.3	0.15	250	-1.0	1.3	100 000	1.0	-	3.0	1.4	B8B	60
DH150	(DD)	6.3	0.225	250	-3.0	1.0	54 000	1.3	-	-	-	B8A	9
LN152	(TP)	6.3	0.3	100	-2.3	4.0	12 500	1.4	2.0	0.3	0.9	B9A	13
B152	(DT)	6.3	0.3†	170	-1.5	7.0	12 000	4.8	2.2	0.4	1.5	B9A	1
DH142	(DD)	14.0	0.1	170	-1.6	1.5	48 000	1.65	2.75	1.5	1.3	B8A	9
DL145	(DD)	15.0	0.1	250	-5.9	5.0	13 500	3.4	3.6	3.7	1.5	B8A	9
Marconi and GEC													
DH81	(DD)	6.3	0.3	250	-0.68	1.0	58 000	1.2	2.4	1.4	1.7	B8B	12
DL63	(DD)	6.3	0.3	250	-3.0	4.2	22 500	1.6	1.5	3.5	2.3	IO	29
H63	(DD, VM)	6.3	0.3	250	-2.0	1.0	66 000	1.5	2.3	3.7	2.5	IO	18
DL82	(DD)	6.3	0.3	250	-3.0	5.0	17 000	1.4	2.0	1.5	2.0	B8B	12
DH76	(DD)	13.0	0.16	250	-3.0	1.1	5800	1.2	1.5	5.0	1.5	IO	29
DH101	(DD)	19.0	0.1	250	-3.0	1.0	58 000	1.2	2.4	1.4	1.7	B8B	12
Mazda													
6F13	(P)	6.3	0.35	200	-1.8	12.6	5300	11.3	-	-	-	B8A	8
6LD20	(DD)	6.3	0.25	250	-5.9	5.0	13 500	2.3	3.6	3.7	1.5	B8A	9
HL133DD	(DD)	13.0	0.2	250	-5.4	6.0	14 000	2.3	3.5	4.5	3.5	MO	10
10LD11	(DD)	15.0	0.1	250	-5.9	5.0	13 500	2.3	3.6	3.7	1.5	B8A	9
6/30L2	(DT)	6.3	0.3	200	-7.9	10.0	5300	3.4	2.5	2.1	2.5	B9A	39
6F1	(P)	6.3	0.35	200	-1.8	12.6	5300	11.3	-	-	-	B8A	17
6F11	(P)	6.3	0.2	100	-1.8	5.75	9000	2.85	-	-	-	B8A	8
6F12	(P)	6.3	0.3	250	-2.0	12.6	8000	9.4	-	-	-	B7G	21
6L1	(DT)	6.3	0.4	250	-11.5	10.0	6200	2.8	2.8	2.3	2.7	B8A	13
6L18		6.3	0.5	250	-13.3	12.0	3000	5.5	4.6	5.8	2.2	B8A	6
6L19	(DT)	6.3	0.4	250	-3.1	4.0	20 000	2.75	2.9	2.5	2.5	B8A	13
6L34		6.3	0.3	250	-1.5	10.0	10 500	8.5	5.1	0.1	3.6	B7G	24
6LD3	(DD)	6.3	0.23	150	-3.0	1.0	54 000	1.3	3.0	1.9	1.3	B8A	9
30L1	(DT)	7.0	0.3	150	-3.4	16.0	3800	6.6	2.3	0.5	1.1	B9A	28
30F5	(P)	7.3	0.3	170	-1.85	12.6	-	11.0	-	-	-	B9A	10
30FL1	(T, BT)	9.4	0.3	200	-7.9	10.0	5300	3.4	3.6	2.6	2.7	B9A	49
20L1	(DT)	12.6	0.2	250	-11.5	10.0	6200	2.8	2.8	2.3	2.7	B8A	13
10LD3	(DD)	13.0	0.1	250	-3.0	1.0	54 000	1.3	3.0	1.9	1.3	B8A	9
30PL1	(T, BT)	13.0	0.3	200	-7.9	10.0	5300	3.4	2.6	2.0	2.4	B9A	27
10L1		19.0	0.1	250	-1.5	10.0	10 500	8.5	5.1	0.1	3.6	B7G	24
10F1	(P)	22.0	0.1	200	-1.8	12.6	5300	11.3	-	-	-	B8A	17

Mullard														
DAC32		(SD)	1.4*	0.05	90	0	0.15	240 000	0.275	1.3	6.0	1.0	IO	91
DCC90		(DT)	1.4*	0.22†	90	–2.5	3.7	8300	1.8	0.9	1.0	3.2	B7G	8
KBC32		(DD)	2.0*	0.05	100	0	2.4	21 000	1.2	1.9	7.0	3.1	IO	88
EAC91 M8097	(SQ)	(SD)	6.3	0.3	200	–3.2	7.5	12 800	2.8	1.7	0.4	1.6	B7G	23
EBC41		(DD)	6.3	0.23	250	–3.0	1.0	54 000	1.3	2.75	1.5	1.3	B8A	9
EC31			6.3	0.65	250	–16.0	20.0	3300	3.2	–	–	–	IO	20
EC53			6.3	0.25	200	–3.3	7.5	11 400	2.9	1.3	0.13	1.3	B3G	
EC90			6.3	0.15	250	–8.5	10.5	7700	2.2	1.8	1.3	1.6	B7G	15
EC91 M8099	(SQ)		6.3	0.3	250	–1.5	10.0	12 000	8.5	5.3	5.8	2.5	B7G	24
ECC31		(DT)	6.3	0.95	250	–4.6	6.0	14 000	2.3	4.0	1.9	3.4	IO	22
ECC32		(DT)	6.3	0.95	250	–4.6	6.0	14 000	2.3	4.3	2.0	4.3	IO	26
ECC33		(DT)	6.3	0.4	250	–4.0	9.0	9700	3.6	3.5	1.5, 1.2	2.5	IO	26
ECC34		(DT)	6.3	0.95	250	–16.0	10.0	5200	2.2	3.5	1.8	4.0	IO	26
ECC35		(DT)	6.3	0.4	250	–2.5	2.3	34 000	2.0	3.0	1.0, 1.3	2.5, 3.0	IO	26
ECC40		(DT)	6.3	0.6	250	–5.2	6.0	11 000	2.7	3.0, 2.6	1.15	2.6, 2.7	B8A	13
ECC81 M8081	(SQ)	(DT)	6.3	0.3†	170	–1.0	8.5	12 000	5.5	2.2	0.4, 0.5	1.5	B9A	1
ECC85		(DT)	6.3	0.435	250	–2.3	10.0	9700	5.9	3.0	0.18	1.5	B9A	39
ECC91 M8081	(SQ)	(DT)	6.3	0.45	100	–0.85	8.5	7100	5.3	2.2	0.4	1.6	B7G	17
EABC80		(TD)	6.3	0.45	250	–3.0	1.0	50 000	1.4	1.9	1.6	2.2	B9A	2
EBC90		(DD)	6.3	0.3	250	–3.0	1.0	58 000	1.2	2.3	1.1	2.1	B7G	19
ECC82 M8136	(SQ)	(DT)	6.3	0.3†	250	–8.5	10.5	7700	2.2	1.6	0.5	1.5	B9A	1
ECC83 M8137	(SQ)	(DT)	6.3	0.3†	250	–2.0	1.2	62 500	1.6	1.6	0.46	1.7	B9A	1
ECL80		(TP)	6.3	0.3	100	–2.3	4.0	12 500	1.4	2.0	0.3	0.9	B9A	13
UC92			9.5	0.1	170	–1.0	11.5	10 500	6.7	2.6	0.24	1.6	–	–
HBC90		(DD)	12.6	0.15	*Other data as Type EBC90*									
HBC91		(DD)	12.6	0.15	250	–2.0	1.2	62 500	1.6	–	–	–	B7G	19
PCL83		(TP)	12.6	0.3	250	–8.5	10.5	7700	2.2	2.0	0.35	1.6	B9A	27
UBC41		(DD)	14.0	0.1	170	–1.6	1.5	42 000	1.65	2.75	1.5	1.3	B8A	9
PCL82		(TP)	16.0	0.3	100	0	3.5	28 000	2.5	2.7	4.0	4.0	B9A	37
UCC85		(DT)	26.0	0.1	200	–2.1	10.0	8300	5.8	0.003	0.008	0.008	B9A	39
UABC80		(TDT)	28.0	0.1	200	–2.3	1.0	50 000	1.4	1.9	1.4	2.0	B9A	2
UCL83		(TP)	40.0	0.1	200	–1.5	2.4	34 000	2.5	2.3	0.32	1.6	B9A	27

Valve rectifiers

Table 17. Further valve data – *continued*

Type	Heater		Type of rectification	Input volts (R.M.S.)	Max. rect. current (mA)	Max. reservoir capacitance (μF)	Min. series resistance (Ω)	Base	
	Volts	Amps						Type	Ref.
Cossor									
OM1	30.0	0.2	H.W.	250	120	32	50	IO	55
Marconi and GEC									
U84	4.0*	1.0	F.W.	250–0–250	75	16	100	B8B	24
U81	6.3	1.6	F.W.	500–0–500	150	16	100	B8B	24
U82	6.3	0.6	F.W.	325–0–325	75	4	150	B8B	1
U76	30.0	0.16	H.W.	250	100	32	100	IO	55
U101	50.0	0.1	H.W.	250	100	32	100	B8B	25
U54	5.0	2.8	F.W.	500–0–500	250	16	75	IO	62
U78 / QA2407 (SQ)	6.3	0.6	F.W.	325–0–325	70	8	435	B7G	31
U709	6.3	0.95	F.W.	350–0–350	150	–	270	B9A	31
U319	20.0	0.3	H.W.	250	170	–	55	B9A	18
U31	26.0	0.3	H.W.	250	120	32	100	IO	55
U107	40.0	0.1	H.W.	250	90	12	200	B7G	13
Mazda									
UU10	4.0	2.3	F.W.	500–500	180	8	–	B4	14
U201	20.0	0.2	H.W.	250	f90	16	47	IO	55
U281	28.0	0.2	H.W.	250	120	16	47	IO	55
U404	40.0	0.1	H.W.	250	90	50	180	B8A	f1
Mullard									
AZ31	4.0*	1.1	F.W.	500–0–500	60	60	–	IO	60
GZ32	5.0	2.3	F.W.	500–0–500	125	60	150	IO	62
GZ33	5.0	2.8	F.W.	500–0–500	250	–	–	IO	62
GZ34	5.0	1.9	F.W.	550–0–550	160	60	175	IO	62
EZ35	6.3	0.6	F.W.	325–0–325	70	16	350	IO	54
EZ40	6.3	0.6	F.W.	350–0–350	90	50	300	B8A	14
EZ41	6.3	0.4	F.W.	250–0–250	60	50	325	B8A	14
EZ80	6.3	0.6	F.W.	350–0–350	90	50	300	B9A	31
EZ81	6.3	1.0	F.W.	350–0–350	150	50	240	B9A	31
EZ90 / M8138 (SQ)	6.3	0.6	F.W.	325–0–325	70	8	520	B7G	31
PY31	17.0	0.3	H.W.	250	125	60	175	IO	55
UY41	31.0	0.1	H.W.	250	100	50	210	B8A	22
HY90	35.0	0.15	H.W.	117	100	40	120	B7G	33
UY85	38.0	0.1	H.W.	250	110	100	100	B9A	18

Cathode-ray tuning indicators

Table 17. Further valve data – *continued*

Type	Heater		Target volts	Target current (mA)	Grid voltage change	Base	
	Volts	Amps				Type	Ref.
Cossor							
63ME	6.3	0.3	250	4.5	22	IO	46
64ME	6.3	0.2	250	0.75	2.5 + 16.00	IO	48
65ME	6.3	0.3	250	2 to 2.3	−1.0 to −16.0	B9A	41
EM81	6.3	0.3	250	2 to 2.3	−1.0 to −10.5	B9A	41
Ferranti							
VFT4	4.0	0.5	200–250	0.5	20.0	IO	46
FT4	4.0	0.5	200–250	0.5	6	IO	46
VFT6	6.3	0.3	200	4.5	22	IO	46
Marconi and GEC							
Y65	6.3	0.3	180–250	4.5	11	IO	46
Mazda							
6MI	6.3	0.3	250	1.16	22.5	IO	46
ME91	9.0	0.2	175	2.7	19	MO	21
IOM1	18.0	0.1	250	1.16	22.5	IO	46
IMI	1.4	0.025	90	0.25	13.5	B8D	9
			60	0.12	8.0		
6M2 (dual sensitivity)	6.3	0.2	250	0.46	4	IO	135
					20		
10M2 (dual sensitivity)	12.6	0.1	200	0.4	3	IO	136
					20		
Mullard							
EM34	6.3	0.3	250	0.75	5 & 16	IO	48
UM34	12.6	0.1	250	0.75	5 & 16	IO	48
DM70	1.4*	0.025	85	0.17	10	B8D	9
			60	0.1	7		
EM80	6.3	0.3	250	2.3	13	B9A	41
EM81	6.3	0.3	250	2.3	9.5	B9A	41
UM81	19.0	0.1	200	7.0	13.0	B9A	41

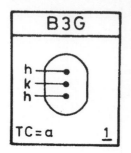

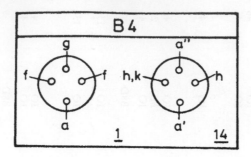

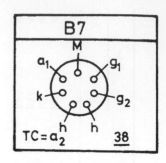

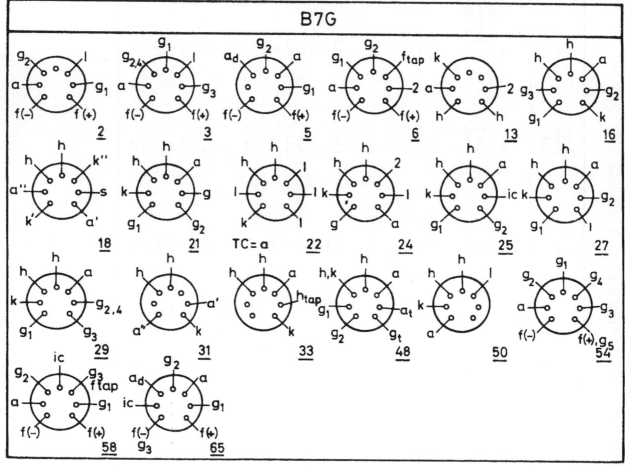

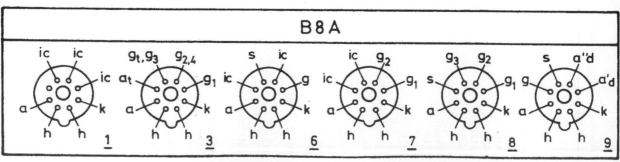

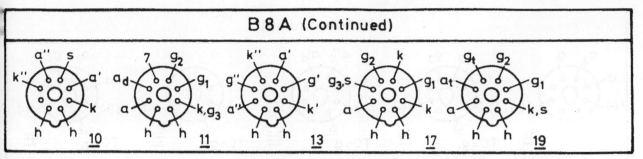

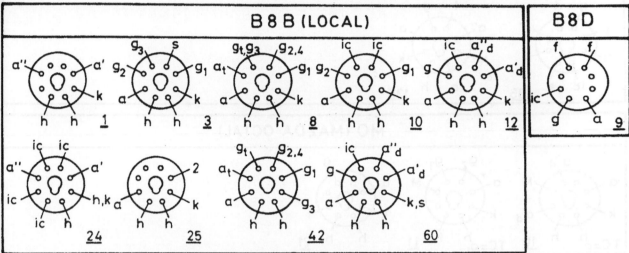

'k and g₅ to centre spigot

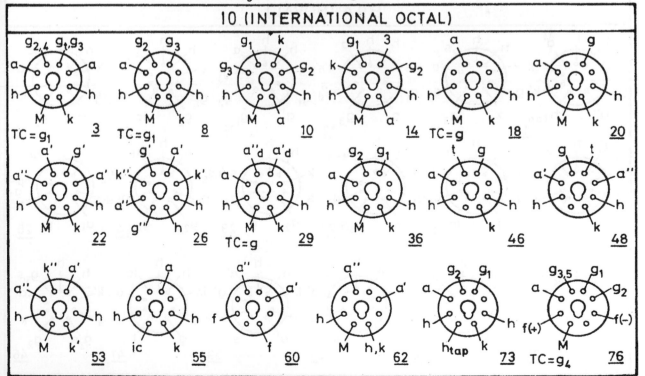

10 (Continued)

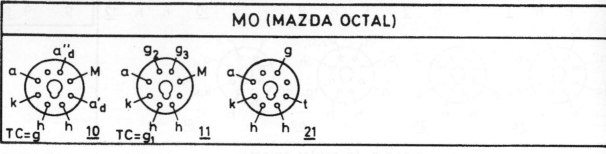

MO (MAZDA OCTAL)

B9A (NOVAL)

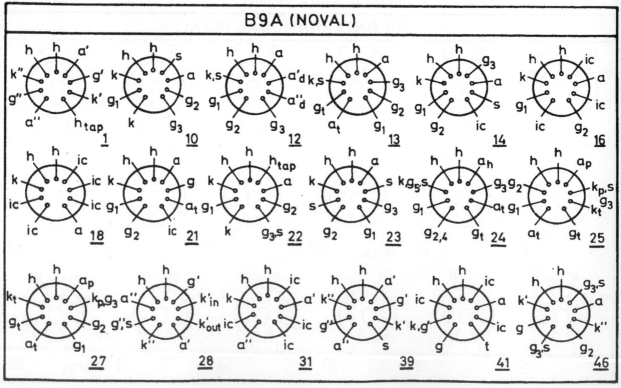

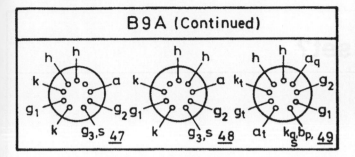

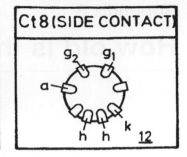

ADDENDUM TO TABLE 17

Reference Mazda type 6C9 (p. 203)

There is also an American valve bearing the same type number but of an entirely different construction. This latter is a double RF pentode for use in VHF/FM tuner units, one section operating as FR amplifier, the other as a self-oscillating mixer. To accommodate the necessary number of electrode connections a special 10-pin base is employed. It has nine pins arranged as for the conventional B9A base, plus another, concentric to the rest. Details of the connections are as follows:

Pin No. 1 G1, first section
            2 G2   „   „
            3 A   „   „
            4 Heater
            5 Heater
            7 K,G3,5, second section
            7 G1   „   „
            8 G2   „   „
            9 A   „   „
        10 (central pin), K,G3,S, first section

Appendix 3

How old is that radio set?

It is possible to make a fairly accurate estimate of the age of a radio set by taking a close look at the dial, because over the years the wavelengths and names of various stations were changed several times. Although the exact dates of the alterations are known, there is an inevitable time gap before all the sets being produced were fitted with appropriately marked dials. As far as the BBC is concerned, the 1930s are split into pre- and post-1935 by the introduction of the Droitwich high power long wave transmitter in October 1934. Five years previously a National and Regional system of broadcasting had been inaugurated when the Brookman's Park (North London) Station had been brought into service. There were six regional stations in the original scheme, all on medium wave, plus the National transmitter on long wave from Daventry. Because this latter was not powerful enough to cover the entire country unaided, there were relay transmitters for London, the North of England and Scotland. Thus, radio dials of 1930–35 will have Daventry shown on 1554.4 m long wave, and the relays at, respectively, 261.6 m, 301.5 m and 288.5 m. The regional programmes for these areas were on 356 m, 480 m and 376.4 m, and for the Midlands from another transmitter at Daventry on 398.9 m.

When Droitwich came into service (now on 1500 m) high power relays for the North and Scotland were no longer required, but that for London was retained. Hence, dials after 1935 no longer showed North National or Scottish National transmitters. The wavelengths for the Regionals were altered and augmented as follows: London 342.1 m, Midlands 296.2 m, North 449.1 m, Scotland 371 m, plus West on 373.1 m and Northern Ireland on 307.1 m. These held good until the ending of the National/Regional system on 1 September 1939. Also to be found during this period are Radio Luxembourg on 1293 m (previously 1190 m) and Fécamp/Radio Normandie on 226.1 m (223 m). These were two of the most popular commercial stations to be heard in Britain at the time.

During the war the BBC's domestic services were known as the Home and Forces programmes. They were transmitted on groups of frequencies in the MW band under a scheme designed to prevent enemy aircraft from 'homing-in' on them. The only receivers in which they were shown on dials were the 'utility' sets. The Droitwich LW station was closed down for a while but returned for broadcasting news, etc., to occupied Europe.

At the end of hostilities a modified Regional system was reintroduced, under the names London Home Service, Midlands Home Service, etc., the wavelengths being: London 342.1 m, Midlands 296.2 m, North 449.1 m, Northern Ireland 285.7 m, Scottish 391.1 m, Welsh 373.1 m and West 307.1 m and 216.8 m. The place of the National programme was taken by the 'Light', again on 1500 m, plus a number of relays on 261.1 m. Post-war sets should thus be easily enough recognised, but beware of, and don't be fooled by, the few that retained the name National for some time (in the case of the Invicta models 200 and 200 W, until 1948!). The above BBC wavelengths were in use until 1950, but to narrow down the field a little, sets from 1947 can be expected to bear the name of the Third programme on 514.6 m and 203.5 m.

Radio Normandie, closed down in early 1940, never returned, but Luxembourg was soon back after the war on 1293 m. In a sense it had never gone away, for it had been commandeered by the Nazis and used for propaganda broadcasts to Britain.

The next landmark was the Copenhagen Plan (named after the venue of an international conference) which came into force in March 1950. The changes on dials after that date were as follows: London 330 m, Midlands 276 m, North 434 m, Northern Ireland 261 m, Scottish 371 m, Welsh 341 m and West 206 m. The Light programme continued on 1500 m, but the relays changed to 247 m. The Third went to 464 m and 194 m for the main and relay stations, respectively. At the same time Luxembourg was given its celebrated 208 m wavelength. There were no major changes after this until 1967, which is just outside our terms of reference for this book, but there was a milestone in May 1955, when the first VHF transmitter (Wrotham) came into full service. Thus any set having FM facilities must date from that time.

Clues inside the set

The sizes of the valves tell a good deal about a set's age, too. In direct contrast to dials, which tended to grow larger through the period under discussion, valves became smaller. The large British-based types were in general use throughout the 1930s, but there are a couple of milestone dates, when other valves were to be found. 'P' base (side-contact) series-heater types were in some sets as early as 1933, but were popular for only a short while, until replaced by a range identical electrically, but fitted with British bases. From this they can be placed pretty firmly as mid-1930s. However, 6.3 V heater 'P' bases were fitted in a number of sets from 1938. In that same year EMI and GEC employed Marconi-Osram octal valves for most of their models, and were followed by other firms. The American UX types had also been employed from the early 1930s, with the International octal base appearing in 1935, initially fitted only to the new all-metal valves but soon spreading to conventional types. The so-called 'British octal' base devised by Mazda appeared in 1938. In the following year came the octal-based equivalents to the 'P' valves and the new all-glass 'loctal'-based types appeared in some sets.

It seems reasonable to assume that had the war not intervened, the older bases would have virtually disappeared by 1940, but in the event supply difficulties resulted in their being retained in many of the models which were produced, although on a restricted scale, in that year. However, the above information, in conjunction with that given for dials, should enable a reasonable dating to be put upon sets of the pre-war period.

When full-scale production resumed again in 1945/46 octals dominated the scene, with Mazda octals finding favour in only a few makes. This situation continued for a few more years, but 1947 brought a renewed interest in 'loctals', of the original American, Continental (B8G) and British (B8B) varieties. The new miniature B8A all-glass valves also began to be employed. The first of the later B9A range appeared in sets towards the end of 1950. Very small portable sets ('personals'), including the unofficially imported American models of the war-time era and those produced in this country just afterwards, used the miniature all-glass B7G valves. These superseded octal 1.4 V types in most portables of any size by 1950 and continued to be used until the end of the valve period. It is useful to bear in mind that from 1954 the low consumption (25 mA) filament versions took over, most easily recognisable in Mullard form by the serial numbers which all terminated in –96, as opposed to the earlier –91 and –92.

In general, it is safe to conclude that any set using exclusively octal valves was made before 1952, and for loctals, 1956, but note that a few models, notably those with high quality output stages, continued to employ perhaps one or two octals among a majority of miniatures for many years after these dates.

If you wish to pin down the age of a set more closely than the approximations made possible from the above, it pays to examine the smoothing condensers, which were in many cases date stamped by the makers. By the law of natural perversity you will more than likely have to undo fixing clamps, etc., to be able to find the date, but it is well worth a try. Obviously, there will be some interval between the time that the condenser was made and when it was used by the radio manufacturers, but this probably would not exceed a few months. Again, replacements that were fitted long after the set was produced will be easily spotted.

The Radiophile

A Happy Christmas to All Our Readers
THE LEADING MAGAZINE FOR ALL VINTAGE RADIO ENTHUSIASTS

Combined Issues Nos. 77 and 78, Christmas and Hogmanay 1998/9. Edited by Chas. E. Miller

CONTENTS INCLUDE:

*Receiver Profile - The "Baby" Hale *In Our Workshop * Designing the Murphy A8, by Frank Murphy *Monty McNab, Short Story *A Handy Home-Built Signal Generator *My World of Wireless, by Jaymar * Improvements in Loudspeakers, by Albert Porter *The Bill Smith Chronicles - Postscript *
* And Much More in Your Special Double Issue Radiophile *

Published by Subscription; for rates see page 24. All Enquiries re Subscriptions to The Admin Office, "Larkhill", Newport Road, Woodseaves, Stafford, ST20 0NP.

Index

Presented Free with Popular Wireless, October 19th, 1929.

THE "POPULAR WIRELESS"

6d

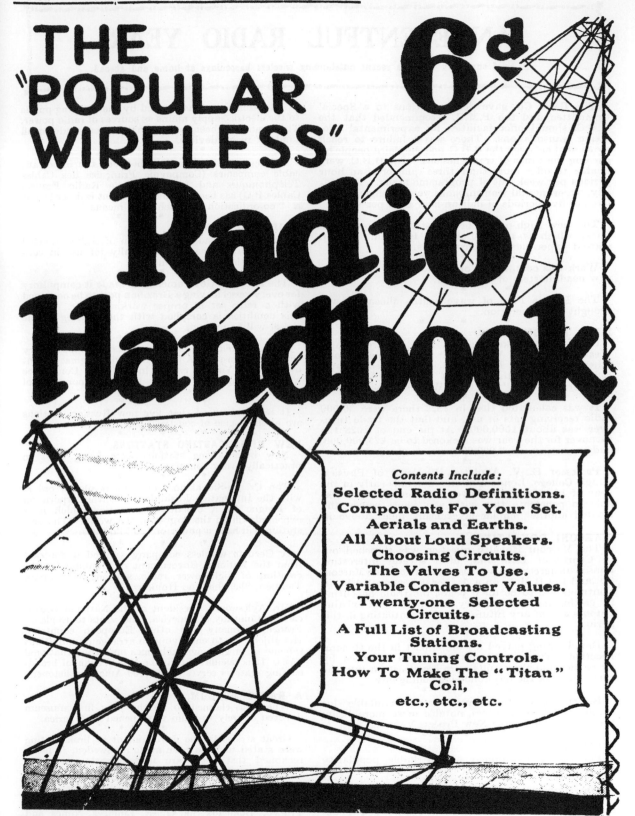

Radio Handbook

Contents Include:

Selected Radio Definitions.
Components For Your Set.
Aerials and Earths.
All About Loud Speakers.
Choosing Circuits.
The Valves To Use.
Variable Condenser Values.
Twenty-one Selected Circuits.
A Full List of Broadcasting Stations.
Your Tuning Controls.
How To Make The "Titan" Coil,
etc., etc., etc.

AN EVENTFUL RADIO YEAR

A brief summary of some recent outstanding wireless happenings at home and abroad.

The Baird Co. gave demonstrations to a Special Committee and the P.M.G. recommended that the B.B.C. should offer facilities for experimental television transmissions. There was a failure to reach agreement between the B.B.C. and the Baird people regarding the times of transmissions. The B.B.C. were unable to offer more than three quarter-of-an hour periods per week out of programme hours. Eventually, experimental transmissions were arranged on the basis of five periods of half-an-hour each per week.

The new high-power broadcaster at Brookman's Park, which is to be London's Regional Station, started experimental transmissions.

Work was commenced on Broadcasting House, the new headquarters of the B.B.C.

The Prague Plan of wave-lengths allocation was brought into operation.

B.B.C. RESIGNATIONS.

Capt. P. P. Eckersley, Mr. R. E. Jeffrey, Capt. West, Mr. K. A. Wright, Mr. Eric Dunstan, and other well known B.B.C. officials resigned. Capt. Eckersley joined the staffs of " Popular Wireless," " Modern Wireless " and the " Wireless Constructor."

It was calculated that in 1922 there were 30,000 radio receiving sets in use, and that the trade turnover was about £500,000. At the end of 1928 the turnover for the year was reckoned to be £25,000,000, and the sets in use 3,000,000.

Professor E. V. Appleton, Professor of Physics, King's College, London, disclosed details early in the year of his Heaviside Layer experiments. He was able to penetrate this with very short waves and located another similar layer many miles above it.

MARCONI ROYALTIES.

The Marconi Royalty question was thrashed out in Courts of Law and, after several interesting phases, an agreement was reached between the Marconi Co. and the R.M.A. on a 5s. per valve royalty basis. Contracts are to be for 5 years, the trade benefiting by getting the use of a very large number of existing patents and of any future patents eventuating in this period.

Broadcasting played a leading part in the General Election. There were pre-Dissolution speeches by all parties and the results were broadcast far into the early hours of the day following polling day.

Better valves than ever became available for listeners and constructors, notable newcomers being the remarkable Cossor New Process Valves.

PORTABLE SETS.

Portable sets attained a remarkable popularity which at the time of writing shows no signs of waning.

The combination of electric gramophone and radio set in the one instrument has created great interest, and manufacturers selling them report heavy demands.

Attention was at last turned by the listening public to the electric-supply mains as sources of radio power. The result has been exceptionally good trade in all kinds of mains devices.

A long-standing difference between the important cable companies (Compagnie Française des Cables Téléphoniques and La Compagnie Radio France, Cables P Q) has been removed by what is described as a " co-ordination of technical means."

RAILROAD RADIO.

The Federal Radio Commission of U.S.A. allotted five short-wave channels especially for use in connection with railway trains.

The Brazilian Government has made it compulsory for every vessel leaving a Brazilian port to be equipped with a radio set in proper working order. Unless this condition is complied with the clearing of the vessel will be refused.

Radio developments started taking place very actively in Russia. A party of Soviet engineers went over to New York to engage in Technical Conference with the engineers of the Radio Corporation of America.

It was announced that five new Russian stations are in the course of construction.

NEW BROADCASTING STATIONS.

New broadcasting stations were being built in practically every country.

The Czecho-Slovakian Government placed orders with the International Standard Electric Corporation of Prague for a broadcasting station which it is expected will be the largest in the world. The new station will have a power of 120 kilowatts.

A German wireless company secured a contract from the National Government of Nanking for the erection of high-power radio stations in Canton, Tientsin, Shanghai and Hankow.

Mr. Aylesworth, President of the National Broadcasting Company of America, paid visits to England, France, Germany and other European countries. His object was to endeavour to arrange an ambitious scheme of programme exchanges on an international basis. It is calculated that the N.B.C. chain of broadcasting stations serves 50,000,000 American listeners.

A RADIO MUSEUM.

What was claimed to be the very first museum devoted entirely to radio was opened in America.

Great strides in the development of broadcasting were stated to have been made in Sweden, it being reckoned that there are now 66 sets per 1,000 inhabitants.

Radio-Paris and Radio Toulouse were voted by French listeners as the two most popular stations in France. Bordeaux Sud Ouest, Limoges, Nimes and Toulouse P.T.T. were found to be of little interest.

SELECTED RADIO DEFINITIONS

Here are practically all the wireless names and terms you are likely to meet in " P.W." together with concise and lucid definitions. This list was specially compiled for this book.

By THE " P.W." TECHNICAL STAFF.

AERIAL CIRCUIT.—The circuit containing the aerial, and all apparatus connected between it and the earth, usually comprising a variable condenser and tuning coil.

AERIAL TUNING INDUCTANCE (A.T.I.).—The variable inductance or tuning coil in the aerial circuit, by means of which the wave-length of the circuit can be adjusted. The greater the wave-length required, the greater must be the size or portion of the A.T.I. included in the circuit.

ALTERNATING CURRENT (A.C.).—A current which reverses its direction of flow a definite number of times per second.

AMPERE.—The unit of electric current, being the current that can be driven through a resistance of 1 ohm by a pressure of 1 volt.

AMPLITUDE.—The maximum value which an alternating current or voltage attains in either direction. The current rushes produced by spark signals have an amplitude which decreases towards the end of the wave train.

ANODE.—The terminal by which current enters electrical apparatus. Thus the Plate of the Valve is its anode according to standard electrical practice—not the electron theory.

ANODE-BEND RECTIFICATION.—Detection or Rectification using the bend in the anode current—grid volts characteristic of a valve.

ANODE CURRENT.—The current which is driven, by the high-tension battery round the anode circuit and through the valve.

AUTO-TRANSFORMERS.—A transformer in which the one winding is tapped off the other winding, the connections being exactly the same as those for a potentiometer. The difference lies in the fact that the magnetic effect in the auto-transformer ensures that the primary and secondary currents are in the inverse ratio of the voltages, which is not the case in the potentiometer.

BACK E.M.F.—A voltage acting in opposition to a normal flow of current.

BAFFLE or BAFFLE-BOARD.—A wooden or other screen used in conjunction with a loud speaker to prevent the interaction of the sound waves emanating from the front and back of the instrument.

BALANCED ARMATURE.—A type of movement figuring in loud-speaker units. A piece of soft iron acting as an armature is balanced between the poles of a permanent magnet. The balance of this is upset by the operating energy and a diaphragm or cone moved accordingly.

BLASTING.—The distortion which follows the overloading of microphones, valves, loud speakers, etc.

CAPACITY.—The property which enables apparatus to store a quantity of static electricity when electrical pressure is applied. Capacity is measured by the quantity of electricity that can be forced into the apparatus by a pressure of 1 volt. The unit of capacity is called a " Farad."
The capacity of an accumulator is measured in ampere hours. In this case the term is used in a somewhat different sense. Ignition capacity is double actual capacity. One ampere hour means capacity to deliver a current of one ampere for one hour.

CARRIER-WAVE.—The steady H.F. oscillations emitted by a wireless telephony transmitter. These oscillations are varied or modulated by the speech and music.

CATHODE.—The terminal by which current leaves any piece of electrical apparatus. In a thermionic valve the filament is the cathode. In a cell supplying electricity, the cathode is the negative terminal (according to standard electrical practice).

CHOKE.—A coil of wire which offers considerable opposition to varying and alternating currents, but which may have low direct-current resistance.

COUPLING.—The connection by means of which electrical energy is transferred from one circuit to another. The transference may be brought about by means of condensers (capacity coupling), by electro-magnetic induction, as in the transformer (inductance coupling), or by connections similar to those in an auto-transformer (direct coupling).

CURRENT.—An electric current is a movement of negative electrons, driven by an electro-motive force. A current cannot flow unless there is an electro-motive force to drive it, and a conducting path for it to flow along. The unit of electric current is the ampere.

CUT OFF.—The point in the frequency scale at which apparatus, such as a loud speaker, ceases to operate.

CYCLE.—A complete alternating current or voltage wave, extending from one maximum value to the next maximum value in the same direction.

DAMPING.—Loss caused by energy absorption in mechanical or electrical apparatus.

D.C.—Direct Current.

D.C.C.—Double Cotton Covered.

DIELECTRIC.—A substance which will allow practically no electric current to flow through it—i.e. a nearly perfect insulator. The term is usually applied to the insulating material in a condenser.

EDDY CURRENTS.—Currents induced in metal by adjacent varying magnetic fields.

ELECTRODE.—A part of a valve or of a battery.

ELECTROLYTE.—The solution or paste used in a battery.

ELECTRO-MOTIVE FORCE (E.M.F.).—The force which is necessary to produce an electric current, and upon the value of which depends the amount of current, measured in amperes, in any particular circuit. Electro-motive force is measured in volts.

ELECTRON.—The ultimate particle of matter, consisting of an indivisible negative electric charge. A stream of negative electrons constitutes an electric current.

EMISSION.—The stream of electrons thrown off by the heated filament of a valve.

ETHER.—The all-pervading medium through which radio waves are presumed to vibrate.

FARAD.—The unit of electrical capacity, being the capacity of a condenser which will store 1 coulomb of electricity at a pressure of 1 volt, or which will take 1 second to be charged to a pressure of 1 volt by a current of 1 ampere.

FILTER.—An arrangement of inductances and condensers which will pass, or prevent from passing, varying currents of certain frequencies.

FREQUENCY.—The number of times per second that an alternating current or voltage attains its maximum value in one direction; the number of complete wireless waves received per second.

FULL-WAVE RECTIFICATION.—The rectification of alternating current so that both half-cycles are used.

GRID BIAS.—A voltage applied to the grid of a valve in order to bring its operating characteristic to a certain desired condition.

GRID CIRCUIT.—The circuit which externally connects the filament and grid of a valve, and is completed internally by the electron stream between them.

HALF-WAVE RECTIFICATION.—The rectification of alternating current so that only one half-cycle is used.

HARMONICS.—Frequencies which are multiples of other frequencies.

HEAVISIDE LAYER.—An upper layer of atmosphere (60 to 200 miles above the earth) which is presumed to affect the transmission and reception of radio by acting as a reflecting or absorbing screen.

HENRY.—The unit of self-inductance, being that inductance which will so retard any change in the value of a current that it takes 1 second for 1 volt to raise the current in a circuit by 1 ampere.

HIGH FREQUENCY.—A term applied to alternations or waves which occur at frequencies too high for audibility; sometimes called " Radio Frequency." High frequency may be taken to include all frequencies above 20,000.

HIGH-FREQUENCY RESISTANCE.—The resistance which a conducting path offers to high-frequency currents. Skin effect renders this higher than the resistance that would be offered by the same path to a continuous or low-frequency current.

HYSTERESIS.—The lagging effect observed in the magnetising of iron.

IMPEDANCE.—The total opposition offered by a circuit, or a piece of apparatus, to a varying or alternating current, being made up of the combined effects of resistance and reactance.

INDUCTANCE.—The property of a circuit which operates and retards any change in the value of the current flowing. Inductance has the same effect upon an electric current as inertia or momentum has upon a moving body.

INDUCTION.—The production of an electromotive force in an electric circuit through the agency of another circuit, without any direct electrical connection between the two. Induction may be brought about by lines of electric force (electrostatic induction), or by lines of magnetic force (electro-magnetic induction). Upon the latter depends the working of the transformer and the loose coupler.

KILOCYCLE.—One thousand cycles.

KILOWATT (K.W.).—The unit used for measuring large amounts of electric power, being equal to 1,000 watts or 1⅓ horse power.

LAMINATIONS.—Layers or thin sheets. A laminated core in a transformer or choke is a core built up of thin sheets of iron.

LINES OF FORCE.—The paths along which acts the force due to a magnet or electrically-charged body.

LOOSE COUPLING.—The fairly weak magnetic or electric linkage between two coils or circuits.

LOW FREQUENCY.—Frequencies up to about 20,000 cycles.

MAGNETIC FIELD.—The space surrounding a magnet, extending as far as its magnetic influence is appreciable. Any space pervaded by lines of magnetic force is a magnetic field.

MANSBRIDGE CONDENSER.—A type of large capacity fixed condenser.

MEGOHM.—The unit used for measuring high resistance, being equal to 1,000,000 ohms.

METAL RECTIFIER.—A rectifier, consisting of two plates of metal in contact, used mostly in mains units.

MICROFARAD (MFD.).—The practical unit of capacity equal to one millionth of a farad.

MICROHENRY (MH.).—The practical unit of inductance equal to one millionth of a henry.

MILLIAMPERE.—Thousandth part of an ampere.

MOVING-COIL LOUD SPEAKER.—An instrument whose diaphragm is operated by a small coil of wire suspended in a strong magnetic field. The input energy is fed into this Moving Coil.

NATURAL FREQUENCY.—The frequency at which a circuit containing inductance and capacity will most readily oscillate.

NEGATIVE POLE.—A pole that is at a lower potential relatively to another, the positive pole.

NON-INDUCTIVE.—A wire resistance wound so that it has negligible inductance is so described.

OHM'S LAW.—The law which states the relations existing in any circuit between current, voltage, and resistance. These relations are as follow: Amperes = Volts ÷ Ohms. Volts = Amperes × Ohms, and Ohms = Volts ÷ Amperes. Thus, for example, 36 volts are required to send a current of 4 amperes through a resistance of 9 ohms.

OPEN CIRCUIT.—A broken circuit; a circuit through which current cannot flow.

OSCILLATORY CIRCUIT.—A circuit having inductance and capacity and comparatively low resistance.

PARALLEL.—Two or more conductors or pieces of apparatus are in parallel when they are so connected that the current in the circuit divides, and part goes through each of them. Cells are connected in parallel when the required current is equal to the sum of the currents which can be given by each individual cell, and the voltage required is that of a single cell.

PLATE CIRCUIT.—The circuit which externally connects the filament and plate of a valve and is completed internally by the electron stream between them.

PLATE OR ANODE IMPEDANCE.—The internal impedance of a valve.

PLATE OR ANODE VOLTAGE.—The potential difference (voltage) existing between the plate (anode) of a valve and its filament.

POSITIVE POLE.—A pole having a relatively higher potential than another pole, the negative.

POSITIVE AND NEGATIVE (+ and —).—Names given to distinguish the terminals of a source of electric supply. Current is assumed to flow round a circuit from positive to negative, although it actually consists of an electron stream flowing from negative to positive. Positive and negative terminals are often distinguished by the colours red and blue or black respectively.

POTENTIAL DIFFERENCE (P.D.).—The difference of potential or electrical pressure between two points is the electro-motive force trying to send current from one point to the other. (Voltage.)

POWER VALVE.—A valve designed to handle relatively large inputs. A power valve does not necessarily give greater amplification—frequently its amplifying properties are comparatively low.

PRIMARY CELL.—A cell which produces current by chemical activity and which cannot be recharged like an accumulator (secondary cell).

PRIMARY CIRCUIT.—A circuit which hands on applied energy to another (secondary) circuit.

REACTION.—A system of feeding-back the amplified energy from the anode circuit of a valve to its grid circuit for further amplification. Indiscriminately applied, this scheme may result in a receiving set radiating interfering energy.

RESISTANCE.—The opposition offered by an electrical path to the passage of current when no reaction is present. Except in the case of high-frequency currents, resistance depends purely upon the conducting path and is independent of changes in the value of the current. The resistance of any path depends upon the length and sectional area of the path, and upon the material of which it is composed.

RESONANCE.—A circuit is in resonance when its frequency corresponds to that of the applied energy. Resonance in a loud speaker indicates a tendency to vibrate more readily on certain notes.

SATURATION.—In a valve—where further increases in anode voltage produce no corresponding increases in anode current. The maximum amounts in such cases are known as saturation currents.

Magnetic saturation indicates a point when further increases in the magnetising force fail to cause increases in flux density. This condition is met with in L.F. chokes and transformers.

SCREENING.—The separation of components or circuits by metal partitions or boxes in order to prevent coupling effects.

SERIES.—Two or more conductors or pieces of apparatus are in series when the whole of the current in the circuit has to pass through them one after the other. In a simple receiving circuit, for instance, the telephones and crystal detector are connected in series. Cells are connected in series when the required voltage is equal to the sum of the voltages of the individual cells.

SHUNT.—When two portions of a circuit or pieces of apparatus are connected in parallel, one is said to shunt the other.

SIDE - BANDS.—A number of high-frequency waves above and below the frequency of the Carrier-Wave produced by Modulation.

SMOOTHING CIRCUIT.—An arrangement of chokes and condensers designed to suppress the irregularities in a current supply.

SOFT VALVE.—A valve in which a little gas remains. The anode current is carried partly by this gas as well as by the electron stream from the filament.

SOLENOID.—A coil of wire wound in a long spiral, for the purpose of producing a magneto-motive force along its axis.

STALLOY.—A special steel used widely in the construction of L.F. chokes and transformers.

STEP-DOWN TRANSFORMER.—A transformer which steps down voltage—the current in the secondary being proportionately greater.

STEP-UP TRANSFORMER.—This steps up the voltage.

SULPHATING.—A chemical effect in an accumulator caused by neglect or age and which impedes the action of the device.

TIGHT COUPLING.—The strong magnetic or electric linkage between two coils or circuits.

TRICKLE CHARGER.—An accumulator charger which has a very low output. Batteries can be left connected to such a device throughout the day and night.

UNTUNED (or APERIODIC).—A term applied to an aerial and H.F. transformer or coupling coil indicating that such is not tuned to any one particular frequency.

VOLT.—The unit of electro-motive force or electrical pressure, being that pressure which will drive a current of 1 ampere through a resistance of 1 ohm. The electro-motive force of a single accumulator cell is about 2 volts.

VOLTAGE.—Potential difference in volts.

VOLTAGE DROP.—The voltage used in driving current through a circuit or across a piece of apparatus. Voltage Drop across a 10-ohm resistor passing ½ ampere would be 5 volts.

WATT.—The unit of electrical power, being the power exerted by a current of 1 ampere flowing under a pressure of 1 volt. 746 watts are equivalent to 1 horse-power.

WAVE-LENGTH.—The distance travelled by a wireless wave, while it increases from zero to its maximum value in one direction, reverses, attains its maximum value in the other direction, and falls to zero again.

WAVE TRAP.—A device designed to eliminate an interfering station.

WET BATTERY.—A term that used to be applied to an accumulator, but is now more widely applied to the small Leclanché cell batteries employed for H.T. purposes.

Aerial Interaction

Cases of this cannot, as a rule, be cured by one party only, and the co-operation of both parties concerned is desirable.

The trouble generally arises from the use of small sets which necessitate the employment of large amounts of reaction, and, as a rule, the only certain cure is for both parties to build larger sets, preferably employing neutralised H.F. stages.

Sometimes a cure can be effected by making sure that a different earth is used by each receiver, and, again, the use of a small condenser in series with each earth lead will sometimes improve matters.

Aerials That Swing

There is no need to have the aerial absolutely tight unless short-wave work is being indulged in, when it is an advantage, because a swinging aerial may then affect reception. The advantage of a certain amount of "give" in the aerial wire is that there is less strain upon the mast, supporting stays, etc., particularly with a high aerial, and for ordinary reception a certain amount of give is in no way detrimental. It is, however, important to see that the aerial does not kink anywhere, and if this happens it is sure to break sooner or later, which may give rise to noises in the set.

SELECTED "P.W." CIRCUITS

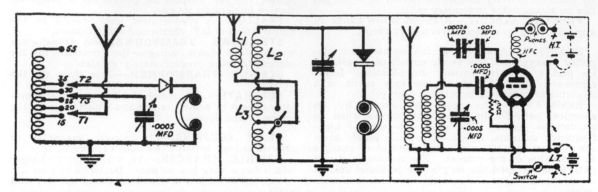

Left, a sensitive and selective type of crystal circuit. Tapping T_3 is normally placed on the 55-turn point, T_1 and T_2 being adjusted for best results. Numbers indicate turns of No. 24 D.C.C. wire on a 3-inch tube. Centre, a standard form of wavechange crystal circuit. L_1, No. 25 or 35, L_2, No. 60, L_3, No. 150 centre-tapped. Right, the popular Reinartz circuit as a single-valver arranged for plug-in coils.

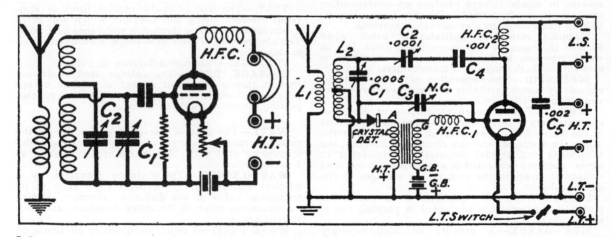

Left, a typical circuit of Reinartz type in which the reaction condenser is placed at the lower end of the reaction coil, thereby reducing hand-capacity effects. Right, a recent development of the " Trinadyne " circuit, employing a crystal detector and a valve which gives reaction and L.F. amplification. L_2 is a centre-tapped coil, C_3 a condenser of the neutrodyne type, and the valve an L.F. or small power. (H.F. type for headphone work.)

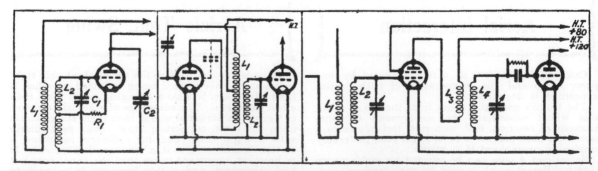

Some typical methods of H.F. inter-valve coupling. Left, split-secondary neutralising circuit. Centre, split-primary neutralising circuit in which the dotted condenser represents the plate-to-grid capacity of the valve. Right, plain transformer coupling as often used with screened grid-valves.

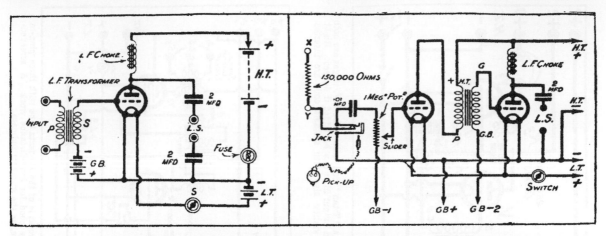

Left, a single-valve L.F. amplifier incorporating a safety fuse and an output filter properly arranged to reduce battery coupling effects and to isolate the loud speaker completely when a mains unit is used. Right, a two-valve L.F. amplifier for use with any set not already incorporating L.F. stages, and with a jack for the insertion of a plug connected to a gramophone pick-up. Note the volume control marked " 1 meg. Potr."

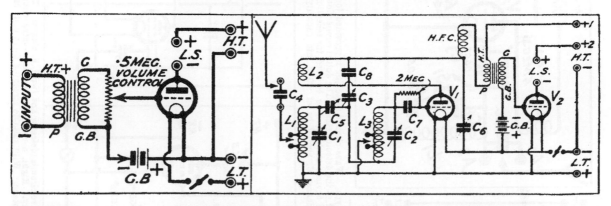

Left, a single-valve L.F. amplifier similar to the one above, but including a volume control and omitting the output filter. Right, an exceptionally high selectivity two-valve circuit for the Regional scheme. Coils L_1 and L_3 are of the " X " type (60 for low waves and 250 for long waves). Note the special reaction scheme : C_3 (·0001 mfd) is the main control, C_5 (·0001 mfd) is a compression-type adjustable condenser giving a preliminary control of reaction. C_5 is a neutrodyne-type condenser providing a small capacity to couple the two circuits together.

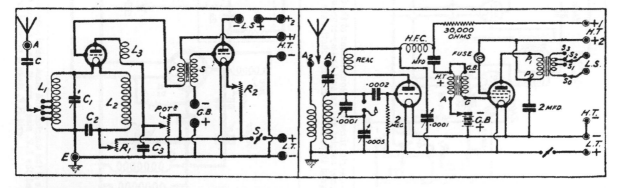

Left, a good example of the Filadyne circuit in one of its later forms. Note the division of the tuning coil into portions so that separate filament chokes are no longer necessary. The potentiometer provides the control of reaction. Right, a good modern form of two-valve short-wave circuit, using a pentode in the L.F. stage (note the special output transformer). The resistance-feed method is used for the L.F. transformer connections, and the set can be used on the broadcast waves by inserting suitable coils.

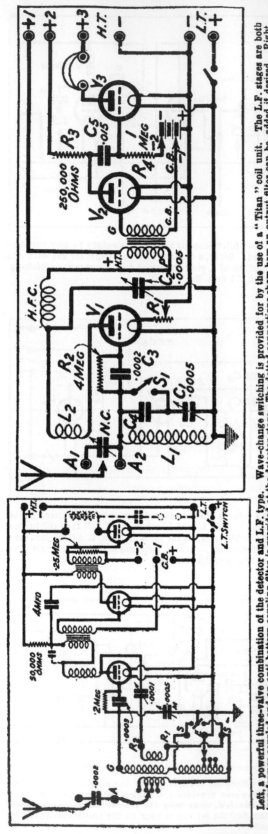

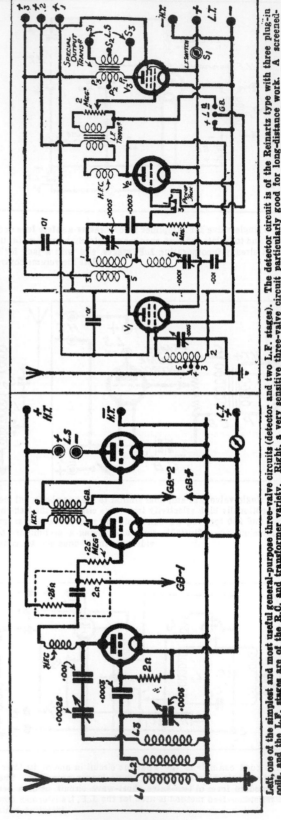

Left, a powerful three-valve combination of the detector and L.F. type. Wave-change switching is provided for by the use of a "Titan" coil unit. The L.F. stages are both transformer coupled and an anti-battery-coupling filter is provided at the detector stage. The dotted connections show how an output filter can be added if desired. **Right** a very good three-valve circuit for short-wave work with special devices enabling it to be used for reception of the local station also when suitable coils are inserted. Condenser C_1 should be of ·0002 mfd capacity.

Left, one of the simplest and most useful general-purpose three-valve circuits (detector and two L.F. stages). The detector circuit is of the Reinartz type with three plug-in coils, and the L.F. stages are of the R.C. and transformer variety. **Right,** a very sensitive three-valve circuit particularly good for long-distance work. A screened-grid H.F. stage is provided and a pentode L.F. stage. With suitable screening 6-pin coils can be used here, as the numbers indicate.

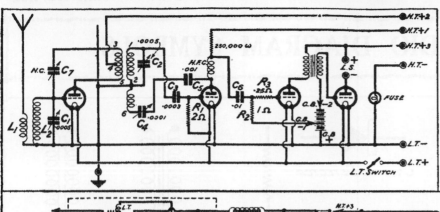

A good general-purpose four-valve circuit for 6-pin coils. The H.F. stage is of the split-primary neutralised type, followed by a leaky-grid detector, a resistance-coupled L.F. stage, and then a final transformer-coupled stage. A little screening is required for the best results.

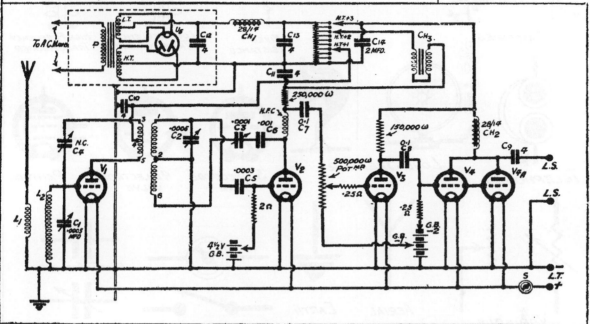

A five-valve circuit designed to give very fine quality of reproduction and a large output with suitable valves.   Note the built-in mains H.T. circuits (alternating-current mains).

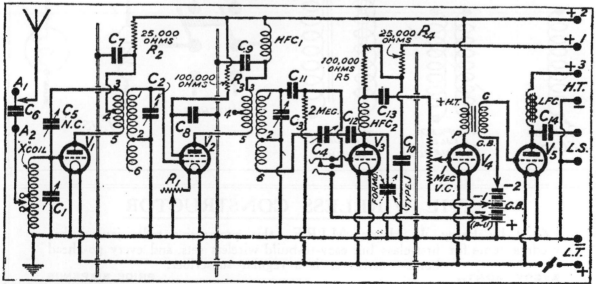

A particularly good five-valve combination consisting of a neutralised H.F. stage, a screened-grid H.F. stage, detector (note the jack for gramophone pick-up), one R.C. coupled L.F. stage and one transformer stage.

RADIO DIAGRAM SYMBOLS

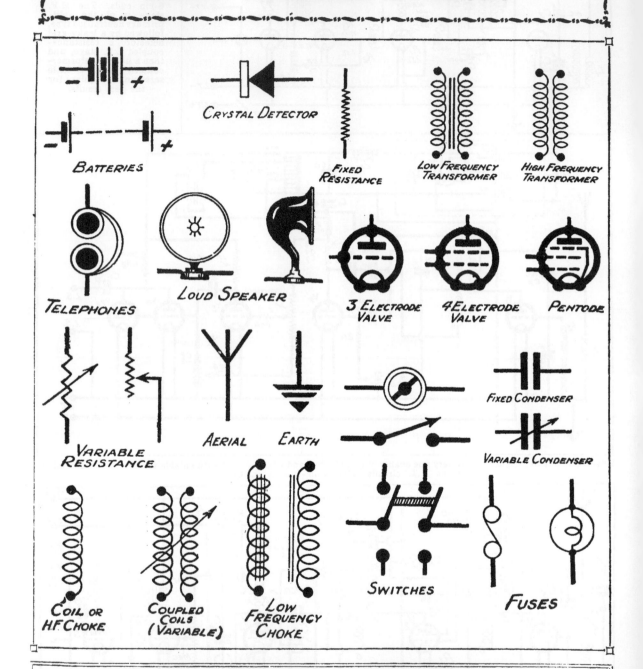

BATTERIES

CRYSTAL DETECTOR

FIXED RESISTANCE

LOW FREQUENCY TRANSFORMER

HIGH FREQUENCY TRANSFORMER

TELEPHONES

LOUD SPEAKER

3 ELECTRODE VALVE

4 ELECTRODE VALVE

PENTODE

VARIABLE RESISTANCE

AERIAL

EARTH

FIXED CONDENSER

VARIABLE CONDENSER

COIL OR H.F. CHOKE

COUPLED COILS (VARIABLE)

LOW FREQUENCY CHOKE

SWITCHES

FUSES

THE WIRELESS CONSTRUCTOR

Edited by Percy W. Harris, M.I.R.E., this progressive radio journal has made a name for first-class but easy-to-build wireless sets, and every go-ahead home-constructor is a regular subscriber.

EVERY MONTH   **PRICE SIXPENCE**

Printed by The Amalgamated Press, Ltd., Printing Works, Sumner Street, London; S.E.1.

Printed and bound by CPI Group (UK) Ltd, Croydon, CR0 4YY

03/10/2024

01040336-0008